Applying Ecology

Alan Beeby

Department of Biotechnology
South Bank University
UK

CHAPMAN & HALL
London · Glasgow · New York · Tokyo · Melbourne · Madras

Published by Chapman & Hall, 2-6 Boundary Row, London SE1 8HN

Chapman & Hall, 2-6 Boundary Row, London SE1 8HN, UK

Blackie Academic & Professional, Wester Cleddens Road, Bishopbriggs, Glasgow G64 2NZ, UK

Chapman & Hall GmbH, Pappelallee 3, 69469 Weinheim, Germany

Chapman & Hall USA., One Penn Plaza, 41st Floor, New York, NY10119, USA

Chapman & Hall Japan, ITP - Japan, Kyowa Building, 3F, 2-2-1 Hirakawacho, Chiyoda-ku, Tokyo 102, Japan

Chapman & Hall Australia, Thomas Nelson Australia, 102 Dodds Street, South Melbourne, Victoria 3205, Australia

Chapman & Hall India, R. Seshadri, 32 Second Main Road, CIT East, Madras 600 035, India

First edition 1993
Reprinted 1994

© 1993 Alan Beeby

Typeset in 10/12pt Palatino by Type Study, Scarborough
Printed in Great Britain by The Alden Press, Osney Mead, Oxford

ISBN 0 412 44420 4

A Catalogue record for this book is available from the British Library

Library of Congress Cataloging-in-Publication Data available

∞ Printed on acid-free paper, manufactured in accordance with ANSI/NISO Z39.48-1992 and ANSI/NISO Z39.48-1984 (Permanence of Paper)

Contents

Preface   ix

Acknowledgements   xi

1 Applying ecology   1
1.1 Thinking about ecology   2
1.2 Scientific method and ecological modelling   2
1.3 Levels of organization   5
1.4 The problem of scale in ecology   6
1.5 Ecological methodologies   8
1.6 Ecological stability   17
1.7 Stress   21
1.8 Theory and application   24

2 Measuring the effects of pollution   29
2.1 The nature of pollution   30
2.2 Adaptation to the environment   32
2.3 Allocating energy   34
2.4 The accumulation of pollutants in individuals   38
2.5 Measuring toxicity   42
2.6 Biological monitoring   46
2.7 Measuring toxic effects in soil communities   56
2.8 Biochemical measures of stress in whole organisms   61
2.9 Scope for growth and protein turnover   64

3 Exploiting variability   71
3.1 Variation and adaptation   72
3.2 Specialization and generalization   80
3.3 Using tolerance as a pollution indicator   84
3.4 Metal tolerance in plants   88
3.5 The use of tolerant microorganisms in soil remediation   94
3.6 The use of GEMs in bioremediation   102
3.7 Recombinant technology and the release of GEMs   106

4 Managing populations   115
  4.1 Population growth   116
  4.2 The limits to population growth   118
  4.3 Determinants of stability   121
  4.4 Exploiting a population   122
  4.5 Managing forests for maximum sustainable yield   135
  4.6 Conserving populations   138
  4.7 Loss of genetic variation   147

5 Managing pests   157
  5.1 Defining the pest   159
  5.2 The functional response of a predator   170
  5.3 The numerical response of a predator   173
  5.4 The use of models in biological control   179
  5.5 Finding natural enemies   181
  5.6 The ecology of the pest   185
  5.7 Other forms of biological control   186
  5.8 Chemical control and integrated pest management 194

6 Conserving communities   201
  6.1 The distribution of rare species   202
  6.2 Fragmentation and insularization   206
  6.3 Species – area relationships   207
  6.4 Insularization and reserve design   216
  6.5 Insularization and species composition   220
  6.6 Selecting nature reserves   228
  6.7 Reserve management   239

7 Establishing ecosystems   243
  7.1 Succession and disturbance   245
  7.2 The principles of reclamation and restoration   262
  7.3 Reclamation   263
  7.4 Restoration and soil erosion   278
  7.5 The restoration of semi-natural habitats   280

8 Modelling ecosystems   287
  8.1 Principles of modelling   289
  8.2 Hubbard Brook   295
  8.3 Nutrient cycling and the impact of pollutants   297
  8.4 Eutrophication in freshwater ecosystems   300
  8.5 Microcosms as experimental models of larger systems 310
  8.6 Bioaccumulation   314

9 Exploiting ecosystems 327
  9.1 Simple process control 329
  9.2 Sewage treatment 332
  9.3 Sludge disposal problems 348
  9.4 Alternative methods of treating sewage effluent 350
  9.5 Landfill ecology 357
  9.6 Composting 364

10 Assessing large-scale ecological change 369
  10.1 Extending ecological theory to the larger scales 372
  10.2 Ecosystem stability 373
  10.3 Stability and diversity 375
  10.4 Regional diversity 382
  10.5 Landscape and regional ecology 385
  10.6 Evaluating global ecological change 398

References 403

Index 425

Preface

Undergraduates often distinguish between textbooks and reference books according to their thickness. The latter are to be dipped into, whereas textbooks have a greater chance of being read in their entirety. I hope this book is short enough to encourage students to read it from cover to cover – ecology is one science that requires an overview of the whole discipline. The book attempts to review the theory behind a range of applications of ecology, and a more complete understanding emerges from a broad view of the subject.

The central aims of this text are student understanding of these main concepts and to provide a critique of their current applications. In each chapter the relevant ecological theory is developed before considering how it is applied to particular environmental problems. For the most part, examples or specific details of a technique are confined to the figures. This allows the concepts to follow sequentially, enabling students to structure their thinking without undue clutter. The development of some of these ideas does not end with a single chapter but is followed up with applications in different areas, so that some themes run throughout the book. Chapter 1 provides an overview of ecological thinking, and introduces some of the general ideas that recur throughout the text.

Next there is a conventional development of ecological theory – following the normal hierarchy from the level of the individual organism through to ecosystem ecology. This is not an exhaustive survey of applied ecology, and much has had to be pruned away to keep it to a reasonable size. To aid this, and to give coherence to each subject, specific habitats or groups of organisms are considered in each chapter. Nevertheless, I have enjoyed using contrasting examples in most chapters, and hope that this serves to highlight the theory.

Overall, the main concepts, the major groups of organisms, the significant pollutants and the important environmental issues are covered, in some combination. The recommended texts at the end of each chapter are a guide to where this material can be followed up. The main concepts are highlighted in bold type where they first occur, and a series of cross-references are used for direction to related topics elsewhere in the text.

This is not a handbook of practical ecology, nor is it a complete account of ecological theory. Instead it concentrates on the interface between theory and application, attempting to show how each can illuminate the other. For many people, applied ecology is about conserving habitats, endangered species or solving pollution problems, but ecologists also make contributions in such diverse areas as agriculture, resource and waste management, environmental planning, pest control and so on.

In most respects applied ecology is not one subject but several, although their collective spirit is certainly ecological. The theory underpinning them becomes useful when it makes reasonable predictions about ecological systems, including those which have been badly degraded. In applying this theory, applied ecologists can throw new light onto these biological processes, helping us to define what should be best practice in our attempts to restore or conserve them.

My past students have unwittingly shaped this book, and I am grateful for all they have taught me; my thanks also to my patient postgraduate students, Larry Richmond and Richard Post. It is a pleasure to thank the several people who read and commented on various chapters – Charles Banks, Elliot Gingold, Paul Hart, Sharon Hulm and Chris Platt. Ian Spellerberg kindly reviewed it all, and I am indebted to him for his many valuable suggestions. My thanks also to Bob Carling, Clem Earle, Susan Hodgson and Anne Waddingham at Chapman & Hall for their patient help. Jackie Beeby read the whole book and improved most of it. I hope Ralph and Kate will eventually approve of what their parents have produced.

It is also my pleasure to record my gratitude to the friends who have helped in the past – my mentor, John Bullock, and also Frank Clark, Brian Beeby, Garry Dyer, Steve Carnachan and Ges Sturgess.

This book is my thank you to Brian Wilson.

Acknowledgements

My grateful thanks go to the following for their permission, and in several cases, the original artwork for figures used in the book.

Figure 2.9 Dr Peter Donkin and Dr John Widdows and *Chemistry and Industry*, **21**, 733 (1986) Trade Publications Ltd

Figure 2.14 Dr Thomas Borchardt and *Marine Ecology Progress Series*, **42**, 27 (1988) Inter-Research

Figure 3.6 Prof. Dennis D. Focht and *Applied Environmental Microbiology*, **50**, 1060 (1985) American Society for Microbiology

Figure 4.7 The Editor, *J. Fish Res. Board Can.*, **30**, 2230 (1973)

Figure 4.18 Prof. Gary E. Belovsky and *Viable Populations for Conservation* (M. E. Soulé, Ed.) Cambridge University Press

Figure 5.5 Prof. Sir Richard Southwood and *Journal of Animal Ecology*, **45**, 957 (1976) Blackwell Scientific Publications Ltd

Figure 5.8 Dr Jeff Waage and *Parasitology*, **84**, 247 (1982) Cambridge University Press Ltd

Figure 5.9 Prof. Robert M. May and Prof. M.P. Hassell and *Philosophical Transactions of the Royal Society*, **B318**, 132, (1988) The Royal Society

Figure 6.5 M.E. Soulé and *Biological Conservation*, **15**, 269 (1979) Elsevier Applied Science Publishers Ltd

Figure 6.8 Prof. James F. Quinn and *Oecologia*, **75**, 136 (1988) Springer-Verlag

Figure 6.11 The late Dr J. G. Denholm and Dr Ian Denholm and *Journal of Applied Ecology*, **22**, 232 (1985) Blackwell Scientific Publications Ltd

Figure 7.6 Prof. Peter Vitousek and *Colonization, Succession and Stability* (A.J. Gray *et al.*, eds.) Blackwell Scientific Publications Ltd

Figure 7.7 Prof. William A. Reiners and *Ecological Bulletin*, **33**, 520 (1981)

Figure 9.8 Prof. Hans Brix and *Ambio*, **18**, 101 (1989) Royal Swedish Academy of Sciences

Figure 9.9 R. D. Zweig and *Ambio*, **14**, 72 (1985) Royal Swedish Academy of Sciences

Frank Clark kindly provided the photograph of the humpback whale at the beginning of Chapter 4.

Myakka River State Park, Sarasota County, Florida. Its wetlands are home to a large variety of birds and reptiles. Elsewhere, in the drier areas, there are also remnants of the native hammock vegetation, dominated by evergreen hardwoods.

1

Applying ecology

This is a book about ecological theory and how it can be applied to environmental problems. It reviews the principles that underlie ecological processes and the extent to which these are used in environmental management. This cannot be a survey of all the applications of ecology, but it attempts instead to map the common ground shared by theory and practice. Ecological concepts underpin most efforts at assessing and managing environmental problems, although at the same time solving these problems provides an important test of theory. In this way, both theory and practice should benefit from the insights of the other.

Ecosystems are immensely complicated systems. We are all impressed by the elaborate architecture and diversity of life found in a mature woodland or a coral reef. Yet despite the very different species found in these habitats, similar processes and mechanisms organize each one. Our understanding of these systems, how they function and how they are threatened, will determine our success in protecting them. This first chapter sets out some of the themes and concepts that recur throughout the book, and introduces some basic ideas about ecological methodology. To appreciate their relevance, we first review some particular aspects of analysing ecological systems.

The science of **ecology** attempts to explain the relationship between organisms and their environment. It describes the distribution and abundance of species in space and time, seeking to explain how populations grow or become extinct, how communities are organized and change in the face of disturbance. Many of its basic principles can also be applied to a range of human activities – from the microbiology of the sewage works to the setting of fishing quotas. Similarly, our efforts to conserve the major ecosystems and their diversity depends upon our understanding of ecological processes from the smallest to the largest biological scales.

This text provides an overview of the problems which applied ecology addresses, rather than the detail of each method. There can be no manual or handbook of techniques for the applied ecologist: there are few methods which the manager of a fishery or a sewage works share, or which are common across the whole range of the subject. Here the central aim is to provide a theoretical backbone to help organize the student's thinking and understanding, and to highlight the principles that unite different areas of the subject. In this way, we can identify those theories that have the greatest utility and have survived the test of being applied.

1.1 THINKING ABOUT ECOLOGY

Ecology has no all-embracing, universally accepted theory binding the subject together. Indeed, there is relatively little agreement amongst ecologists on which concepts are the most important (Cherrett, 1989). The central theory of biology is the evolution of species by natural selection in which the individual is the unit of selection. While evolution theory is pivotal to much of ecological thinking, ecology covers those domains above the level of the single organism, at the population and community levels, where natural selection can only operate indirectly.

Equally, applied ecology has no general theory, nor is it likely to develop one, if only because of the range of disciplines which it encompasses (Slobodkin, 1988). Nevertheless, there are several concepts which will prove useful throughout this book and four of these – levels of organization, problems of scale, regulation and stress – are introduced below. There is much debate about the meaning of each of these, and they are given different emphases according to the nature of the study and the background of the ecologist. Similarly, there are several approaches to studying ecological processes and we briefly consider their essential features later on. One general principle, the formal method of testing of scientific hypotheses and the derivation of models, is common to all. It is worth making a brief diversion to consider the principles of the scientific method and its relationship to the use of models in ecology.

1.2 SCIENTIFIC METHOD AND ECOLOGICAL MODELLING

We all use theories to help us to make sense of the world – mental constructs that enable us to predict its behaviour and to judge appropriate courses of action. Most of us will revise these theories when they fail, or when we learn of a more accurate view. Scientific theories too, are retained only as long as they remain useful. Science proceeds by a formal method of creating hypotheses, statements that can be tested, or more accurately, falsified. This requires that the terms of the statement are precisely defined. The hypothesis that 'all leaves are green' does this – it specifies the domain of interest – all leaves – and a property that they share – greenness. This is a general statement and, by itself, is relatively trivial, but it can be tested and potentially falsified: in this case, simply by finding a leaf which is not green, say from a copper beech tree.

A prediction may be tested through observation (the next leaf we find will be green) or experimentation (by growing a plant in the absence of light). Such tests help to refine our view of the world and make our theories more accurate descriptions of it. Properly formulated hypotheses serve to focus the attention of the investigator on what is essential in a problem. To be testable a statement has to be reduced to its simplest form, removing conditions to which it does not apply – for example 'all *living* leaves are green'. We then look for data that will falsify the hypothesis, the non-green leaf, or we speed the process of falsification by experimentation. Properly designed experiments can give an insight into a problem, indicating the conditions under which a hypothesis fails and the cause of particular results. Unfortunately, the scale and complexity of many ecological processes can make such experimentation impractical or meaningless. At the global or ecosystem level we have to rely on observation and correlations between observations to identify the most significant factors in ecological processes. The problem of extending ecological theory to these larger scales is considered in the final chapter.

All science works by reducing a problem to a series of testable hypotheses, but different subjects focus their attention at different levels, and on different scales. Experimentation requires us to define the domain of interest very precisely. A hypothesis that is not precise will be compromised by the circumstances where it doesn't hold good. Even the most robust and useful theories can be made conditional by the scale of an observation, becoming invalid at a different scale. Euclidean geometry is a good example.

Part of the problem in ecology is deciding whether the chosen scale is appropriate. Ecosystems are perhaps the most complex systems that any scientist attempts to understand, and ecologists have to partition them into smaller, quantifiable parts, such as the total number of species present, or the rate of energy transfer. Although these divisions can be useful for a specific purpose, they can give us little insight into the integration and functioning of the whole system. Instead we have to construct models of the system, summaries of the interrelationships between these various components. A **model** represents a distillation of the features of interest, with the extraneous 'noise' removed. A road map is a familiar example of a model – emphasizing the roads at the expense of other information.

Much of ecology is about constructing models, from the simple one-line equations used to describe population growth or pollutant uptake, to the multicompartment models used to set fishing quotas or describe nutrient transfer within an ecosystem. These are mostly numerical models, residing only in the memory of a computer as an algorithm. They consist of a series of statements, or hypotheses, linked together, describing the relationship between its parts – 'if A then B'. These can be refined by altering the values of each parameter or the terms defining their relationship, either in the light of further observations, or of experiments to test particular parts of the model. Their complexity depends upon the number of hypotheses which are combined to form the model, but each of these are falsifiable in the same way as any other scientific hypothesis. The process of model building is described in more detail in section 8.1.

The most useful models are the simplest – these correspond to the most general theories, and are more likely to be applicable over a range of circumstances. However, a compromise has to be struck between simplification and the complexity needed to describe a system adequately, particularly in applied ecology. Simplification is necessary for the process of modelling to begin at all, but the model should make predictions which are realistic, testable and not trivial. Its accuracy can be improved by defining each of its terms more precisely, or the range over which they vary.

Both general and specific models benefit from this sort of review and development. This is especially important where we are attempting to simulate accurately the behaviour of a particular ecosystem, say the North Sea. Then detail is essential if the model is to be used to make predictions about the future behaviour of the system and the impact of any changes that we might make to it. The simulation serves as an aid to decision-making – either before some development, or following some disturbance, such as an oil spill. Such models are commonly derived from a large database of observations collected over a range of conditions and time, allowing us to measure the variation in each of its terms and to quantify the uncertainty or stochasticity, of the model.

A **stochastic model** incorporates any variability in each of its terms and thereby allows for a variety of outputs. In contrast a simple **deterministic model** fixes the relationship between its parameters and only one outcome is possible for a given set of conditions and inputs. Deterministic models are sometimes described as analytical; useful for understanding the principal factors governing the behaviour of the system. However, stochastic models may also be analytical, and are not confined to being used for simulations. An example of the contrast between the two approaches is given in section 5.4.

Analytical models attempt to isolate general principles that can be applied across a range of populations and communities. Ecologists still argue whether there are repeatable patterns in different ecosystems and the extent to which these indicate the same organizing mechanisms, some process common to all communities. As we see below, part of this discussion revolves around the scale at which the hypothesis and the observation is made.

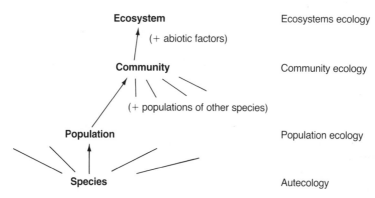

Figure 1.1 The levels of organization in ecology and their associated disciplines.

1.3 LEVELS OF ORGANIZATION

One feature typical of both the woodland and the coral reef is their large number of species, with each species adapted to the particular conditions of its habitat. Describing the adaptation of an individual to its environment is the first in a hierarchy of subjects that are grouped under the umbrella title of ecology (Figure 1.1). At the species level ecology shares common ground with other areas of biology, including genetics, biochemistry, physiology and ethology. Together these can describe the nature of an organism's adaptation and explain its distribution.

A **species** is composed of a number of populations which can actually, or potentially, breed with each other to produce viable, fertile offspring. **Communities** comprise collections of different species, interacting with each other; the **ecosystem** represents the community set in its non-living or abiotic environment. These levels of organization divide ecology into its three classical areas of study:

- **Population ecology:** the dynamics of populations of single species or of several species competing with each other. Applied aspects include the management of populations for exploitation (Chapter 4), for conservation and control (Chapter 5);
- **Community ecology:** the interrelations between species gathered together in one area. Understanding community processes is central to the conservation of habitats (Chapter 6) and their restoration (Chapter 7);
- **Ecosystem ecology:** the relationship between the community and the non-living components of its environment. Models of nutrient and energy flow help to explain pollution impact (Chapter 8). There are also functional properties of ecosystems that we can exploit in natural and artificial ecosystems (Chapter 9).

In addition there are those areas of the subject that operate over larger spatial scales (the landscape, the biome) and temporal scales (palaeo-ecology, quaternary ecology). The subject also merges with the humanities and some ecologists create higher levels of organization to accomodate this – an ecosystem in a political, social or economic context is termed a 'noosystem' by Barrett (1985).

At each level we can use the response of the living components to measure of the scale of some environmental change. A number of responses of individuals are used to monitor pollution in the short term (Chapter 2) or using population adaptations over the longer term (Chapter 3). In the same way, we can use changes in the structure and composition of communities to assess impacts on a larger spatial scale (Chapter 10). Indeed, Odum (1985) has suggested that the appearance of ecosystem-level effects of disturbance should cause alarm because this indicates that regulatory and homeostatic process at the lower levels of this hierarchy, in individuals and populations, may have broken down.

1.4 THE PROBLEM OF SCALE IN ECOLOGY

One obvious feature of these levels of organization is the change of scale implied as we move from the species to the ecosystem. The temporal and spatial scale of any study will be critical in determining what it can detect, its level of resolution. For example, sampling the population of a small mammal may be confined to a small area, with a time scale set by the length of its reproductive cycle. In contrast, the northern coniferous forest runs almost completely around the globe and has a history going back thousands of years.

For much ecological work, the domain of interest is defined as much by the practicalities of the study, as by its biological scale. For this to be valid, the ecologist has to be aware of the way in which observations will be compromised by the limitations of the sampling process. Data collected over the lifetime of the average ecologist may give little indication of the long-term processes operating in the coniferous woodland. Populations and communities vary in time and space and the frequency with which each dimension is sampled determines the validity of our observations. Properly designed, a sampling programme will avoid bias and correctly quantify the spatial and temporal variability in the data (Green, 1979). Where the practicalities of a survey mean that only a partial view can be obtained, this has to be acknowledged from the outset.

A well-designed experiment attempts to control variables and isolate the significant factors governing an effect. In ecology, the scale and complexity of many ecological processes may limit our capacity to manipulate many of these variables. Small, rapidly growing organisms may be easily controlled in the laboratory but not so the communities from which they came, even though this is the context in which we seek to understand the organism's activity.

Ecologists have tried to mimic complex communities in experiments using microcosms and mesocosms (Chapter 8). Applied ecology also offers some scope for larger scale experimentation on ecosystems, using controlled experiments in the restoration of degraded habitats (Chapter 7). However, much ecological experimentation is limited by the constraints of scale, and ecologists then need to rely on observational data to test hypotheses.

1.4.1 The scale of disturbance

One common method of experimentation is to perturb or 'challenge' a system and by watching its response, to gain some insight into the factors that govern its behaviour. Disturbance is a natural feature of many ecosystems, a prime cause of many spatial and temporal patterns of species distributions (Chapters 7 and 10). The magnitude and frequency of disturbance is a factor structuring a number of ecosystems: to have an effect, the disturbance must be large enough to cause the loss of some individuals or some species, but a disruption that is too frequent or too severe may destroy the whole system (Chapter 6). Similarly, there has to be sufficient time between disturbances for the system to recover a maximum number of species. The stability of the community may depend upon the frequency of these changes and the time marked by the interactions between its component species (May, 1974).

Natural ecosystems may become adapted to regular disturbances of a particular scale and the species composition may then reflect the likelihood of disturbance (Rapport *et al.*, 1985). Some established woodlands for example, are dominated by plant species adapted to frequent, small forest fires. Similarly, the life cycles of individual organisms are commonly adapted to seasonal changes which they can anticipate, and also for some long-term and less predictable changes. The frequency of these disturbances, their predictability and magnitude need to be considered in a sampling programme.

Some species rely upon the high probability of a disturbance for their continued survival. These **fugitive** species maintain their population by moving from one gap to another. Within each gap they are eventually lost as more competitive species begin to colonize the area. For example, the wild cherry (*Prunus pennsylvanica*) maintains its abundance by rapidly invading gaps in the forest, relying on the frequency of tree falls or forest fires to create new spaces for it to use. In each case it will eventually lose out to more shade-tolerant competitors, but it 'wins the war, though it loses every battle' (Urban *et al.*, 1987).

1.4.2 The effect of scale on management

Decisions taken about the management of ecological resources are closely linked to the scale of a problem. A plant found to be a locally rare, although it is abundant elsewhere, may be given a low priority for conservation, compared to others which are globally rare (Chapter 6). Similarly, action to control a pollutant may only follow if a large proportion of a community is affected, rather than a single species.

The management of a woodland can also reflect the natural scale of disturbance in these ecosystems. Clear-cutting small areas of forest is accepted as the best management strategy in many commercial, large-scale woodlands. This mimics the natural process of gap formation following tree fall, from which the forest will rapidly regenerate itself, either from the surrounding vegetation or with active re-planting. Similarly, well-meaning attempts to prevent forest fires are now known to be deleterious in many temperate forests: a regular series of small fires prevents the build up of excessive fuel (litter), and frequent disturbance encourages the regeneration of the ground flora. A small, isolated forest may be encouraged to burn in several stages so that its capacity to recover is not reduced by a single, large burn. Other forms of resource management can also leave patches undisturbed, to speed the recolonization of gaps.

In the same way, some of the effects of overfishing might be avoided if parts of the stock had extended periods without exploitation. This has led to suggestions for some fisheries to have designated areas where no fishing is allowed – a spatial equivalent of an off-season, where recruitment of stocks is unaffected by fishing activity. This is possible in the many species that feed and reproduce in particular areas, before migrating into other waters.

A large number of populations and communities have become reduced in size, rescaled by the influence of man, and our management strategies have increasingly to reflect these changes.

1.5 ECOLOGICAL METHODOLOGIES

Three basic approaches are used in ecological research, differing according to the level at which they define their domain of interest. This necessarily leads to their using different scales of resolution, but these approaches also reflect different perspectives on the principal mechanisms which organize populations and communities.

The crucial question concerns the processes which regulate the interactions between the species, and the extent to which these impart stability to the larger community. At the level of the individual and the species, natural selection is clearly the mechanism determining features of their morphology and physiology. Similarly, for species and populations, selective pressures and competition can limit population growth (Chapter 4). However, it is more difficult to describe the processes by which communities achieve any form of regulation.

The extent to which communities or ecosystems are self-regulating is one of the key issues in theoretical ecology, and one that continues to partition ecologists according to their methodology.

1.5.1 Reductionist ecology

This method attempts to understand the behaviour of a complex system by reducing it to its component parts and defining the relationships between them. Reductionists thus suggest that a community can be understood from the behaviour and interactions between its constituent populations. This method uses two basic approaches:

Population/community ecology

The first approach derives from population ecology. On one hand, ecologists are interested in populations in their own right and have developed a variety of models to examine the regulation of population size. The same methods are used to model the interactions between a small number of species, such as species competing for the same resources, or predators and their prey. The community is seen as a matrix of interacting species and, if these interactions were to be fully quantified, the behaviour of the community, it is argued, would be deducible from these population dynamics. The great strength of this approach is its predictive power, because each term in these models can be quantified: both deterministic and stochastic models can be used to predict individual population sizes and their effect on each other.

The fluctuations of real world populations may be explained simply by a mathematical analysis of their growth rates and the time lags in their interactions with other species and the environment (May, 1974). By looking at particular components of such communities, reductionists argue that this approach allows us to define the important relations and rules governing both population dynamics and community structure, at least for some part of the community. Some suggest that the stability of the larger community is an evolved trait, derived from the intrinsic regulation of its constituent populations.

This obviously begs the question of whether the collection of populations in a community could be described sufficiently to allow us to model the whole system. Other ecologists have highlighted the dangers of applying simple models to highly variable populations in highly variable environments. This is especially a problem with those exploited species whose abundance may be determined by ecosystem-level effects, such as nutrient supply (Hall, 1988).

Systems ecology

The second approach concentrates on the functional connections between the biotic and abiotic components in the community. In doing so, it emphasizes the ecosystem rather than the community (Figure 1.1). This regards the ecosystem rather like a machine, passing nutrients and energy between individual species or between functional groups such as trophic levels. The method uses mathematical models and systems analysis to identify the important parameters controlling each function of the system (Chapter 8). However, some species do not belong in a single functional group and models of energy or nutrient transfer become problematical if a species feeds at several positions in a food web. Defining the ecosystem and the roles of its various members has become criticial as the grouping of the species and their relative significance will change when we model some other process in the ecosystem. By the same token, the properties of the system vary with different spatial and temporal scales.

Both these reductionist methods are analytical, attempting to isolate the significant factors that control a process. Our understanding comes from describing how the component units interact, and both methods aim to predict future behaviour as a result of modelling these interactions. For example, we might be able to predict the size of a predator population from the numbers of prey or the rate of decomposition from the nitrogen content of the soil. They differ in the object of the study – either the populations of individual species or the rates of functional processes. Both methods can also be synthetic – models of significant relationships can be tested by trying to reconstruct parts or the whole of a community. This has been done with populations of small organisms in the laboratory, or by reconstructing parts of a community in a microcosm (Chapter 8). On a larger scale, much of restoration ecology is synthetic in this way (Jordan *et al.*, 1987), as are our attempts to re-establish endangered species in their former habitats or to create the reserves where they might flourish.

1.5.2 Holistic ecology

Some ecologists have argued that these reductionist methods could not fully describe community and ecosystem-level processes. In their view, ecosystems are highly integrated systems, with recurrent patterns that only become apparent from considering the whole system. These are the so-called 'emergent properties'. For example, in most ecosystems, an

Table 1.1 Trends expected in stressed ecosystems (abbreviated from Odum, 1985)

Energetics
1. Community respiration increases
2. Production/respiration ratio becomes unbalanced
3. Exported or unused primary production increases

Nutrient cycling
5. Nutrient cycling increases
6. Nutrient loss increases

Community structure
7. Proportion of *r*-strategists increases
8. Size of organisms decreases
9. Lifespan of organisms or parts of organisms (e.g. leaves) decreases
10. Food chains shorten because of reduced energy flow at higher trophic levels
11. Species diversity decreases

General system-level trends
12. Succession reverts to an earlier stage
13. Efficiency of resource use decreases

This is a list shortened from Odum's original, chosen when the effect is largely self-evident. However, the meaning of some of these terms may only become apparent in subsequent chapters. Odum offers a justification for all of the effects he predicts that may occur in a stressed ecosystem.

emergent property is their relatively fixed trophic structure and the proportion of species in the higher trophic levels. A high degree of integration is also suggested by the very predictable responses of communities to some form of stress. This can disturb the relationship between different species and in turn, it is suggested, lead to a predictable sequence of changes, which may reflect the scale of the impact (Table 1.1). The holistic view has been confused with the mechanistic approach used in systems ecology, but the latter is decidedly reductionist (McIntosh, 1985).

In its most extreme form, holistic ecology views the community as a 'superorganism' with mechanisms of regulation and coordination to maintain it close to an equilibrium, rather like the homeostatic mechanisms of an individual organism. Little evidence has been found for any 'superorganism' effect, and it is widely accepted today that communities are much looser assemblages than these ideas would have us believe. This does not deny that there are features of their organization which are consistent from one community to another (May, 1981a).

Much of the argument centres on the mechanism by which such regulation is meant to operate. In an individual or species it is relatively easy to show how regulatory processes can evolve through natural selection. Through coevolution, the same can be shown for mutualist species interactions, or predator–prey relationships (Wilson, 1988), but this is less easily demonstrated for a whole community. Most particularly, if community stability is a product of the interactions between the living and abiotic components, can the selective pressures operating on individual species generate a self-regulating community? Natural selection chooses between the genes of different individuals and will only affect the nature of the community as far as it determines the relationships between these individuals.

The holistic view of ecosystem integration relies upon two essential features – those of emergent properties and of the regulation of ecosystem processes.

Emergent properties

These are the observed properties of a number of individuals, or species, or of a whole community, which cannot be deduced from the individual components. We have to distinguish these from the collective properties of the system, such as its number of species or its productivity, properties that are simply the sum of the individual contributions (Salt, 1979). In contrast, an emergent property could not be predicted from the individual components: for example, the shoaling behaviour of a herring would not be deduced from observing a single fish. In Salt's view, emergent properties are probably the result of natural selection, operating on collections of individuals or species which are most likely to be found together. For example, some features of a predator–prey interaction, such as particular patterns of behaviour in response to a specific predator, may only be obvious when the two species are observed together (Salt, 1979).

In an ideal world, a reductionist would hope to explain the properties of the community from the behaviour of its components and their interactions: a study of the different cogs in a watch and their relation to each other, would eventually enable us to describe its emergent property of keeping the time. The complexity of natural ecosystems makes this a highly unlikely prospect in ecology. Perfect knowledge of the watch makes us the watchmaker, whereas a perfect knowledge of a tropical forest might make us divine.

There is no natural ecosystem on the planet that we have described to an extent where we can predict its behaviour even for short periods of ecological time. We are unlikely to adequately model more than a small part of most ecosystems, even assuming we adopt the correct methods (Bender *et al.* 1984) and scale (Wiegart, 1988). Ironically perhaps, holistic ecology may take a more practical view of what can be known about communities and ecosystems.

Integration and regulation

Holistic ecologists also suggest that most natural and undisturbed ecosystems are in some form of equilibrium, a product of internal mechanisms of regulation. Behind this is the idea that the interactions between species have evolved and, during its development, a community becomes highly integrated. It then has some capacity to regulate itself in the face of disturbance. Studies that have looked for such stability have, for the most part, failed to produce convincing evidence, either for the stability of populations within a community or its species composition (Connell and Sousa, 1983). Though this has been used as a stick with which to beat the holists, part of the problem is that many studies have failed to make due allowance for an appropriate time scale, or to demonstrate that the system was indeed stable before any disturbance.

In practical terms, many communities show some degree of regularity in a number of parameters, maintaining relatively constant structures and functions over the long term. This allows us to use some community-level indications of disturbance, which may be more appropriate for judging the biological significance of the stress. For example, in toxicology, the reductionist approach of testing a small number of indicator species is considered by Maciorowski (1988) to be of limited value for measuring the impact of a pollutant at the community level. Indeed, in restoring some eutrophic freshwater systems, a holistic approach alone has led to an understanding of the nature of the problem and the most viable long-term solutions (section 8.4).

In addition, all ecological studies have to integrate some components of the lower levels of organization, and are never entirely reductionist. The resolution that an ecologist choses is frequently a matter of practicality, rather than following any concrete biological demarcation. For example, we rarely need to know the details of species' reproductive biochemistry to describe its population growth, yet a description of the hormonal control of ovulation may be essential to conserve an endangered species. Similarly, many population studies treat the soil as a black box, simply measuring inputs and outputs of nutrients and energy. The constituent microorganisms and their interactions are ignored, and their crucial role in making nutrients available to higher plants is taken for granted. Yet there are a number examples where the population dynamics of a herbivore can only be explained by the changes in the nutrient availablity of the soil.

1.5.3 Hierarchy theory

Hierarchy theory suggests that the organization of the ecosystem results not from individuals, species or trophic levels, but from the rates of

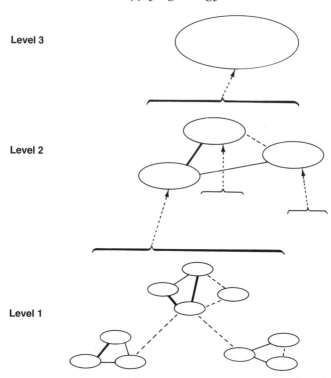

Level 3

Level 2

Level 1

Figure 1.2 A representation of the principle behind hierarchy theory. At each level, components are collected into functional groups, defined by some process for which they are responsible – such as decomposition or photosynthesis. The strength of the connection between each member is indicated by the thickness of the line joining them. The boundary between one level and another is set by a sharp change in the rate of that process. In effect, the lower level groups are nested into those of the higher level groups. The whole hierarchy is determined by the rates of processes between the different levels. Moving up the hierarchy, the spatial and temporal scale increases. (After Urban *et al.*, 1987.)

different processes that drive the ecosystem. This approach collects the components of an ecosystem into functional groups, say the leaves in a forest. Groups are themselves allocated to levels, according to the rate at which some process, such as photosynthesis, proceeds. The groups are thus 'nested' into a hierarchy, so that the next level might include all photosynthetic surfaces, followed perhaps by a level including all primary producers and so on (Figure 1.2). Rather like systems ecology, a different hierarchy has to be constructed to map each process, although many functional groups will remain the same for different processes (Allen and Hoekstra, 1987).

One justification for this new approach is a recognition of the particular problems in describing the complexity of ecosystems. O'Neill *et al.* (1986) suggest it is these methodological difficulties that have left ecology without an all-embracing theory. There are too many components in an ecosystem for a reductionist approach ever to describe fully its behaviour (as in describing a machine) and too few to average their behaviour using the large-number methods of statistics (such as those describing the properties of molecules in a gas). Ecosystems are 'medium-number' systems. The only way to model them is to condense the number of components to a manageable size to allow a small-number approach. Hence the nesting of functional groups in a hierarchy, partitioned according to an appropriate spatial and temporal scale.

The hierarchy will have certain important properties. At the smaller scales the functional groups operate at faster rates and show fluctuations with a high frequency. These fluctuations are damped moving up the hierarchy (Figure 1.3). For example, photosynthetic activity in a leaf changes rapidly with variations in the carbon dioxide content of the air, but the growth of the whole tree responds more slowly, and the 'signal' of the carbon dioxide level is integrated over a much longer time scale. Thus we do not detect these minute-by-minute variations in the annual growth ring of the tree. Change in the species composition of the forest happens at an even slower rate. In effect, the signal is attenuated as it passes through each level: the leaf, the tree, the community and beyond. The boundary between one level and another is defined by a distinct change in the rate of the process (Figure 1.3). In effect, each succeeding level serves to filter out the higher frequencies of the level below.

The rates at the lower levels in the hierarchy are constrained and limited by the slower rates operating at the higher levels. This serves to regulate the ecosystem and give the appearance of constancy at the higher levels – any change is simply happening more slowly at these larger scales. As an alternative example, the abundance of a fugitive species, such as the wild cherry, may be highly variable in one location, but over a large forest and a number of gaps, its abundance is more constant. Its population is limited by the total area of the gaps.

The attentuation of the signal up the hierarchy means that the lower frequencies of a process will be found over the larger spatial and temporal scales. Disturbances with a low frequency will generally operate over larger areas and longer periods, and would only be detectable at these scales. For example, the long-term rise in atmospheric carbon dioxide levels relates directly to the capacity of photosynthesizers on the planet to fix the gas and its impact may only be measurable on the regional or global scale.

The theory also implies that we cannot understand the regulation of some ecological processes without considering how the higher levels in

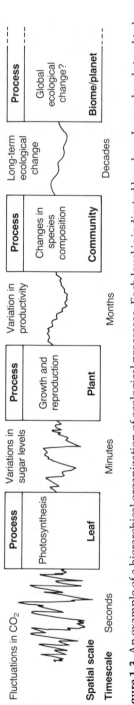

Figure 1.3 An example of a hierarchical organization of ecological processes. Each level is indicated by a box. Lower levels tend to be smaller with faster rates. The higher frequencies of the lower level processes are lost moving up to the higher levels – the spatial and temporal scales become larger, and the 'signal becomes attenuated'. The slower rates at these levels constrain the processes at the lower levels, and give the system its apparent constancy. For example, a slow rate of change in the species composition may limit the growth in biomass by limiting the colonization of new individuals or species.

the hierarchy constrain the lower levels. This is effectively a 'top-down' approach, which begins with the largest functional unit, the ecosystem, and then partitions it, rather than by starting with the parts and attempting to reconstruct the mechanism.

This is one justification for using a 'black-box' approach to modelling some ecological processes, rather than attempting to reduce an ecosystem to its component species. In practice, ecologists frequently do this anyway, using simple input and output terms to model lower levels (section 8.3).

The application of hierarchy theory to ecology is relatively new, although the principle itself has a longer history (Wilson, 1988). Its predictions now need to be tested. For example, a disturbed ecosystem will be freed of the constraints of the higher levels of the hierarchy (Figure 1.3), and some processes may run at a faster rate. An example is the nitrate loss to groundwater in disturbed soils (section 8.3). If the system is to regenerate, the slower frequencies have to be re-established to constrain the system, and prevent the nutrient leaching away.

1.6 ECOLOGICAL STABILITY

All living organisms can regulate their processes and adapt to the most probable environmental changes. They have a range of homeostatic mechanisms to maintain physiological and biochemical stability both within the tissues and within an individual cell (section 2.2). Between individuals, interactions may similarly be regulated and the apparent constancy of many communities is frequently taken as an indication of some mechanisms of control.

The idea of a 'balance of nature', that ecosystems are in some sort of equilibrium, is perhaps as old as man himself. However, ecologists do not agree on its reality: Charles Elton, a theoretical animal ecologist, thought it a myth, though it was very real to Aldo Leopold, a practitioner of game management (McIntosh, 1985). Few, if any, studies have effectively tested the idea (Connell and Sousa, 1983), partly because of its imprecise meaning and the failure to devise a testable hypothesis. Even so, much of modern ecology has an implicit assumption of an equilibrium condition, and it is often an explicit objective in conserving a species or a community (Chapter 6).

Stability is commonly represented as population stability, or to a lesser extent, as a constancy in the species composition of a community. Other measures are possible (Table 1.2) and indeed, the basic principles have been applied to population genetics, physiological ecology and ecosystem ecology. Properly defined, stability theory offers a single paradigm that can be used at all levels of organization, and against which we can measure the success of our efforts at managing populations and communities.

Applying ecology

Table 1.2 The effect of different stressors on five ecosystem level parameters (after Rapport *et al.*, 1985)

Type of stressor	Size of nitrate pool	1° Productivity	Size[H] distribution	Species diversity	System retrogression
Harvesting of renewable reserve, e.g. fishing	–	–	*	*	+
Pollution e.g. acid rain	*	*	–	*	+
Introduction of exotic species, e.g. fish	–	–	*	–	+
Extreme natural event, e.g. earthquake in a tropical forest	–	*	*	*	+

These are the five ecosystem-level parameters considered to be most useful indicators of stress across a range of ecosystems. The table is based on reported effects, summarized by Rapport *et al.*

A dash indicates that no effect was measured; * that the ecosystem was displaced from its original state compared with an unstressed system; + that there is evidence of the system being shifted back to a state comparable with an earlier stage in its succession (autogenic recession).

[H] Size distribution refers to a reduction in average size of biota, if only temporarily.

1.6.1 Types of stability

Stability refers to the response of a system to a disturbance. A system is said to be stable if it resists being deflected by some external factor or returns to its original state after being deflected. Stability can be classified according to the nature of the original state:

- **homeostasis:** the capacity to return to an original steady state after disturbance;
- **homeorhesis:** the capacity to return to an original trajectory or rate of change after disturbance (O'Neill *et al.*, 1986).

In the first case, it has to be established that a system was at equilibrium before disturbance, undergoing no significant change: for example, a population that returned to its predisturbance density after a pollution insult would be stable. In the second case, stability would be indicated if the population returned to its previous rate of change following disturbance, or the amplitude of its fluctuations were the same after insult (Lawton, 1987a).

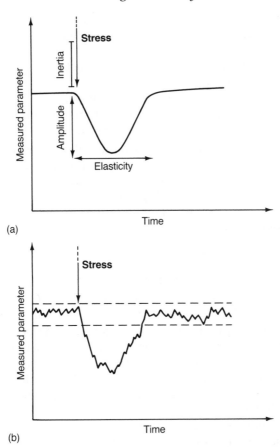

Figure 1.4 (a) The terminology of stability. The same principles may be applied to an individual, a population, a community or some ecosystem process. Inertial stability is the size of the stress needed to deflect the parameter. Adjustment stability has two components: amplitude, which is the maximum deflection of the parameter that allows the system to return to its original state, and elasticity, which is the speed of return to the pre-stress equilibrium. (b) Homeorhetic stability. Although the parameter has no equilibrium position, it does return to the same amplitude of fluctuations after disturbance.

Stability can be divided into two components, measuring the resistance to deflection and the speed of return from a deflection (Figure 1.4). These different forms of stability are given different names by different authors (Connell and Sousa, 1983; Pimm 1984) but we shall largely follow the names and justifications of Westman (1985) and Underwood (1989).

- **Inertial stability:** this is the maximum size of the disturbance which a system can withstand before being deflected. This is also called **resistance**.
- **Adjustment stability:** this refers to the characteristics of the return to equilibrium and is further divided into:
  - **elasticity** (the speed of return to equilibrium);
  - **amplitude** (the maximum deflection from equilibrium which still allows the system to return to its original state).

Others measures of stability have used the scale of a deflection, the shift to a new equilibrium, paths of return to the original state, (Pimm, 1984; Westman, 1985; Underwood, 1989) or the range of disturbance over which á system is stable (May, 1974). A system may also have more than one equilibrium position, each of which will have some degree of inertia and adjustment stability (Chapter 8).

1.6.2 Stability in populations

Connell and Sousa (1983) concluded there was little evidence of homeostatic stability in their survey of previous studies on natural populations, either at a single equilibrium density or at multiple stable states. They also found limited evidence of homeorhetic stability: one pest species (the larch tortrix moth) was found to resume a cycle of population change, one generation after perturbation – in this case, the application of an insecticide. A common problem with many of these studies was their use of an inappropriate time scale over which to measure this stability. Underwood (1989) provides a review of the necessary experimental techniques to demonstrate demographic stability.

Even simple deterministic models of population dynamics can show a great variety of behaviour, including cyclical change or even chaotic fluctuations (May, 1986). It is probably unrealistic to expect natural populations to remain close to a single density in a fluctuating environment (Hall, 1988). Instead a population may be deemed stable if its fluctuations remain within a range under normal conditions, even though it shows no single equilibrium density (Figure 1.4b): stochastic models use the size of the fluctuations as an indication of its stability (May, 1974). Over a series of observations, a population or a community found to have a constant mean, a constant variance and a constant time structure would be stable (Williamson, 1987). If the measurements are taken over too short a period, this stability might never be detected in a population varying in this way (Pimm, 1984).

1.6.3 Stability in communities

The evidence for inertial stability in the community is also equivocal. Theoretical ecologists have modelled this as constancy in the species composition, but there are alternatives (Table 1.1). Some features of the community may be constrained to be structured in a particular way: May (1981a) points out that all communities are constructed around a trophic 'skeleton' independent of which species occupy the roles of primary producer, consumer and decomposer. In freshwater habitats at least, the ratio of predators to prey appears to be fixed (Lawton, 1989), indicating that there is some regulation of the configuration of these communities.

There may also be common responses to disturbance. A range of ecosystem-level responses to stress have been suggested by Rapport *et al.* (1985) (Table 1.2). These are taken to be properties of ecosystem functions and are independent of the nature of the disturbance or the ecosystem itself. As such they may be used as generalized indicators of disturbance, using the principle behind Odum's list (Table 1.1). Admiraal *et al.* (1989) were able to detect human impact on five aquatic ecosystems in the floodplain of the river Rhine based on Rapport's criteria. The elasticity of these ecosystems to natural disturbance suggests they would respond well to attempts to restore them. In this study, adjustment stability has a wholly practical meaning: the long-term species composition of the restored communities is largely unpredictable, but their productivity and overall species richness are expected to increase in response to management.

Many communities undergo cyclical change with different parts of the system in different phases of the cycle. Then individual patches within a community can have very different collections of species. Overall, the mosaic formed by these patches represents an equilibrium for the whole community. Obviously our capacity to detect stability in such a community will depend upon the number of patches sampled, and the spatial and temporal scales over which we sample. More generally, hierarchy theory attributes the greater constancy at the community level to the attenuation of ecological processes between patches.

The patchiness of communities is a crucial factor in the conservation of individual species (Chapter 4) and in the biological control of pests (Chapter 5). The capacity of a community to withstand disturbance is also dependent on area and the frequency of displacement (Chapter 6).

1.7 STRESS

The use of of the term 'stress' should not be confused with the precise meaning it has in physics (Grime, 1989). **Stress** is here taken to be an applied stimulus, measured by its capacity to deflect some living component of the ecosystem. This response can be measured at any level, from the homeostatic responses of an individual to the nutrient cycling of an ecosystem. We divide stress into **disturbance**, which is an unplanned

Applying ecology

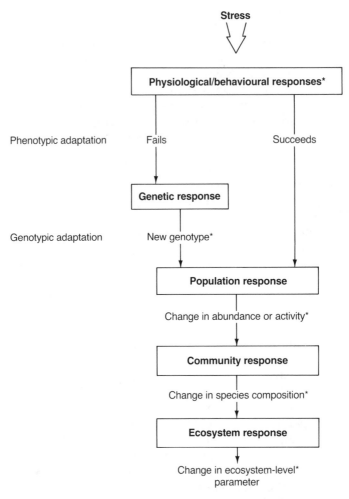

Figure 1.5 The effect of stress at different levels of biological organization. Each box represents a level at which there is some buffering capacity, reducing the impact of the stress on the level above. Asterisks denote points where this can be measured.

stress, and **perturbation**, which is a planned manipulation as part of an experiment (Bender *et al.*, 1984).

Perturbing a system while holding other variables constant is a standard method of experimentation common to all biology, from the study of enzymes to the study of ecosystems (Rosenzweig, 1987). While only a small number of natural ecosystems have been the subject of perturbation experiments, degraded habitats can be found throughout the world, providing a large variety of data on ecological processes. We can learn much from these inadvertant experiments: today we can measure the rate of evolution using plant populations that have adapted to the pollution created 150 years ago (section 3.4).

All levels of organization have some capacity to resist disturbance (Figure 1.5). For an individual, normal homeostatic mechanisms are able to accomodate small-scale variations in its environment. An organism may survive even if these fail, though at some cost to its metabolic or reproductive functions. These are all responses which we can measure (Chapters 2 and 3). At the population level, the scale of a stress may be measured by a reduction in the birth rate or an increase in the death rate. The loss of one or more species or an increase in the abundance of others then becomes a community-level response (Chapter 6). Each level can accomodate some degree of stress and we may detect no displacement of the system at one level because of the buffering of the lower levels. For example, a pollutant would stress individuals within a population, but if this has no impact on their reproductive rate, there would be no obvious change in the population or the community.

The frequency of a disturbance and the elasticity of the system will govern the sampling interval needed to describe any effect. A persistant but low level of stress may require a long sample interval to detect any response: this is part of the difficulty in establishing any ecological change associated with global atmospheric warming. Hierarchy theory suggests that stresses with a low frequency will operate on the larger spatial and temporal scales.

A high frequency of disturbance may select a community of species able to recover quickly from each stress (Figure 1.6). One example is the microbial community that becomes established in the upper layers of a trickling filter of a sewage works (section 9.2). This community is regularly washed away as it grows too large, and thus becomes dominated by rapidly growing species.

Alternatively, if disturbances are too frequent, a displaced system may have no time to recover and may then succumb to further disturbance (Figure 1.6). These and other potential interactions between the duration of a disturbance and the response time of a community can lead to unpredictable results and highly variable communities (Bender *et al.*, 1984). It also highlights again the problem of choosing a suitable sampling interval, determined both by the frequency of disturbance and the elasticity of the system.

In applied ecology, we have to draw a distinction between natural processes of environmental change and those that are artificial. Most communities we observe today are, to varying degrees, still adjusting after the ice ages of 10 000 years ago. Superimposed upon this are the changes brought about by agriculture over the last 4000 years and industrialization in the last 200 years. Today, many plant and animal

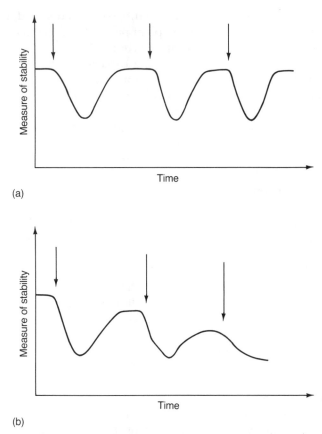

(a)

(b)

Figure 1.6 The effect of the frequency of disturbance on the recovery of some measured parameter. In (a) the system is sufficiently elastic to allow recovery before a subsequent stress. In (b) either the system's return is too slow or the frequency of disturbance is too high for recovery, and the system is gradually degraded. (After Underwood, 1989.)

populations continue to adjust to the reduction in their habitats as tropical woodlands have been destroyed in the last few decades.

1.8 THEORY AND APPLICATION

Most biologists have been taught to expect self-regulation in biological systems and most ecologists accept that natural ecosystems do not need managing. In addition, applied ecologists recognize the capacity of ecosystems for self-repair (Bradshaw, 1987a) even under the most stressed of circumstances. Although the ideas of stability and regulation await verification, they are often an implicit assumption on the part of environmental managers. Their central aim is to produce a population or community able to sustain itself, usually within fairly narrow limits.

In the absence of a general ecological theory, the concept of stability has been used as a general principle applicable to the individual, the population, the community and the ecosystem. Although their mechanisms of regulation differ, their capacity to restore themselves (whether to their original state or not) is a common biological phenomenon. In addition, the notion of adjustment stability offers some scope for quantifying these responses: we can indirectly measure the biological impact of a stress by the displacement it causes. It also allows us to compare stresses and to design experiments that properly test hypotheses about the nature of that impact.

Similarly, the levels of organization are also part of the standard equipment of most ecologists, even if they do contain logical inconsistencies (O'Neill *et al.*, 1986). It is important that the relationships between levels are not forgotten: the implications of a decision taken at one level have to be fully considered at all other levels. The functional grouping of hierarchy theory may help to clarify these relationships. Whatever school of thought they are allied to, ecologists have to be aware of the limits of their predictive powers, above or below the level at which they are working.

This requires the applied ecologist to define the objectives of any management or monitoring exercise from the outset. Ecosystems serve multiple functions and man makes use of these directly and indirectly (Table 1.3).

Both the resources we remove from an ecosystem, or the services it provides have a value and, although the marketplace can determine this for some of these functions, for others it cannot, even though their utility is unquestioned. Environmental management objectives are rarely measured in purely economic, social or in ecological terms. Decisions have to be taken in all three contexts, and the ecologist has to be able to see the whole story. Although there are established frameworks for analysing the social and economic implications of change in the environment, these are largely beyond the scope of this book. A balanced picture would require a review of social and economic theory, and a second volume. This is not to deny the significance of these factors, and the applied ecologist will frequently have to work in a legislative and political framework that tempers wholly ecological objectives. Here the aim is to clarify thinking about ecological problems using ecological theory. In the real world, the options for action may well be constrained by more mundane considerations.

More than one ecologist has suggested that our experiment with the ecology of this planet has provided for a greater understanding of ecological processes. Inadvertantly we have learnt much about adaptation and evolution, extinction, dispersion and colonization. Theoretical

Table 1.3 Three possible objectives for ecosystem management (after Westman, 1985)

Objective	Purpose	Comments

Preservation of ecosystem function	Stabilize:
	  Landform
	  Soil
	  Nutrient release
	  Water flow
	  Populations
	  Energy fixation
		Scale will depend on particular objectives, e.g. control of landslip may require large masses of vegetation

Preservation of integrity of community	Preservation of community and of single species	Stability or at least long-term persistence will depend on the size and heterogeneity of the habitat
Management of ecosystem for exploitation	Yield of resources or of an amenity or service, e.g. fishing, parks, pollution absorption, storage of water	Includes management of population and of ecosystem to maximize return

ecology has been nourished by this supply of data, even if the experimental design has rarely been perfect. Unfortunately, we cannot wait for the quality of the data to improve while this particular experiment continues.

Summary

Ecology deals with the higher levels of biological organization, including the individual, the population, the community and the ecosystem. This requires data to be collected over a range of scales. Most theoretical ecology makes use of models, effectively a sequence of linked hypotheses that can be tested by observation or experimentation. Different approaches are adopted at different levels – reductionist ecology suggest that ecological processes can be understood from a study of the component parts of systems, most especially populations. Holistic ecology suggests there are properties of ecosystems that are not apparent from considering only their components, and that ecosystems show a high degree of regulation and integration. Finally, hierarchy theory groups the components of a system into functional units, separated at different levels according to process rates. According to this view, the regulation of the system is derived from the constraining influence of the higher levels.

Stability is briefly presented as a unifying concept in ecological theory. Two types of stability are identified – inertial stability and adjustment stability, though the evidence for each at the population or community level is equivocal, due in part to the practical difficulties of measuring stability at an appropriate scale. Stress, and the use of perturbation experiments are fundamental to much biology, and can tell us much about the functioning of ecological systems.

Further reading

Cherrett, J.M. (ed.) (1989) *Ecological Concepts*, Blackwell, Oxford.
May, R.M. (ed.) (1981) *Theoretical Ecology*, Blackwell, Oxford.
McIntosh, R.P. (1985) *The Background to Ecology* Cambridge University Press, Cambridge.

The presence of particular forms of lichens has long been used to indicate levels of atmospheric pollution, particularly sulphur dioxide. This foliose species attests to the air quality of the Pyrenees.

2

Measuring the effects of pollution

A toxic pollutant can induce one of two responses in a living organism: it may have some sublethal effect, perhaps causing an impairment of growth, reproduction or metabolism. Alternatively it may, sooner or later, cause the death of the individual. Both lethal and sublethal effects can be quantified and are the most direct methods of judging the biological impact of a pollutant.

Unfortunately, the death of a single organism can tell us relatively little about the ecological significance of a pollutant. Instead we need to measure the response of a large number of individuals to establish the dose likely to produce that response in the majority of the population. Different individuals will respond at different doses, so we measure both the average response and its associated variation. The toxicity test is one obvious way of establishing these dose–response relationships. Ideally, the dose is applied in a form and at levels that the organism is likely to encounter in the environment, allowing us to assess its ecological significance.

A direct toxic effect is usually the result of some interaction between the pollutant and the biochemistry of the cell, impairing the function of an enzyme or altering some structural protein. This causes the cell to function improperly, and the individual then incurs an energetic cost, either in repairing the cell, degrading the poison or isolating it from sensitive tissues. As we shall see later on, each of these responses is used as a measure of the sublethal effects on the individual. The same methods can be extended to the population or the community; in particular the energetic cost of a toxic insult is a common currency, useful in judging the impact of a stress at each level. It also gives an insight into the effect of the pollutant on community processes and the interactions between species within a trophic system.

A characteristic of life is its capacity to accommodate or adapt to change in its environment. Even pristine, unpolluted habitats require their inhabitants to adjust to continually changing conditions. We need to distinguish between these 'background' fluctuations and a response attributable to the pollutant alone. Following some form of natural disturbance for example, many soil processes require over 30 days to recover previous levels of activity (Domsch, 1984). This is a measure of the elasticity of an unpolluted soil community (section 1.6); to be judged as having a significant effect, a pollutant has to extend this recovery period. Similarly, a depression of activity greater than 50% exceeds the normal amplitude of responses from the microbial community, again indicating an impact on the soil (Figure 1.4). With very severe impacts, the system may be unable to resume its predisturbance state and there would then be some long-term displacement of the community. This same principle can be applied to the cell, the individual or the population.

Adaptation is the process that allows organisms to adjust to a changed environment. Within a single cell or a whole organism, homeostatic mechanisms operate to maintain the relatively stable conditions which their metabolism requires. Any stress may be accommodated by a change in the physiology of the organism or avoided by an alteration of its behaviour. These are purely phenotypic responses, and are not passed on to the offspring. Over a number of generations, however, a persistent stress may select particular genotypes (section 3.1). Then differences in the survival or reproductive prospects of individuals have produced genotypic adaptation by the population, with only the fittest surviving to reproduce.

In this chapter we concentrate on the use of phenotypic responses as measures of pollutant impact and leave such genetic effects to Chapter 3. Here our aim is to survey the range of methods which have been used to assess the biological significance of pollution, moving from the individual to the community. We also survey various biochemical measures that have been used to measure responses at both these levels. Two main examples are used – intertidal benthic species and the microbial communities of the soil. These help to introduce the ideas of biological monitoring and community-level indicators of stress.

We begin with a brief survey of the nature of pollution. The emphasis here is on the biological properties of pollutants rather than any detailed discussion of their chemistry. The nature of adaptation and the allocation of energy are then used to introduce the concept of an energetic 'cost' of a pollution burden. Pollutant uptake can also be usefully compared to the rate of energy assimilation and a number of simple models of assimilation are introduced.

2.1 THE NATURE OF POLLUTION

Pollution is any increase in the concentration of matter or energy generated by human activity which degrades a living community or its abiotic environment. A pollutant may affect some aspect of the non-living environment but cause no measurable impact on the organisms living in it. Our concern here is with the biological effects of pollution and, most especially, the relationship between the scale of an impact and the concentration of a pollutant in the environment or an individual.

Both the dispersal of a pollutant and its toxic effects are determined by the nature of its emission and its physical and chemical properties. Emissions may be from a point source, such as the exhaust of a stationary car, or from a line source – a moving car. The dispersal of the aerosol created behind a car is usually very rapid, determined in part by the prevailing weather conditions and local topography. This can spread the toxic metals and gases over a wide area, reducing their impact on the roadside ecosystem. In contrast, the impact of the oil spilled from the *Exxon Valdez* in Prince William Sound was made worse by the low temperatures, calm seas and the shape of this part of the Alaskan coastline, all of which which slowed the dispersal of the slick. Pollutant emissions can be episodic – occurring in short bursts – or continuous. A low level of release may have a large impact on the community if it is sustained for a long period and the pollutant is not readily degraded or lost from the system.

A pollutant shows **persistence** if it remains in the system in a form that continues to have an impact on the biota (section 8.6). The noise from a car backfiring has effectively no persistence, whereas some elemental pollutants may continue to have an impact over many years. Radionuclides are lost through radioactive decay and their persistence will depend upon their half-lives and their residence time in the community. Other pollutants may be weathered or degraded by physical, chemical and biological processes. The hydrocarbons in an oil spill are degraded both chemically and biologically, although some fractions are more persistent than others, depending on their chemistry (Table 3.1).

Most often, a dose–response relationship is used to measure the impact of various concentrations of a pollutant on a population. The aim is usually to specify a **threshold** concentration at which some effect is observed and at which some action might be taken in the field. But such measures represent a great simplification of ecological processes and a threshold defined under one set of conditions may not apply under all circumstances. A healthy and well-fed trout may have a much higher threshold than the average fish in the wild; juveniles of most animals are commonly more susceptible to pollutants than adults. Also, the high concentrations of a pollutant in most ecosystems rarely accumulate rapidly, but arrive over a period: resident species may thus have been exposed to low doses for some time and been able to adjust to them.

These are not the only reasons why simple measures of concentrations in an individual or its habitat are often a poor indicator of pollutant impact. In many cases, the total concentration in the environment will not be the dose to which the biota has been exposed. For example, some toxic metals become isolated from living systems, reducing their toxicity: lead is readily bound by organic matter in soils, reducing its availability for uptake by both plants and animals. Lead is also commonly held in inert tissues such as the skeleton, away from sensitive tissues, so a whole body concentration does not indicate any toxic effect. Pollutants rarely occur singly and their interactions with each other can further complicate their impact. Two pollutants whose combined impact is greater than the sum of each acting independently are said to be **synergistic**: their interaction enhances their toxicities. One pollutant is **antagonistic** to another where it ameliorates the toxicity of the second. Overall these effects mean that the impact of a pollutant may bear little relation to any single measure of its environmental concentration or total body burdens.

Other determinants of a pollutant's impact are the physiology of affected species and their ecology. The response of different species of lichens to sulphur dioxide, for example, depends upon features of their morphology, physiology and the substrate to which they are attached (Hawksworth and Rose, 1976). These collectively determine both the entry of acidic solutions into their cells and their capacity to buffer the cell contents against low pH. These differences mean that some species are more susceptible to acid deposition than others. The types of lichens growing in an area have been used to map average concentrations of SO_2 in the atmosphere in a relatively well-established dose–response relationship.

2.2 ADAPTATION TO THE ENVIRONMENT

The distribution of a species is the product of its evolutionary history, its power of dispersal, its interactions with other species, and the abiotic features of its environment. Generally, we expect a species to be most abundant where conditions for its growth and reproduction are optimum. Outside this range, the biotic or abiotic conditions limit growth and reproduction, and their numbers tail off (Figure 2.1). For example, the distribution of many plants reflects environmental gradients of soil pH or moisture, and a species with growth requirements for particular nutrients will be most abundant where these are optimum. Such effects are readily observed moving up a mountainside, where groups of species are found in well-defined zones, reflecting (amongst other things) the effect of altitude on the annual temperature range.

Outside its optimum range the organism grows less well. It has fewer resources available for growth or reproduction, either because it can only secure smaller amounts of energy or nutrients or because it has to use a large proportion of these resources to maintain its internal environment.

This is a very simplistic account: the performance of a species depends upon a number of factors, biotic and abiotic, and its abundance is

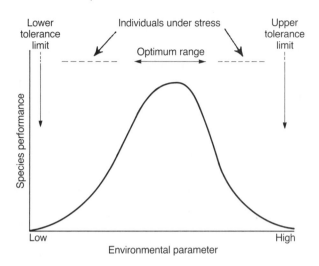

Figure 2.1 Growth, reproduction or some other measure of a species' performance will be maximized within its optimum range for some environmental parameter, such as temperature. Beyond this range, performance declines as more resources have to be directed at surviving the increasingly harsh conditions. At very high or very low temperatures no survival is possible, setting the tolerance limits for this species.

determined by a complex of such factors and their interactions. Additionally, individuals within a population will differ in their tolerance to environmental conditions, increasing the range over which the population is able to survive. This becomes important when we attempt to define the toxicity of a pollutant for a species. There is no single value for a population at which a pollutant is deemed toxic, but rather an average value and some measure of the variability between individuals.

In adapting to a changing environment, organisms may alter their behaviour, morphology, or physiology to accommodate the new conditions. The old dog lying in the sun pants to increase its heat loss, when others get up and move into the shade – both are behavioural responses. The shape of leaves on a plant may change if they grow under conditions of water shortage, while a range of physiological mechanisms attempt to keep the conditions close to the optimum for cellular processes to function efficiently. Protein structure and function are easily denatured by extremes of temperature, acidity, salinity and so on, and this is especially critical for the enzymes driving metabolic activity. More resistant forms of enzymes may be produced and, in some cases, organisms can increase their production of an enzyme or protein to detoxify or bind a poison in their tissues. This however, diverts energy and resources away from normal growth and reproduction.

With short-term stresses, such homeostatic mechanisms may be sufficient to maintain an internal equilibrium, but where these costs have to be met over an extended period, the organism may widen its tolerance limits by modifying its physiology, through **acclimation**. For example, man is able to acclimate to the lower oxygen pressures at high altitude, by increasing the number of red blood cells and by changes in heart function. These changes are reversed when returning to lower altitudes. Where there are regular and predictable cycles of change in the environment, organisms may anticipate the need to acclimatize. By using reliable cues, such as daylength, they can adjust to seasonal cycles in the weather or the diurnal cycle of exposure with the tides on a seashore. Then their activity becomes concentrated into the periods most likely to be productive.

However the scope for such adjustments is limited to relatively small changes in external conditions. Where new conditions extend over more than one generation, acclimatization may be replaced by a more permanent adaptation. **Tolerance** is an inherited, genotypic adaptation to stress. The genetic constitution of a population reflects selective pressures that were operating on its parents and ancestors, those individuals able to reproduce under the prevailing conditions. This inheritance may or may not equip the present generation to survive the environment in which they find themselves: only those who reproduce may be described as 'fit'. For this reason, we judge the fitness of any group by their capacity to produce viable offspring, usually over several generations (section 3.1). A tolerant race has a different optimum range for a particular stress compared with non-adapted members of the species and a variety of plants and animals have evolved such **ecotypes**, tolerant to all of the main classes of pollutants. Indeed, Gray (1989) has suggested that communities with a high proportion of ecotypes may be characteristic of disturbed habitats, an indication of long-term stress.

2.3 ALLOCATING ENERGY

Every organism has to apportion its resources of energy and nutrients between various functions. The details of this allocation is some indication of the condition of the organism: when resources are abundant, that remaining after metabolism can be directed toward growth and reproduction; at other times, a greater proportion would be used in maintenance, perhaps accommodating a stress.

A simple energy budget for an animal is shown in Figure 2.2. Only a small fraction of the energy consumed becomes available for biological processes (M) and the efficiency of energy assimilation varies widely between animals and their diets – between 10 and 50% for herbivores and detritivores and up to 80% for some carnivores. This partly reflects the digestibility of the food, and also its abundance – a detritivore surrounded

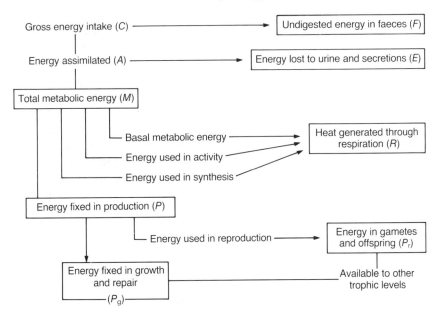

Energy used in metabolism $(M) = R + E$
Scope for growth $(P) = A - M$

Figure 2.2 An energy budget for an individual animal and the derivation of scope for growth. Note that this excludes heat received directly from the sun or the immediate environment – a major source of energy for ectotherms.

by dead and decaying matter does not need to be efficient at digesting its food: a carnivore that kills infrequently has to make better use of what is available.

All organisms have to meet basic metabolic costs for essential mainten-ance of the tissues – their **basal metabolic rate**, a minimum rate of energy consumption. This rate shows a consistent relationship to body weight in animals:

$$\text{basal metabolic rate} = a\,(\text{body weight})^{b} \qquad (2.1)$$

This equation describes the line of this relationship where a is the intercept on the y axis and b is its slope.

For the mammals, a group that ranges from the shrew to the whale, these values are largely constant: $a = 0.68$, $b = 0.75$. More significantly, the value of b is constant over a much wider range of animals. The reasons for this are not fully understood, but it is generally thought that b relates basal metabolic demand for energy to the surface area of the animal, as both are functions of body weight, oxygen absorption and energy loss through heat dissipation (Schmidt-Nielsen, 1979). Although larger animals will require a larger total energy intake, their requirement per unit body weight is actually smaller. However, surface area cannot itself be a factor directly regulating basal metabolic rate – instead it has probably been a crucial selective feature in the evolution of its control mechanism, for all animals (Gordon *et al.*, 1982).

The value of *b* remains more or less the same between endothermic and ectothermic animals. **Endotherms** generate most of their heat internally, derived from their diet, whereas **ectotherms** rely on external sources, most especially the sun. Most plants and animals are ectotherms and only the birds and mammals are wholly endothermic. All endotherms can maintain a relatively constant internal temperature and are therefore termed **homeothermic**; animals without this ability become more inactive as the ambient temperature falls, and are called **poikilothermic**. Note however, that some ectotherms are classed as homeothermic because they can regulate their body temperatures to some extent by altering their behaviour. This distinction is thus neither clear nor rigid.

Not surprisingly, endotherms have higher metabolic costs and need to consume 2–3 times as much energy to maintain their body temperature – up to 99% of the energy assimilated by a large mammal may be used in respiration alone (*R*), compared with just 60–90% in ectothermic insects. This leaves a small proportion of the energy assimilated available for growth and reproduction (Figure 2.2); as a result, a mammal can devote a smaller proportion of its energy to reproduction than an equivalent sized reptile (Colinvaux, 1986).

As we shall see below, the relationship of metabolism to body size has also been implicated in the uptake and retention of pollutants.

2.3.1 Community energetics

Only the energy fixed in the tissues can be consumed by other organisms (Figure 2.2). This fraction (*P*) of the energy received by an organism (*C*) decreases from primary producer to herbivore to carnivore, that is, with each successive trophic level. This progressive loss of energy from the system limits both the length of the food chain and the size of populations at the higher trophic levels.

The principles (and notation) used to construct an energy budget for an individual can be used in the same way for each trophic level, or even the whole community. Similarly, we can measure the efficiency of energy assimilation and transfers, and the rate at which energy is transferred from one compartment to the next. The **assimilation efficiency** for a trophic level (or an individual) is the energy assimilated (*A*) as a percentage of that consumed (*C*):

assimilation efficiency

$$= \frac{\text{energy assimilated at trophic level } n}{\text{energy ingested in trophic level } n} \times 100. \tag{2.2}$$

As we discussed above for different organisms, the assimilation efficiency varies widely between different trophic levels. This equation has its counterpart when we measure the rate at which a pollutant moves between individuals or trophic levels.

A large proportion of the energy assimilated is used in metabolism (M) and lost as heat generated by respiration (R). The remainder can be used to build tissues or produce offspring. The **production efficiency** is the energy fixed in tissues (P = growth + reproduction) as a proportion of that assimilated (A) by that trophic level:

production efficiency
$$= \frac{\text{energy fixed in the tissues of trophic level } n}{\text{energy assimilated by trophic level } n} \times 100. \qquad (2.3)$$

Again, the same equation is valid for an individual or a community. This efficiency also varies greatly with both the trophic level and with the type of organism. Some insect carnivores can achieve values of over 55% but for many species, the production efficiency is much lower, especially amongst endotherms. In contrast, microorganisms, important decomposers in all ecosystems, have very high production efficiencies, probably due to their short lifespan and rapid population growth (Begon *et al.*, 1990). Production efficiencies often drop in organisms and communities subject to some form of stress.

Energy is lost in its transfer from one trophic level to the next. The rate at which energy is passed between levels is measured as the **exploitation efficiency**:

exploitation efficiency
$$= \frac{\text{energy ingested in trophic level } n + 1}{\text{energy fixed in trophic level } n} \times 100 \qquad (2.4)$$

where trophic level $n + 1$ is feeding on the lower trophic level, n.

The energetics and productivity of the whole ecosystem can be assessed in a similar way. For very complex communities, such as the soil, few attempts are made to describe the impact of a pollutant for individual species or populations: instead, the energetics of the whole community are used to gauge the impact of a stress.

These simple equations of energy transfer can be used in a comparable way to measure the assimilation and mobility of pollutants. In addition, a reduction in production efficiency is one of a range of responses that has been used to assess the impact of a pollutant on individuals.

Others include changes in behaviour, morphology, reproductive success, developmental biology and competitive ability. Each may have implications for the larger community and can therefore lead to significant impacts at the population and community levels. Sheehan (1984) gives an extensive review of the methods used to assess community and ecosystem change under pollution insult.

2.4 THE ACCUMULATION OF POLLUTANTS IN INDIVIDUALS

Bioaccumulation is a general term used to describe the retention of a pollutant in any living component of a community. Estimating the rate at which a pollutant is retained by successive trophic levels is important for assessing the hazard it represents for the higher trophic levels of a food chain.

A number of terms have been used to describe pollutant accumulation in organisms, often with little consistency. Here bioaccumulation is divided into two components according to the route of entry into the tissues. The **concentration factor (CF)** measures the proportion assimilated from the diet:

$$CF = \frac{\text{concentration of pollutant in organism}}{\text{concentration of pollutant in the diet}}. \tag{2.5}$$

This can represent the accumulation in a single individual or a trophic level and is equivalent to the assimilation efficiency described earlier for energy (equation 2.2).

Pollutants may also be assimilated across the body wall or respiratory surfaces, that is, from non-dietary sources. The **accumulation factor (AF)** allows for this, along with dietary uptake:

$$AF = \frac{\text{concentration of pollutant in organism}}{\text{concentration in surrounding water or air}}. \tag{2.6}$$

A pollutant may be assimilated by this additional route through passive uptake, say from a fine aerosol entering the lungs or by adsorption on the surface of gills.

A simple model can be used to demonstrate the movement of a pollutant into the tissues under passive uptake, and the balance achieved when the rate of input is matched by the rate of loss (Figure 2.3). This is most easily conceived for an aquatic animal living in contaminated water, when passive uptake may take place across the body wall, across the respiratory surfaces, or by assimilation from the gut. The balance between the concentration in the water and in the animal will depend upon the affinity of the pollutant for the tissues.

At equilibrium, the rate of loss is balanced by the rate of uptake and the accumulation factor is the same as the balance between the two (Figure 2.3a):

$$\frac{q}{W} = \frac{k_1}{k_0}. \tag{2.7}$$

(a) Single compartment model

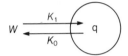

(b) Three compartment model

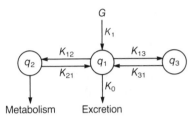

Metabolism   Excretion

Figure 2.3 Simple compartment models for the uptake and loss of pollutants by either plants or animals. (a) Single compartment model assuming an aquatic animal: W = concentration of the pollutant in the water; q = concentration of the pollutant in the whole animal; K_1 = rate constant for pollutant uptake; K_0 = rate constant for pollutant loss. (b) Assumes a terrestrial animal assimilating the pollutant from its gut and partitioning this between three compartments: q_1 = blood or haemolymph that interfaces with the sources of the pollutant (gut wall or respiratory surface); q_2 = liver or digestive gland, the principal site of metabolism; q_3 = other compartment(s), including storage; G = the concentration of the pollutant at the interface. Rate constants are shown for the movement of the pollutants between the various compartments. In both models, rate constants are taken to be simple proportions of the concentration of the pollutant in each compartment. More elaborate models are necessary if these rates vary. In practice, the rates may be derived from experimental observations, using regression analysis to find lines of best fit. (After Moriarty, 1988; Walker, 1990.)

This represents the **partition coefficient** between the animal and the water, for that pollutant. Some have a high affinity for fats and oils and tend not to be soluble in a polar solvent like water. These 'lipophilic' pollutants include a variety of hydrocarbons and organic pesticides with a high partition coefficient, indicating their affinity for lipid-rich tissues.

These same compartment models can be used to estimate the concentration factors for pollutants by different members of a community. Janssen *et al.* (1991) fitted one-compartment models to their experimental data on cadmium assimilation by four species of soil invertebrates. Two arthropods (a springtail insect and an oribatid mite) fed on contaminated algae had a much lower CF (0.09–0.17) than the two carnivores (a carabid beetle and a pseudoscorpion) which fed, in turn, on the springtail (0.35–0.59). From the model it appears that the springtail, *Orchesella cincta*, maintains a low body burden by a low assimilation rate and a high excretion rate. Janssen and his co-workers suggest that in this species the equilibrium concentration of cadmium is probably more influenced by the rate of excretion than by the rate of assimilation.

Using the same principle, more detailed models can be developed by dividing the organism into two or three compartments. These apply particularly to terrestrial animals where the gut is the main route of uptake, and transfers are between this, the blood and some site of storage (Figure 2.3b). Different transfer rates can be calculated for each compartment, giving a detailed description of the movement of the pollutant through the organism. More recent methods allow for the variability in the rates of uptake and loss, and transfer between compartments. These stochastic models give a range of possible values for the level of pollutant in each compartment and also its mean residence time (Matis *et al.*, 1991). Further elaboration allows the rate constants to be varied with seasonal and physiological changes in assimilation. Similarly, the models may include the metabolic degradation of organic pollutants within the tissues or the decay of different radionuclides with their various half-lives.

In principle, multicompartment models could be developed to describe the trophic mobility of pollutants through a community, but there are a number of complicating factors, most especially defining the trophic interactions between species. One difficulty is assigning some species to a single trophic level: an omnivore, for example, may feed on several trophic levels within a food chain. Similarly, the rates of transfer of pollutants from one trophic level to the next have to allow for losses through dispersal and tissues not consumed (Walker, 1990). Some pollutants, particularly persistent and lipophilic ones, may achieve higher concentrations in successive trophic levels, though accumulation in the upper trophic levels may simply be due to these animals tending to be larger. The question of the trophic mobility of pollutants is discussed more completely in section 8.6.

2.4.1 Factors governing pollutant assimilation

Both the chemistry of a pollutant and the biology of the organism determine rates of assimilation and loss. For example, the uptake of toxic metals by bivalve molluscs may be linked to their to metabolic rate. Boyden (1974) was able to relate total metal content to body weight for six marine species by an equation equivalent to that for basal metabolism (equation 2.1). Many metals show a relationship of content (µg) to body weight with a *b* value close to 0.77 – this was found for copper in all six species. Boyden suggested that uptake may thus be some function of metabolic activity in these molluscs. He also acknowledged an alternative possibility, that it was due to some other property linked to body size, such as gill area. In some species, the content of particular metals can be related directly to body weight ($b = 1$). Boyden attributed this to a simple relationship between body weight and binding sites for the metals, though Fagerstrom (1977) argues that this may be more indicative of whole-body metabolism, at least where the animal is in equilibrium with its surroundings. Either way, this illustrates how pollutant assimilation may be closely linked to body size, reflecting not only physiological processes but perhaps a range of other factors which change with body size (Newman and Heagler, 1991).

The pattern of toxic metal assimilation is also determined by homeostatic mechanisms for regulating essential metal uptake. Over a range of animals, zinc, cadmium, lead and other elements are bound in concretions or granules within particular cells, a device that helps regulate some essential metals such as calcium, iron and copper (Hopkin, 1989). Additional sites of storage can include the skeleton, the exoskeleton or the shell: the route followed by a toxic metal depends upon its coordination chemistry and the existing pathways for essential metal uptake and loss.

The uptake of organic chemicals (and some metal compounds) depends upon their relative affinity for either water or non-polar (lipophilic) solvents. A good estimate of the rate of uptake and the toxic effect of organic compounds can be derived from their partition co-efficient (equation 2.7). In practice, this is measured as the proportion of a compound passing into a non-polar solvent (usually n-octanol) from water. This term, log K_{OW}, shows a close relationship to the rate at which the pollutant is accumulated in the tissues (Figure 2.4), a relationship that appears to hold particularly well for molluscs such as *Mytilus*. This is perhaps because of the limited capacity of the mussel to metabolize the more intractable organic pollutants (Walker, 1990).

Generally the more lipophilic pollutants also have higher toxicities. Lipophilic compounds exert an effect by simply occupying a volume in the cell or in its lipid membrane, altering the permeability of the membrane. The toxicity is largely a function of this displacement rather than the chemical reactivity of the pollutant and this is thought to account for the general narcotic effect of anaesthetics: the simple relationship of K_{OW} to the toxicity of such pollutants in water suggests that their effect is indeed non-specific. For the same reason, two organic compounds together in the cell are likely to have an additive effect, rather than show any synergism or antagonism (Donkin and Widdows, 1986).

Variations from this pattern may be explained by features of the chemistry of the pollutant, using quantitative structure–activity relation-ships (**QSARs**). This is a simple empirical approach which plots the

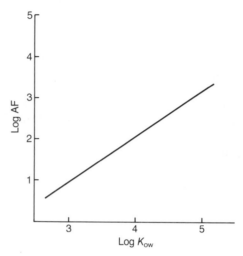

Figure 2.4 The relationship between the log n-octanol partition coefficient (log K_{OW}) of lipophilic organic compounds and the rate at which they are accumulated in the tissues of *Mytilus edulis* (after Donkin *et al.*, 1988, cited by Bayne, 1989).

toxicity of a compound against log K_{OW} and then fits the best line. Different relationships are found for different classes of organic chemicals. Deviations from the K_{OW} toxicity value can be related to the number and position of functional groups on the molecule, or the degree of halogenation (Veith *et al.* 1983). For example, compounds that are easily metabolized will have a lower toxicity than that predicted by their partition coefficent, whereas the larger and more hydrophobic molecules will be accumulated. Toxicity begins to decline at very high molecular weights, possibly because compounds with several aromatic rings are not readily absorbed.

QSARs are a useful method for predicting the toxic effect of complex organic pollutants based on their chemistry alone and, along with the partition coefficient, gives some indication of how the compound is held in the tissues. It also indicates their capacity to accumulate upwards in food chains (section 8.6).

2.5 MEASURING TOXICITY

What is a toxic dose for one individual may be harmless for another. These variations within a population mean that we have to measure the average toxicity of a pollutant and the likelihood of any one individual succumbing. Equally, the different susceptibilities between species make it impossible to define a toxic threshold for the whole community.

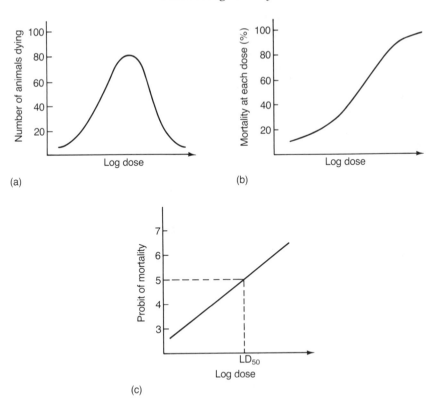

Figure 2.5 The derivation of an LD_{50} – the dose that is lethal to 50% of the population. A number of animals are exposed to the pollutant over a range of concentrations, usually expressed as log dose for convenience. If this range is properly selected, a more or less normal distribution of responses will occur (a), with few animals responding at very low or very high doses. Most die at some intermediate level. If this data is redrawn by plotting the proportion of individuals at each dose (the percentage response at each dose), a sigmoidal dose–response curve results (b). This shape is characteristic of the normal distribution. The final step is to straighten this line. This can be done by expressing the deviation from the mean as a probability unit or **probit**. A probit scale scores each successive standard deviation from the mean as one. The mean is set to 5, so 1 SD above the mean is 6 and 2 SD below the mean is 3. The score for the percent response for each dose is read from tables of probit values. The LD_{50} is simply read off the line; the rate of response (the slope of line) can be compared between treatments or between species.

The **dose** of a poison is the product of its concentration and the duration of its exposure to an organism. A dose is described as **acute** if a large concentration is administered in a short time, whereas a **chronic** dose is a small concentration applied over a longer period. Both an acute and a chronic dose can ultimately lead to the same exposure albeit over different periods. An **acute toxic effect** is a rapid response to the poison often, but not always, leading to death. **Chronic toxic effects** are those that appear over a prolonged period of exposure. This might eventually lead to death, but also includes sublethal effects, such as a reduction in the reproductive potential of a species, changes in behaviour, physiology or morphology (Sheehan, 1984).

Most of these chronic effects are termed **graded responses** because the response can take a range of values. A variety of changes can be measured in relation to dose – the length of time mussels close their shells, the ventilation rate of fish, the incidence of skeletal abnormalities in developing embryoes, or the reduction in their growth rate. When it is necessary to specify a toxic threshold, it may be more appropriate to measure a **quantal response**, which the organism either does or does not show. The most obvious is death, but others include paralysis, the presence of tissue damage, or the absence of a heartbeat. In much environmental toxicology, quantal responses are measured in an acute test, the **LD_{50}** toxicity test. This measures the dose at which 50% of the population are killed within a specified period.

The basic principles of an LD_{50} test are described in Figure 2.5. A range of responses to a poison will be found within a population and, as with many other environmental parameters, this is likely to follow a normal distribution (Figure 2.1). This means that a small number of individuals die at very low concentrations and only a few survive at very high concentrations. The majority die at over a range of intermediate doses (Figure 2.5a). An appropriate range of concentrations has to be found that will give a normal distribution, and a plot of percentage mortality will then follow a characteristic sigmoidal curve (Figure 2.5b). Several methods are available to transform this to a straight line, enabling its slope (the rate of response to increasing dose) to be easily calculated. This also simplifies the calculation of confidence limits and a standard error term for the LD_{50} (Wardlaw, 1985). The dose at which 50% of the population are killed can be simply read off the line. Typically an environmental standard is based on the LD_{50} by applying a safety margin, simply multiplying the value by some factor, such as 0.1.

This value will only be good for the conditions of that test. The result should specify the duration of the test – for example, a 96-hour LD_{50} for pyrethrum upon adult *Daphnia pulex*. There are other qualifications (Table 2.1) that result from the simplification imposed by the test conditions. Some of these problems can be addressed by standardization and comparison with a standard toxicant (such as sodium dodecyl sulphate) which has well-established toxic thresholds.

The LD_{50} method was originally devised to measure the toxicity of a range of drugs for human consumption at a time when no effective assay

Table 2.1 The major limitations of the LD$_{50}$ toxicity test

- Based upon a limited period of exposure
- Based upon a single species
- A single test can only measure the impact on one sex or one stage of the life cycle
- Makes no allowance for the effect of a toxic burden on interactions with other species
- Makes no allowance for synergistic or antagonistic interactions with other pollutants
- The test cannot easily measure the impact on the reproductive potential of a species
- The test is insensitive to chronic toxic effects that may subsequently affect survival in the long term
- As a method of assessing the ecological impact of a pollutant, the test relies upon the correct choice of a sensitive species

Qualifications to the term LD$_{50}$

LD$_{50}$ = lethal median dose – assumes all of dose is assimilated
LC$_{50}$ = lethal median concentration – the concentration present in the surrounding medium causing 50% mortality
LT$_{50}$ = median lethal time – the time taken to kill half the population at a particular level of exposure
ED$_{50}$ = median ecological dose – the environmental level of exposure causing 50% mortality
EC$_{50}$ = median effective concentration – the concentration that depresses some function, such as growth or respiration, by 50%
TL$_{m}$ = median tolerance limit – the concentration where 50% survival is observed
NEC = no-effect concentration – the highest concentration at which a known effect of a pollutant is not observed

was available. Even though it is widely used to assess the environmental toxicity of many pollutants, it has been extensively criticized for being too simplistic (Moriarty, 1988). The procedure can rarely mimic natural conditions of exposure, or properly account for all factors which might affect the impact of the pollutant. In practice the test conditions are a compromise between what is desirable and what is feasible. For example, easily handled adult animals are usually tested, even though younger forms are often more sensitive: the growth rate of larval crabs is the most sensitive indicator of zinc and copper toxicity to these crustaceans (Thorpe and Costlow, 1989), but these rates are not easily or rapidly measured.

The methodology of the test is also crucial to its validity. Pascoe (1983) describes the range of factors that have to be considered to provide viable results. Not only does the organism need to be standardized for size, sex and health, it also has to be acclimated to the test conditions before exposure to the pollutant. Exposure may be of the pulsed or continuous form, by providing a single dose at the beginning of the test, or a continuous exposure using a flow-through system. The test will also specify particular conditions of temperature, alkalinity or other significant factors which have been held constant. In the real world, pollutants occurring together may have antagonistic or synergistic effects on each other and to fully quantify all possible combinations may require a large number of tests. Measuring the LD_{50} of an effluent or river water directly may be a partial answer although this may not indicate the causative agent. In practice a limited number of trials are performed on the more sensitive species in a community and these are then used to set a standard, or threshold, which triggers remedial action. Even so, this threshold gives little insight into community-level processes (section 8.5).

The nature of the response may change with time. Haanstra *et al.* (1985) described dose–response curves for soil respiration, a graded response to various concentrations of nickel over different periods of exposure. At 19 days, the dose-response was linear and only became sigmoidal after 84 days. Using the latter curve, they derived an ED_{50} – the ecological dose at which microbial respiration is depressed by 50% compared with unstressed soils. They also measured the 'ecological dose range' (EDR), the concentrations at which soil respiration was depressed down to 90% and 10% of the undisturbed activity, allowing the rate of decline in respiration to be calculated. The 84-day ED_{50} for nickel in the soil was $450 \, mg.kg^{-1}$, with an EDR of 3090 and $65 \, mg.kg^{-1}$. Notably, both statistics changed with the duration of the test and Haanstra *et al.* suggest that the EDR is preferable to the ED_{50} because it gives a distinct response to changes in the pollutant concentration. Thus the response is more likely to be due to the pollutant rather than with some time-dependent effect, such as a gradual decline in soil nutrients, that would confound a simple toxicity test.

2.6 BIOLOGICAL MONITORING

Simple physical or chemical determinations can be used to measure the rates of pollutant input, distribution and dispersal in the environment, and their assimilation into living tissues. But the total concentration measured in an individual can easily overestimate its biological significance: high levels of surface contamination, or the binding of the pollutant at inert sites may mean that the effective dose is much lower. Biological monitoring aims to assess the significance of a pollutant for an organism in its habitat and for other members of its community.

There are two basic approaches to measuring this impact – the use of monitor and indicator species (Martin and Coughtrey, 1982). Monitor

Table 2.2 Preferred characteristics of species used in biological monitoring programmes

---

- Abundant – allows large samples to be repeatedly taken without significantly changing the population characteristics of the species
- Widely distributed – a single species can then be used to monitor a large area and to compare regions
- Long-lived – can integrate the pollutant over more than one year
- Size – the organism is large enough to provide sufficient tissue for analysis
- Easy to identify and to collect at all ages, throughout the year
- Easy to age, isolating this source of variability

According to the aim of the sampling programme

Sentinel species	**Monitor species**
• Sedentary – measures pollutant at one point only	• Sedentary
• Achieves rapid equilibrium with environmental levels	• Body concentrations reflect pollutant availability to that part of the community
Bioavailability	
• Simple correlation between tissue and environmental levels	• Easy to culture in laboratory to establish nature of toxic effect
• Method of accumulation should measure main source of pollutant, into ecosystem (e.g. aerial deposition)	• An established body of knowledge on the biology of the species
	• Sensitive indicator of likely impact on the rest of the community

---

species are organisms whose ability to accumulate pollutants is used to assess the scale and distribution of a pollution insult. They are generally insensitive or tolerant of the stress. In contrast, **indicator species** are susceptible to the pollutant and their presence or absence is taken to indicate a significant level of contamination.

A monitor species may be used to simply map the distribution and scale of a pollution problem, perhaps as an aid to establishing emission regulations. When the principal concern is not the biological impact of the pollutant, the organism is being used to integrate the pollution 'signal', and is more properly referred to as a **sentinel species**. Our choice of sentinel species is largely determined by their reliability as accumulators of a pollutant, ideally showing a constant (or at least well-defined) relationship between tissue concentrations and ambient levels. An effective sentinel species can increase the sensitivity of our detection methods for particular pollutants: various metal and organic pollutants occur at very low concentrations in both air and water and some species can accumulate the pollutant to measurable levels, accumulating the pollutant over a long period of time. Lichens have been used in this way to map atmospheric flouride distribution. In some cases more than one species may be needed to fully map a pollutant in all components of an ecosystem.

Where the specific aim is to assess the biological impact of a pollutant, we may again sample several species. A pollutant can have several routes of entry into the biota, and become concentrated in different parts of the community. Additionally, its impact may extend throughout the community, by changing competitive or symbiotic relations between species. A monitor species is chosen to represent the major impacts upon a community, either by virtue of its ecology or because of its essential role in that community.

The desirable features of sentinel and monitor species are summarized in Table 2.2. Here we concentrate especially on the use of marine molluscs to monitor metal pollution, although many of the principles can be applied to other organisms used in other environments.

2.6.1 Sentinel species

The ideal sentinel species achieves a rapid equilibrium between the concentration of pollutant in its tissues and that of its habitat; in addition, the levels in the tissues are always a simple reflection of ambient concentrations. This implies that the rate of uptake will be uniform over a range of environmental conditions and between different physiological states of the organism. In contrast, an organism able to regulate its pollutant burden will produce results that are not easily interpreted: for example, the mussel (*Mytilus edulis*), a common bivalve mollusc, is not favoured as a sentinel species for either copper or zinc because of its ability to regulate both metals, although this capacity ceases at high concentrations (Bryan *et al.*, 1985).

Early work on marine molluscs suggested they may be useful as sentinel species for toxic metals: their total body burden appeared to be a function of body weight, perhaps of basal metabolic rate. This is now known to be an oversimplification (section 2.4). Metal uptake in many molluscs is affected by salinity, reproductive activity, interactions with other metals and a range of other factors (Phillips, 1980). Cadmium uptake by marine molluscs, for example, declines with salinity due to the formation of chloride complexes (Bryan *et al.*, 1985). Some of these effects are summarized in Table 2.3.

No one species of marine organism, plant or animal, has been found to be useful as a universal sentinel species for all toxic metals. Bryan *et al.* (1985) recommend that different estuarine species are used according to the metal and its source (Table 2.4).

Table 2.3 Factors affecting metal accumulation in marine organisms

---

Physical/chemical factors
Physical form of metal and its coordination chemistry
Interactions with other metals
Presence of organic compounds chelating metals
Salinity
pH
Temperature/season

Features of organism
Age
Sex
Metabolic rate/body weight
Feeding habits and exposure to non-dietary sources
Location
Essential metal requirements
Sites of metal storage in tissues e.g. shell, metallothioneins
Genetic factors

Sources of stress affecting accumulation
Reproductive activity
Health (condition)
Food quality and quantity
Demand for essential metals
Presence of other pollutants

Table 2.4 A range of sentinel species recommended for monitoring three metal pollutants in estuarine habitats (from Bryan *et al.*, 1985)

	Source of metal uptake		
Metal	*Dissolved in water*	*Dissolved in water and particulates*	*Sediments only*
Cd	*Ascophyllum nodosum*[a] *Fucus vesiculosus*[a] *Littorina littoralis*[c] *L. littorea*[c] *Patella vulgata*[b] *Nucella lapillus*[b]	*Mytilus edulis*[b]	*Scorbicularia plana*[b]
Cu	*A. nodosum* *F. vesiculosus* *P. vulgata* *N. lapillus*	*Ostrea edulis*[b]	*Nereis diversicolor*[d] *Nephthys hombergi*[d]
Pb	*A. nodosum*	*M. edulis*	*S. plana*

[a] Algae; [b] mollusc; [c] crustacean; [d] annelid.

The capacity to achieve a rapid steady state between tissue and environmental concentrations is an important feature of a sentinel species, especially if organisms are to be transplanted from uncontaminated areas to measure concentrations elsewhere. A species that takes a long time to adjust to new ambient levels will give misleading values. The deposit-feeding clam *Scrobicularia plana* will equilibrate to copper within two months, but may take up to a year for lead and cadmium (Bryan *et al.*, 1985). Equilibration is usually faster at higher temperatures, and where there is only a small difference between tissue and source concentrations. In reality, few species will actually reach a steady state with their environment, as both the level of the pollutant in the environment and its rates of emission will be fluctuating with a higher frequency than the tissues of the organism. In effect, the tissue concentrations will represent a moving average of recent ambient concentrations, rather than any single, immediate value.

In the North Sea, mussels have been used to measure the distribution of toxic metals in estuaries and the open waters, using transplanted beds (Borchardt *et al.*, 1988). *Mytilus* was found to accumulate high levels of cadmium and lead around the Jade region of northern Germany, associated with a local lead smelting industry (Figure 2.6). More importantly, the mussels transplanted to the open sea accumulated lead and cadmium, confirming the widespread availability of these metals in these waters. This is due to a high proportion of dissolved metals, to re-mobilization from the sediments under the oxygen-poor conditions of the summer, and a current bringing pollution from the UK.

2.6.2 Monitor species

Monitor species are used to detect biologically significant pollution effects, especially where it is important to establish the proportion of a pollutant available for assimilation. The seaweed *Ascophyllum nodusum* has been favoured for monitoring soluble metals because it is less prone to

---

Figure 2.6 The distribution of six toxic metals in the southeastern North Sea, using the mussel *Mytilus edulis* as a sentinel species. The plots for the open sea represent transplanted mussels retained in suspended bags. Two plots are shown for each site – a simple mean and also a normalized mean (based on multiple regression analysis) which allows for differences in the size of the mussels, times when they were sampled and similar variables. Thus, these normalized values allow more direct comparison between sites. Not only do these sentinel species detect the main metal sources around the Jade estuary, but also the significant pollution in the open sea, previously not suspected. (With permission, from Borchardt *et al.*, 1988.)

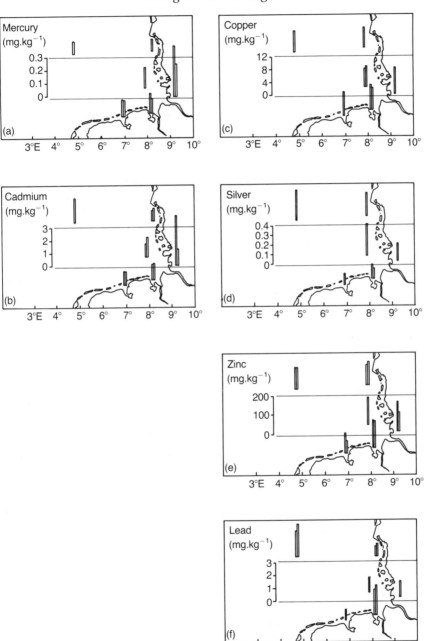

particulate contamination of its surface (Bryan *et al.*, 1985). As a primary producer, *Ascophyllum* thus measures one route of entry into a food chain and the concentrations of metals in the diet of its herbivores. From this we could derive their concentration factor, although we would have to include all other possible routes of uptake to calculate the accumulation factor.

The different propensities for uptake between species, and their exposure to different sources means that a number of species are required to monitor the whole community. Cadmium uptake by *Mytilus edulis*, for example, is accelerated by the metal forming a complex with organic acids, especially the humic acids associated with river and estuarine waters. This could be of major significance for filter feeders (Phillips, 1980). In contrast, copper uptake by the oyster *Crassostrea virginica* is reduced by binding with a range of organic complexing agents. Copper is assimilated mainly as cupric ions, and even where high total copper concentrations are measured, complexation can reduce the biologically available component for this species to very low levels.

The movement of metallic pollutants into the tissues may well follow routes of essential metal metabolism. Some toxic metals may act as analogues for an essential metal, shadowing its uptake and loss. This has been proposed as a method of allowing for many sources of variability, using the homeostatic regulation of the essential metal. Simkiss and Taylor (1981) suggested expressing the contaminant as a proportion of the essential metal concentration, to give a more realistic measure of its biological availability. The regulation of the essential metal should keep this ratio constant – the ratio should vary only if environmental concentrations of the toxic metal change. For example, cadmium will bind to zinc-binding proteins and although zinc and cadmium uptake by *Mytilus* are both affected by variations in the salinity at different sites, the ratio of the bound metals remains virtually unchanged. Polluted sites can then be identified, irrespective of their salinity, where the ratio of cadmium to zinc is significantly elevated. The same technique may be applied to particular tissues, such as the digestive gland, kidney or shell, to improve sensitivity (Simkiss and Taylor, 1981) (Figure 2.7).

Interactions between metals can also alter their availability. A good example is the work of Luoma and Bryan (cited in Bryan *et al.*, 1985). In a survey of the clam *Scrobicularia plana* from a number of estuaries, the total lead content of the sediments upon which it feeds was a poor indicator of the soft tissue concentration. Instead the lead to iron ratio was found to be a better predictor, probably because iron oxyhydroxides in the sediments were significant binding sites for lead. This also illustrates the importance of sources of the pollutant for different members of the community. Only by reviewing several species can we gain a complete picture of the impact of a pollutant on a community. For example, Phillips and Rainbow (1988), compared the relative availabilities of several metal pollutants to four species of barnacle and one species of mussel in Victoria Harbour, Hong Kong. These are all sedentary filter feeders, yet there were major

Figure 2.7 Molluscs have been used in both terrestrial and marine habitats as biological monitors for a range of metallic pollutants. The shells of the common garden snail, *Helix aspersa*, collected over the last century have also been used as an archive of historical metal pollution in Sweden (Gardenfors *et al.*, 1988).

differences between the crustaceans and the mollusc, especially in their cadmium uptake. In this case, a single monitor species would give a misleading impression of the impact of cadmium on the whole biota, or its availability to other parts of this food chain.

Even so, single-species monitor programmes are widely used in this way. *Mytilus edulis* has been the subject of a global monitoring effort for toxic metals and hydrocarbons, sponsored by the United Nations Environment Programme, the 'Mussel Watch' (Bayne, 1989). One major advantage of this species is the large amount of information available on its response to various pollutants. Indeed, because of its global distribution and abundance, its sedentary nature and longevity, it can claim to be the marine animal closest to the ideal sentinel species, at least for some metals and for many hydrocarbons (Table 2.2).

2.6.3 Indicator species

The idea that some species can be used as indicators of the quality of the environment is older than the formal science of ecology. Plant types were used to classify rangelands in the US Stock Raising Act of 1916 and lichens were suggested to be indicators of Parisian air quality in the middle of the nineteenth century. Using the presence or absence of sensitive species to indicate habitat change has been most widely applied to aquatic and, in particular, freshwater ecosystems. At the turn of the century, the impact of sewage discharges on rivers was measured using indicator species that responded to nutrient enrichment and oxygen depletion (Moriarty, 1988). A range of freshwater *biotic indices* have since developed from this first 'saprobity index' (Washington, 1984). These commonly use a number of indicator species, usually invertebrates, to assess the scale of a stress. Few of these indices are universally applicable; those currently used to assess water quality in European rivers are reviewed by Metcalfe (1989).

The principle behind these methods is certainly universal. A stress causes changes in the number or relative abundance of species in a community, which the index measures, often against some scale of community structure or composition. The ideal index shows a simple response with an increasing pollution insult. Some species will be lost following a disturbance simply because they are low in number, but others will disappear because of a sensitivity to the pollutant. Any loss of species creates space and resources for other species, and most communities have a number of 'opportunist' species that will rapidly invade and dominate where disturbance is frequent (Figure 2.8). The marine polychaete *Capitella capitata* is a good example, flourishing in sediments enriched with organic matter. It shares several characteristics with other opportunist species – small organisms with short generation times and a high reproductive capacity. These are sometimes termed *r*-selected species (section 5.2). Larger species are less abundant and more readily lost from stressed habitats, if only by chance (section 4.3). These will thus be unsuitable as indicator species; for this reason, Gray (1989) argues that moderately abundant species are perhaps the most appropriate indicator species.

Again, these effects have to be distinguished from the continual variations of an undisturbed community. Many of these methods suffer from our lack of understanding of the factors governing community composition as well as a shortage of baseline data against which to judge an impact. This is especially so in marine habitats. The density of *Capitella*, for example, is known to respond to the stability of its mudflat habitat, rather than just the level of organic pollution. One alternative is to look at a range of species. Platt *et al.* (1984) suggested that changes in the dominance of benthic species in intertidal habitats could be used to detect disturbance by pollutants. This idea has been developed by Warwick *et al.* (1987) who compared the numerical distribution of individuals amongst species with the division of biomass between them. This is termed the abundance/biomass comparison or **ABC method**.

Under some form of stress, a community is expected to lose its larger species, with smaller-sized species becoming numerically more dominant. The average weight of individuals in the community thus declines.

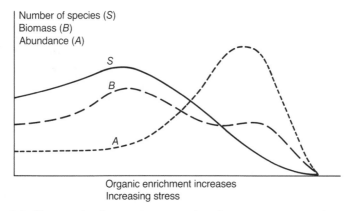

Figure 2.8 Changes in the marine macrobenthic community with increasing levels of organic enrichment. The number of species initially rises (as does the total biomass) but, as the larger, less tolerant species disappear, biomass then declines. The abundance of the more opportunistic species will rise as they exploit the available resources, allowing a second peak in biomass. Eventually, these species also succumb to very high levels of pollution. (After Pearson and Rosenberg, 1978, cited in Gray, 1989.)

A stressed community would then show a more rapid increase in the abundance of species compared with the increase in their biomass (Figure 2.9). A transect of samples across a sewage dumping site off the Ayrshire coast does indeed show this switch between plots of numbers and biomass. Applying the ABC method to the enriched sediments beneath fish farm cages indicated that it could detect such community changes in the space of a few weeks (Ritz *et al.*, 1989).

Other ecologists however, using more extensive data sets, have cast some doubt on the ability of the ABC method to distinguish polluted from uncontaminated sites. Beukema (1988) has shown the method to be extremely sensitive to the presence and abundance of just one or two species. Using 13 years of data from the tidal mudflats of the Wadden Sea in the Netherlands, Beukema found that the ABC plots often depended on the status of the gastropod *Hydrobia ulvae* and the crustacean *Corophium volutator*. These are small, rapidly invading species and, like true opportunists, their numbers fluctuated widely even in unpolluted and relatively unstressed habitats. Such sites would thus be scored as polluted by the ABC method. Additionally, individual sites were not scored consistently over the extended sampling period and the method could not pick up a possible trend of organic enrichment during this time. Meire and Dereu (1990) also show that the ABC method scores unpolluted estuaries with long periods of exposure at low tide as polluted, with the outcome dominated by the abundance of just one species,

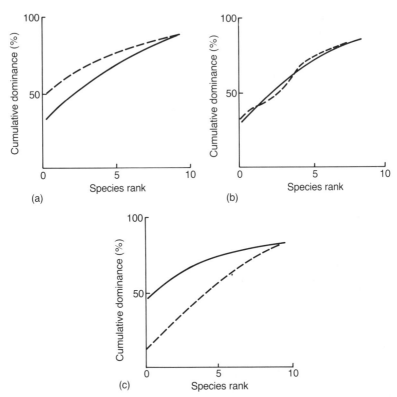

Figure 2.9 Generalized plots of species abundance (continuous line) and biomass (broken line) as cumulative percentages against a ranking of species, where the most abundant species is scored 1. These plots are used to detect areas of organic pollution in marine benthic communities, using the ABC method. Three of these 'K-dominance' curves are shown: (a) shows the pattern in an unstressed system dominated by larger organisms. Then the cumulative biomass curve is always above the plot of species abundance. In a moderately stressed system (b) and a highly stressed system (c) these curves are reversed – the loss of the larger species and the dominance of small-bodied organisms means that cumulative species abundance rises faster than the plot of biomass.

Hydrobia ulvae. Although the method may be a useful indicator of general stress, effective baseline comparisons are needed for it to be a reliable measure of pollution stress alone. These would need to include the natural frequency of disturbance (Ritz *et al.*, 1989).

2.7 MEASURING TOXIC EFFECTS IN SOIL COMMUNITIES

One central problem in toxicity testing is measuring the response of the community to some impact. This is especially true in ecosystems that are difficult to investigate and are poorly described. The soil has a complex community, not readily partitioned into easily counted populations. Thus most studies of toxic stress in the soil have measured some community-level or ecosystem-level parameter (section 8.3). Very direct methods have been used, often using soils *in situ* – in one example, total bacterial numbers showed a sigmoidal response to metal concentrations in the soils around a brass foundary, since 'dose' declines with distance from the smelter (Nordgren *et al.*, 1986).

We know microorganisms largely by their actions, by the substrates they attack, and we measure their response to stress by changes in their activity. These include rates of substrate use, decomposition or respiration, each, in effect, measuring the energetic balance of the community. Similarly, assays of soil enzymic activity have also been used extensively to assess impact. In the next section we also review methods of estimating the size of the soil microbial community by measuring its biomass.

Distinguishing the impact of a pollutant on a soil requires baseline measurements of the frequency and amplitude of natural variations. This means measuring the variation of some soil process over the range of temperature and dehydration that the soil might ordinarily experience, to provide a standard against which to compare any response to a toxic stress. The amplitude of these natural stresses may frequently depress cell numbers and biomass more than 50% (Greaves *et al.*, 1980) and recovery from the stress may still be incomplete after 30 days. Consequently, to establish an impact by a pollutant, Domsch (1984) argued that the response had to exceed this normal range – the amplitude of a response needed to be greater than 50% and the response should still be detectable 60 days after the stress occurred.

A variety of toxicity assays have been used in soil ecosystems but relatively few were recommended by Greaves *et al.* (1980). The favoured methods included measuring the rates of respiration (as CO_2 production), litter decomposition, symbiotic nitrogen fixation and of nitrogen mineralization. Together these measure the most important functions of the soil community – decomposition, nutrient capture and release (section 7.1). A standard protocol for determining the impact of pesticides on particular soil microorganisms or on the activity of certain soil enzymes was also suggested. All of these measures are sensitive to the conditions of a particular soil, and the interactions between the pollutant, the abiotic environment and the living community. For example, lead is readily bound by soil organic matter and clay particles and this lowers its toxic impact on the soil biota (Tyler, 1981). Similarly, the lower solubility of metals in neutral and alkaline soils further reduces their toxicity. For these reasons, the impact of a particular level of pollution may differ considerably between soils with different ecologies, making direct comparisons difficult.

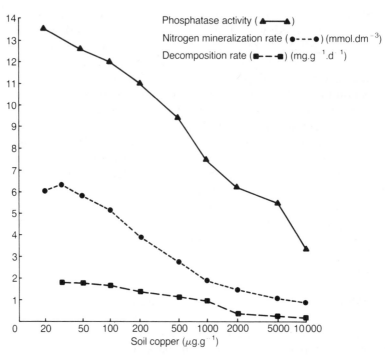

Figure 2.10 Three indicators of soil processes depressed by elevated copper levels in the topsoil of a forest near the brassworks at Gusum, Sweden. Notice that very high metal levels are needed for significant depression of enzyme activity, decomposition or nutrient mineralization rates. (After Tyler, 1984.)

One common effect of toxic metals is to depress rates of soil decomposition (Figure 2.10) and a number of methods have been developed to measure this. These are generally based on weight loss or colour loss from organic materials that are buried and recovered later (Khan and Frankland, 1984). Decomposition often slows in the latter stages, probably because some metals form stable complexes with the humic and fulvic acids that are generated (Tyler, 1981), and the assays need to allow for this.

Carbon dioxide production may also decline with these slower rates of decompositon, although aerobic respiration may itself be impaired by the pollution levels in the ecosystem (Figure 2.11). The response of the microbial community will depend upon its past exposure to toxic metals (Duxbury, 1985), and a site with a long history of pollution may have a community resistant to the stress. In effect, this requires toxicity tests to use properly controlled experiments, in which the conditions of the assay and of the soil are fully specified.

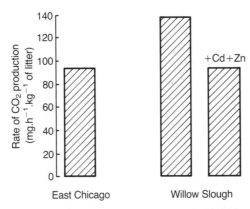

Figure 2.11 The use of carbon dioxide production as an indicator of stress in two forest soils, one of which was contaminated (East Chicago). One set of the control soils (Willow Slough) was also contaminated with $10\,\mu\mathrm{g.g}^{-1}$ cadmium and $1000\,\mu\mathrm{g.g}^{-1}$ zinc. The results record the loss of CO_2 over 23 days as a proportion of the litter present. The treatment of the control soil lowered soil respiration to that of the contaminated site. (After Chaney *et al.*, 1978.)

The same is true of assays relying on enzymic activity: less sensitive forms of some enzymes are produced by microorganisms growing in stressed environments, as part of the tolerance mechanism of these strains. They thus show less response to a given stress than unexposed microorganisms (Klein *et al.*, 1985). A number of soil enzymes are inhibited by high metal levels (Figure 2.10), and especially sensitive are those whose active site contains two sulphydryl (-SH) groups. The two negative charges have a great affinity for metals and form very stable complexes which inactivate the enzyme. The activity of urease, for example, is negatively correlated with the total (copper + zinc) concentration in the soil around a brassworks (Tyler, 1981) and rates of nitrogen mineralization decline in copper-rich soils (Figure 2.10). The results may not always produce a consistent threshold between soils – Tyler (1984) reports a depression of phosphatase activity at high levels of copper in soils, but subsequent work on other contaminated soils found the enzyme to be unaffected by the metal (Nordgren *et al.*, 1986).

There are additional problems of relating soil enzymic activity to biological activity within the soil. Amongst the enzymes responsible for releasing nitrogen from organic compounds, 'urease' actually consists of a group of enzymes derived from bacteria, fungi and higher plants, and can survive long periods outside the cell. Thus, although urease is useful as an indicator of the nitrogen mineralization capacity of the soil, it is a poor measure of the size of its microbial populations. The viability of extracellular enzymes depends on several soil parameters, particularly their clay content. Enzymes bound to clays may be protected from denaturation and loss through leaching (Klein *et al.*, 1985), but their activities may decline by virtue of this binding. Dehydrogenase assays are taken to be more indicative of biological activity because enzymes of this group are only active within the cell. They have been used extensively to assess the impact of pesticides, though the levels of the enzyme do not show any simple correlation with microbial numbers, biomass or respiration (Greaves *et al.*, 1980).

2.7.1 The size of the microbial biomass as an indicator of stress

Various methods have been used to assess the size of the total microbial biomass in the soil and its response to different stresses. For the most part these are gross measures with little, if any, distinction between the various species within the microbial community. Plate counts can be used to provide selective measures of the populations of particular genera but, more frequently, some indirect measure of biomass is used. One example is the fumigation–incubation technique, where the microbial community is first killed using chloroform and the flush of carbon dioxide produced by its subsequent decomposition during an incubation period, after inoculation, is used as a measure of the original microbial biomass (Vance *et al.*, 1987).

A different approach is to measure the adenosine triphosphate (ATP) content of the soil. ATP is the primary energy storage molecule in living cells, and directly reflects the amount of microbial carbon and thus the size and activity of the microbial community. ATP can be readily measured by the light emitted when the enzyme luciferase promotes the degradation of its substrate luciferin, with energy supplied by the ATP of the soil (Jenkinson and Oades, 1979). ATP estimation is very sensitive to disturbance in the soil and has to be measured after a pre-treatment period which standardizes conditions in the soil. ATP will be rapidly broken down during extraction and can be difficult to extract from soils with a high clay content. Visser (1985) shows how ATP measurements are depressed in a mine spoil compared with an nearby undisturbed soil, although both sites were highly variable.

Microorganisms divert more energy to cell maintenance under osmotic stress, leaving less scope for growth (Figure 2.2). Based on this, Killham (1987) assessed the ratio of respired carbon to biomass carbon as an indicator of stress in soils and litters subject to a range of stresses. This ratio, and the allocation of energy that it implies, was a more sensitive measure of a stress in the soil community than either dehydrogenase activity or the respiration rate. In a similar way, the available metabolic energy in the soil community can be measured as the

adenylate energy charge (AEC). This is derived as a ratio of the adenine nucleotides:

$$AEC = \frac{(ATP + 0.5\,ADP)}{(ATP + ADP + AMP)}. \tag{2.8}$$

ATP releases energy when its bonds with the phosphate groups are hydrolysed first to adenosine diphosphate (ADP) and then adenosine monophosphate (AMP). AEC is greatest when this energy is unused and the ATP remains unhydrolysed. Active cells typically have an AEC of 0.8–0.9 but, where their growth is limited, this drops to around 0.5–0.7. As AEC approaches zero (all AMP) so metabolic activity ceases and the cell, the organism or the community, is in decline. A typical grassland soil has a AEC of 0.85, and many soils can maintain these levels for long periods, even in the absence of significant carbon inputs (Brookes *et al.*, 1987). However, a severe stress, such as air drying the soil, will reduce the AEC to 0.45.

The AEC principle was first applied to specify tissues within animals and has been used to measure the effect of stress on the energetic status of whole organisms. Some animals tolerant of temperature and other stresses have higher AECs under stressed conditions, probably because they can reduce their metabolic rate when stressed (Hoffman and Parsons, 1989).

2.8 BIOCHEMICAL MEASURES OF STRESS IN WHOLE ORGANISMS

Enzymic measures of stress have also been used in pollution studies on whole organisms. Cellulolytic activity in the digestive tract of the clam *Corbicula* sp. was used to measure the impact of effluents rich in copper and zinc. At an outfall downstream of a power station on the Clinch River, Virginia, the activity of the enzyme was just 9% of that of uncontaminated clams from upstream (Farris *et al.*, 1988). This was a significant impairment of the mollusc's digestive capability, even though this water met USA quality criteria for copper and zinc and there were no detectable impacts on the macroinvertebrates at the level of the community.

There are thus good reasons for assessing the impact of a pollutant before pathological effects become apparent. A variety of methods have been developed for measuring responses at the biochemical level, where the pollutant first exerts its effect. These can be detected by changes in the cell morphology or impairment of function. Of the biochemical effects that follow a pollution insult, we examine two measurable responses which have been used in the past – the production of enzymes used to degrade organic pollutants and of proteins induced to bind toxic metals.

2.8.1 Enzymes induced by organic pollutants

Mixed-function oxidase enzymes (MFOs) are part of the natural mechanism of a cell to degrade steroids, hormones and some vitamins. They are also induced by high concentrations of a number of organic chemicals, including a variety of drugs, pesticides and hydrocarbons. The enzymes detoxify these chemicals by a series of oxidative reactions, converting an insoluble organic compound to a water-soluble metabolite which can then be excreted (Payne *et al.*, 1987). Levels of MFOs in the tissues of both fish and invertebrates have been used to assess the effect of oil spills in marine habitats. The activity of these enzymes is measured by incubation of a tissue extract (or whole animal extract) with a suitable substrate. Different substrates may be used to quantify and characterize MFO activity, such as benzo(a)pyrene. The enzyme is then defined by its functional activity, for example benzo(a)pyrene hydroxylase (B(a)PH). Alternatively, concentrations of **cytochrome P-450**, which serves as a terminal oxidase, binding oxygen for these reactions, can also be used as a general measure of activity.

A range of potent chemicals will induce MFO activity, including polycyclic aromatic hydrocarbons (PAHs), derived from oil pollution, and polychlorinated biphenyls (PCBs) found in industrial effluents. The oxidation of these compounds may not always effectively detoxify the pollutant for the animal, and sometimes the metabolites produced are more toxic than the primary pollutant. For example, people with a high susceptibility to carcinogenesis tend to produce MFOs more readily: smokers receive high doses of PAHs and the risk of cancer from smoking may be linked with the metabolites from PAHs and with the rate of production of MFOs.

PAHs cause lysosomes to bind with other components of the cell where they become abnormally large. As the lysosome is the main store of hydrolytic enzymes within the cell, this swelling indicates some increase in autophagy and protein breakdown (Moore *et al.*, 1987). PAHs appear to reduce the response time of the lysosome, allowing for faster induction of the enzymes. The latency period of the lysosome has been used as a general measure of cell condition, being shorter in cells under stress (Bayne *et al.*, 1976).

An MFO assay is a very sensitive technique for quantifying the impact of a pollutant. For example, Suteau *et al.* (1988) were able to detect B(a)PH levels in mussels when PAH concentrations in the surrounding water were as low as $130 \, ng.l^{-1}$. Distinguishing an enhanced response from the background variation presents some difficulties, but the method appears to detect effects before other responses: caged rainbow trout held in the effluent from a oil refinery showed elevated MFO activity after 21 days, before any other measures of stress suggested an effect on the indicator species (Payne *et al.*, 1987). A particular value of measuring MFO activity is its significance for higher-level functions, including growth and reproduction. For example, egg viability in some species of flatfish has been found to decline as MFO induction increases.

2.8.2 The induction of proteins to bind toxic metals

Where a pollutant is non-degradable, some other mechanism of detoxification is needed, a method of isolating the poison from sensitive sites in the cell. One example is the range of proteins produced in response to certain toxic metals, and which can bind the metal, preventing damage to other tissues. These metallothioneins were first described for vertebrates and then invertebrates, but have their counterpart in plants as well (section 3.4). They have a low molecular weight with a high cysteine content, an amino acid rich in sulphydryl groups that enable the protein to bind metals.

Ordinarily, metallothioneins are part of the homeostatic mechanism for the assimilation and regulation of two trace elements, zinc and copper. If these elements are in excess, additional metallothionein is produced to prevent damage to cellular processes. These proteins can also be induced by the toxic metals cadmium and mercury and, in some invertebrates, variants have evolved which are more specific for each toxic metal. The measurement of metallothioneins in particular tissues is a sensitive indicator of toxic metal stress.

According to the 'spillover' hypothesis, the toxic effect of a metal may only become apparent when the binding capacity of the metallothionein has been exceeded. The metal then 'leaks' to other parts of the cell, combining with other cell fractions. This idea was tested by dosing brook trout (*Salvelinus fontinalis*) with a range of cadmium doses and was found to be something of a simplification (Hamilton *et al.*, 1987). Although metallothionein production was elevated in the kidney and liver at all doses, the level did not reflect the dose of cadmium. Nor was all of the cadmium in either organ bound wholly to metallothionein. The situation is complicated by the presence of other metals. The trace elements copper and zinc are at much higher concentrations than cadmium in the kidney and liver of vertebrates, and will be bound preferentially by the protein. Metallothionein production by the trout responded to the level of cadmium when its concentration, relative to copper or zinc, was high. The same is true in invertebrates: rates of zinc and cadmium accumulation in crabs and shrimps reflect the balance between the two metals competing for metallothionein sites (Thorpe and Costlow, 1989).

Cadmium will bind to metallothionein at all concentrations, but there is always likely to be some fraction that is free to induce some form of pathology. The proportion that is free depends upon the concentration of trace metal levels competing for binding sites to the protein. This is one reason why Hamilton *et al.* (1987) concluded that metallothionein levels in the tissues are not a useful early warning indicator of metal pollution. They may be of more value as a long-term measure of severe exposure, indicating that at least some members of the population are having to accommodate to a stress by diverting resources into metallothionein production.

2.9 SCOPE FOR GROWTH AND PROTEIN TURNOVER

One other form of graded response has been used to assess the energetic costs of a high pollution burden directly. This method measures the **scope for growth** *(SfG)* of an organism under stress, based on the energy budget of an individual (Figure 2.2). This has been used extensively with mussels.

Of the energy assimilated (*A*), that which is not metabolized (*M*) is available for the production of new tissues or offspring (*P*):

$$\text{scope for growth } (SfG) = A - M. \tag{2.9}$$

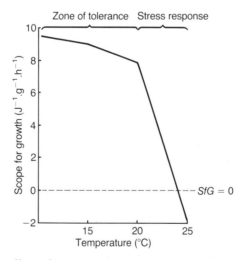

Figure 2.12 The effect of increasing temperature on the scope for growth of *Mytilus edulis*. Up to about 20°C, homeostatic mechanisms can compensate, causing only slight reductions in *SfG*; beyond this level the animal is stressed and *SfG* declines rapidly. This can be seen as one half of the optimum range curve of Figure 2.1. (After Bayne *et al.*, 1983, cited in Koehn and Bayne, 1989.)

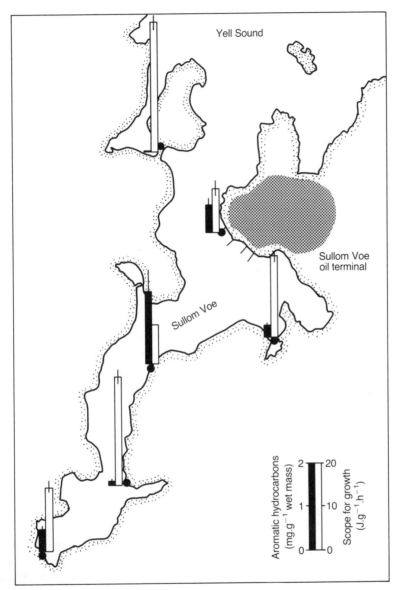

Figure 2.13 Scope for growth and the level of aromatic hydrocarbons in the tissues of *Mytilus edulis* around the Sullom Voe oil terminal in Shetland (summer, 1985). (With permission, from Donkin and Widdows, 1986.)

If we know the amount of energy consumed and the rate of loss through excretion and respiration, we can determine *SfG* as the difference (Figure 2.2). It is important that these terms are measured when the organism is in steady state, rather than when it is adjusting to changing conditions.

A stress will engage homeostatic mechanisms which attempt to restore the equilibrium of the tissues (Figure 2.12) and this alters the balance between these terms, increasing metabolic costs, M. Without an equivalent rise in A, SfG will fall (Koehn and Bayne, 1989).

The growth of an organism integrates a range of physiological, biochemical and cellular processes, and consequently SfG should be a sensitive indicator of any toxic impact. Indeed, responses are usually rapid – within days and weeks. SfG often declines in a predictable way as concentrations in the tissues rise (Donkin and Widdows, 1986) and this has been used to map ambient concentrations of hydrocarbon pollution (Figure 2.13). The oxygen consumption of *Mytilus* is also higher at locations polluted with both hydrocarbons and toxic metals (Sheehan, 1984).

Bayne (1989) describes how a number of measurable impacts combine to produce the reduction in SfG (Figure 2.14). In this diagram, copper is used as an example, though much of this would apply to a range of other pollutants. Copper is a trace element required in small concentrations by the mussel and for which it has a regulatory mechanism. After entering the tissues, copper is bound by a metallothionein. As the concentration of the metal rises so does production of the protein. This starts an escalation of costs which continues as available metal concentrations rise – when sufficient metallothionein cannot be produced, the free copper in the cytosol begins to combine with other proteins and enzymes, impairing their function. Lysosomal activity increases as these proteins are now broken down. Additionally, the metal may be bound by the lysosomal system itself, altering the permeability of the lysosome membrane, causing a more rapid release of its enzymes. This can be detected as a reduction in the period of latency of these enzymes, itself a measure of pollutant impact (Bayne *et al.*, 1976). Lysosomes are commonly swollen under pollution insult as proteins are more rapidly degraded. As much as 20% of the energy used by a mussel may be devoted to protein turnover (Koehn and Bayne, 1989) reducing the SfG.

Detection of SfG or measuring lysosomal stability require comparison with an unpolluted control population, to allow for the seasonal, reproductive and other variations which occur in mussels. When food is scarce in the winter, for example, SfG may be negative, even without a pollutant stress (Bayne, 1989). Used in the field, it is a non-specific response and it may be difficult to attribute the major causes of any reduction. Its sensitivity to particular pollutants has to be determined in the laboratory, although as with other toxicity tests, these values may change with a range of other pollutants (Moriarty, 1988).

One value of SfG is that it directly measures the impact of a pollutant at the population level – any reduction lowers the reproductive potential of an individual, either by reducing the number of gametes or by extending

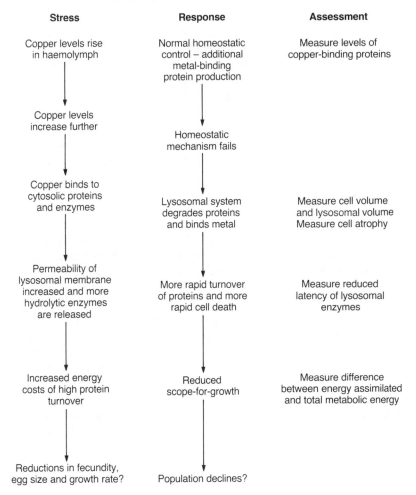

Stress	Response	Assessment
Copper levels rise in haemolymph	Normal homeostatic control – additional metal-binding protein production	Measure levels of copper-binding proteins
Copper levels increase further	Homeostatic mechanism fails	
Copper binds to cytosolic proteins and enzymes	Lysosomal system degrades proteins and binds metal	Measure cell volume and lysosomal volume Measure cell atrophy
Permeability of lysosomal membrane increased and more hydrolytic enzymes are released	More rapid turnover of proteins and more rapid cell death	Measure reduced latency of lysosomal enzymes
Increased energy costs of high protein turnover	Reduced scope-for-growth	Measure difference between energy assimilated and total metabolic energy
Reductions in fecundity, egg size and growth rate?	Population declines?	

Figure 2.14 The response of *Mytilus edulis* to a rise in tissue levels of copper, and the assessment of the stress response.

the time interval between reproductive events. Calow and Sibley (1990) have developed a model to describe how a lower *SfG* might affect reproductive capacity. This will only lead to a reduction in the rate of population growth if the energy expended on total metabolism remains constant. If, on the other hand, metabolic energy is used to 'escape' the stress, a reduction in *SfG* may cause no impact on the population growth rate. A species able to move or adopt some stress-resisting action can then maintain its reproductive potential.

In addition, where stressed conditions persist or cannot be avoided an organism able to reduce its *SfG* would have better survival prospects. If this alone determined reproductive success, this tolerant genotype would then become dominant in the population. On the other hand, meeting the cost of this resistance in the absence of the stress would reduce the competitiveness of the tolerant race, leading to its demise. Indeed, there is evidence that non-tolerant individuals tend to win competitive battles with tolerant individuals when the stress is absent (section 3.4).

This is a wholly physiological model of population dynamics, and excludes any interactions with other species – factors that may be critical for the survival prospects of an individual. Overall *SfG* provides a cause and effect relationship between events at the biochemical and population levels and represents a useful concept and measure of fitness, amenable to further testing.

Measures of stress based on populations are typically long-term assessments, and tell us relatively little about the mode of action of a pollutant. A more recent technique based on indicator species may change this. Felkner *et al.* (1989) describe a method for measuring the impact of a pollutant on a population of the bacterium *Bacillus subtilis* within an hour (two generations) of the population being exposed. Felkner *et al.* shine laser light through a suspension of the bacteria and analyse the pattern of scattering to measure the number, size, shape and distribution of particles. Any change in cell volume, expanding or shrinking in response to the pollutant, is detected. By using 19 strains of the bacteria, some differing by a single gene only, the response can be related to the particular physiological characteristics of each mutant. From this is derived a toxic profile, based on these different responses, and by which a pollutant may be identified. If proven, this method could allow rapid detection of pollutants at biologically significant levels and at the same time provide some insight into their toxic effect.

Summary

The initial toxic impact of a pollutant is at the biochemical level, but responses can be detected beyond the individual, at the population and the community levels. Simple models for the uptake of a pollutant can be used to measure concentration factors and the mobility of a pollutant along a food chain. A large pollutant burden incurs costs for the affected organism – either by impairing some cellular function, as with the narcotic effect of many organic pollutants, or through the use of specific detoxification mechanisms. Both the induction of enzymes to degrade hydrocarbons, or of proteins to bind toxic metals, can exert high energetic costs which can lower the scope for growth of an organism. This reduction in productivity may, in some circumstances, lead to a reduction in population growth rate. All of these responses have been used to measure the impact of a pollutant, but one common feature is the need to distinguish such effects from the background variation in undisturbed populations and communities.

The variation between individuals within a population has to be part of any measure of a pollutant's toxicity. The simple LD_{50} toxicity test can provide this, although it may be a poor indicator of the pollutant's impact in a real ecosystem. Soil pollution monitoring has favoured enzymic measures of stress, or community-level parameters, such as impacts on rates of decomposition or respiration. More direct measures of microbial biomass or their energy status have also been used.

Biological monitoring aims to resolve some of these problems by using particular species to assess the pollutant's impact. Sentinel species are used to map the scale and distribution of a pollutant; monitor species attempt to measure the biological impact for the larger community. Indicator species have been used as a summary of community-level changes, mainly in the form of biotic indices.

Further reading

Calow, P. and Berry, R.J. (eds) (1989) *Evolution, Ecology and Environmental Stress*, Academic Press, London.
Moriarty, F. (1988) *Ecotoxicology*, Academic Press, London.
Sheehan, P.J., Miller, D.R., Butler, G.C. and Bourdeau, Ph. (eds) (1984) *Effects of Pollutants at the Ecosystem Level*, SCOPE/John Wiley, New York.

Parys Mountain in Anglesey, North Wales. Here a number of metal-tolerant grasses have evolved to survive the very high level of toxic metals that occurs naturally in the soil and in the spoil produced from the extraction.

3

Exploiting variability

The immense diversity of life on this planet is perhaps its most valuable resource. We have exploited this variety throughout our history but only now are we beginning to recognize its importance for our future. The origin of this diversity has been understood for just 150 years, although the mechanisms that generate variation and drive evolutionary change have only been described in the last few decades. Yet during this time the destruction of complete ecosystems has accelerated, with many of their species being lost, unrecorded.

Each species has a unique genetic code, and today we are able to read the great library of information stored in their chromosomes. This code controls the efficient biochemical factory within each cell, and its capacity to synthesize a range of organic compounds and catalyse the degradation of others. Much of our economic activity relies upon these powers of anabolism and catabolism. We have unwittingly used this library since the earliest times, breeding domesticated plants and animals, selecting desirable (and occasionally not so desirable) characteristics. Now genetic engineering allows us to choose not only the character but also the organism which carries it. Variation can be transferred between species and new information written into the genetic code of different organisms. Now we can equip microorganisms with the enzymes to degrade pesticides; sheep and cattle with the genes to produce human proteins in their milk; and, by turning native genes off, prevent tomatoes from ripening or rotting prematurely. One way or another, human activity continues to be one of the major selective forces in the environment of many species.

The sum of a species' adaptations to its natural environment represents its **niche**. All environmental factors, biotic and abiotic, generate selective pressures, although population ecologists have long considered the interactions between organisms to be the most significant. These interactions define the position of a species relative to the rest of its community and, in combination with abiotic factors, define its niche. Niche theory is one of the central ideas in evolutionary ecology, matching an organism with the selective forces of its habitat. As we see later on, it also helps to explain why the variation in a species is an essential feature of its ecology.

In this chapter we examine how the adaptive capacity of living organisms has been used to monitor and reclaim degraded habitats. We begin by considering the sources and types of variation and the evidence that stress itself will induce genetic variation. The concept of ecological niche is introduced to help identify those species most likely to adapt to a stress and develop tolerance. The presence of tolerant enzymes or tolerant species can be used to measure the biological impact of pollutants. Similarly, strain selection and the new technologies of genetic manipulation have been used to adapt microorganisms for use in waste treatment and reclamation. Finally, we consider the ecological implications of genetic engineering and, most especially, the potential problems of releasing modified organisms into the environment.

3.1 VARIATION AND ADAPTATION

The totality of the genetic information in an organism is its **genotype**, although note that this term may also be used to refer to just a single character. The genotype for a species represents the totality of genetic information stored in all living individuals, sometimes termed its **genome**. Only some part of the information in the genotype of an individual will be expressed as the **phenotype**, the observed characters.

It is these expressed traits that govern the chance of an individual surviving to reproduce. Carrying the genetic code for a large muscular body, for example, will be of no advantage if those genes are not expressed and the owner loses every encounter with a competitor. In evolutionary terms, fitness is measured by the number of viable offspring produced, that is, the proportion of genes contributed to the total gene pool of the next generation: in this way we measure the **relative fitness** of a particular genotype. Only some fraction of the population will pass their genotype on to the next generation and their progeny will thus represent a limited 'sample' of the total gene pool in the parent generation.

The variation that we observe between individuals within a population derives from two sources – from the code written in the genes and differences induced directly by the environment. Those characters determined solely by the genotype have been inherited from the parents. This is not to say, however, that all of the information stored in the chromosomes is present because of its adaptive significance. In eukaryotes (the higher plants and animals) a large amount of the information in the genetic code appears to be redundant, with little or no selective advantage. This is termed **neutral variation**.

One way we can measure the variation in a population is by counting the varieties of particular proteins. Then we often find that a particular form or polymorphism of one protein may have no obvious selective advantage and its frequency in the population will rise and fall by chance alone, rather than due to selection. Much of this observed variation appears to have little adaptive significance and thus begs the question of whether such neutral variation confers some other, less obvious benefit. Biologists are still arguing about how much of this variation is strictly neutral and whether variation by itself may contribute to the 'vigour' of an individual (section 4.7).

Perhaps only 10% of the DNA in a eukaryote cell actually codes for proteins which adapt the organism to its environment. This is inherited or **genotypic adaptation**, and results from the selective pressures acting on the previous generations. Offspring carry a genetic code determined by their parent's environment, not their own. In the same way, the prevailing environmental conditions will select which genotypes survive into the following generation.

Other forms of adaptation are not inherited, only the ability to change. These include the capacity to adapt some physiological, behavioural or morphological character to accomodate change in the environment – **phenotypic adaptations**. These are often homeostatic responses, short-term changes, confined to the individual. Then it is the capacity to adapt that is written into the genetic code. The colour of a chameleon, for example, will vary with its background but its offspring inherit only the mechanism of change, not a particular colour. In contrast, eye colour in humans is determined solely by the genotype, whereas our height is an result of both the propensity for growth (determined by a number of genes) and the environmental conditions, including our nutritional status (the phenotypic response).

A **gene** is a small piece of DNA, occupying a particular position on a chromosome, or **locus**, and coding for the construction of a polypeptide, the building blocks of proteins. In diploid organisms with paired chromosomes, there are two genes, one on each chromosome, each coding for that character. Within the population there may be many alternate forms, or **alleles** of that gene, but any diploid individual can only have two. If both genes are the same, the organism is **homozygous**, whereas a **heterozygous** individual has two different alleles at that locus. For somebody to have blue eyes, for example, each allele must be of the blue type; the presence of a brown type will dominate the expression of eye colour and the phenotype will be brown. In other cases, dominance of one allele over another may only be partial, or alternatively neither allele may dominate and they are said to be codominant. The human blood groups A and B are coded by codominant alleles and both are dominant over O, giving the four possible phenotypes A, B, AB and O. Many traits are determined by a number of genes and the variation between individuals will depend upon the extent to which each is expressed.

Exploiting variability

Conjugation

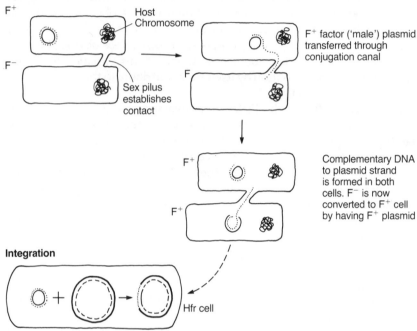

Figure 3.1 Derivation of genetic variation in prokayotes. **Figure 3.1a Conjugation** in bacteria allows direct transfer of DNA between cells. Transposable DNA is genetic material that is transferred directly. One type is the small circular plasmid which may carry several important characters in bacteria, including metal tolerance and drug resistance. In the example above, the F factor plasmid codes for the production of the sex pilus and the formation of a cytoplasmic bridge between the two bacteria, allowing DNA transfer. The F factor converts the recipient cell into an F^+ cell, able thereafter to form a sex pilus itself. Both cells reconstitute the plasmid after transfer by replicating the DNA. Occasionally the plasmid may become integrated into the bacterial chromosome producing a Hfr cell (high frequency of recombination). The F^+ is then replicated into each daughter cell at binary fission. Most importantly, the whole bacterial chromosome will move through the sex pilus at conjugation in Hfr cells, allowing the maximum transfer of genetic information.

Figure 3.1b Bacteria may take up DNA without direct contact with other cells. In **transformation**, DNA in solution, perhaps from lysed cells, is taken up by a second cell. In transduction DNA moves between bacterial cells using bacteriophages as vectors. In **general transduction**, random parts of the bacterial chromosome are incorporated into replicate phages and then transferred to new cells. With **restricted transduction** the division of the prophage DNA from the the host chromosome carries some of the host DNA with it. This is replicated and appears in every phage released from the bacterial cell. In both forms of transduction the foreign DNA can become incorporated into the chromosome of the new cell and is then expressed in this and subsequent generations.

Transduction

General

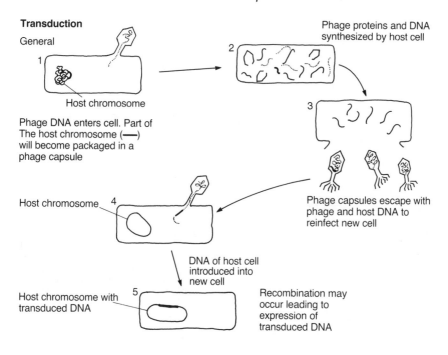

1 Host chromosome

Phage DNA enters cell. Part of
The host chromosome (—)
will become packaged in a
phage capsule

2 Phage proteins and DNA
synthesized by host cell

3

Phage capsules escape with
phage and host DNA to
reinfect new cell

4 Host chromosome

DNA of host cell
introduced into
new cell

5 Host chromosome with
transduced DNA

Recombination may
occur leading to
expression of
transduced DNA

Restricted

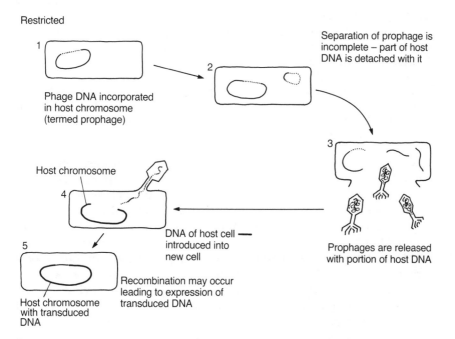

1 Phage DNA incorporated
in host chromosome
(termed prophage)

2 Separation of prophage is
incomplete – part of host
DNA is detached with it

3 Prophages are released
with portion of host DNA

4 Host chromosome

DNA of host cell —
introduced into
new cell

5 Host chromosome
with transduced
DNA

Recombination may occur
leading to expression of
transduced DNA

Transposition

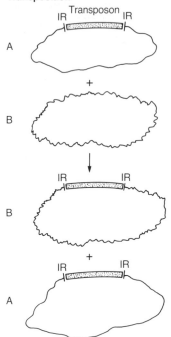

A   Transposon is incorporated into DNA which facilitates its replication – a host chromosome, a plasmid or a phage

B   Bacterial chromosome

B   With replication, the transposon is capable of inserting itself at a large number of sites in the bacterial chromosome using transposase. They thus facilitate the transfer of genes from one bacterium to another, and into the chromosomes of higher organisms.

Figure 3.1c A transposon consists of a mobile section of DNA which can insert itself into the host chromosome, in a form of recombination. In the process of **transposition**, the transposon is duplicated. The transposon consists of an insertion sequence – a segment of DNA that codes for the enzyme transposase to catalyse its insertion into a host chromosome. At either end of this segment are 'inverted repeats' (IR), sequences of base pairs that aid recognition by the transposase enzyme, and which allow recombination between the host chromosome and the transposon. More complex tranposons can insert additional genetic code and aid the insertion of plasmids. Transposons are known also from eukaryotes, including higher plants and animals.

Recessive alleles may not be subject to selection but they can still be passed on to the next generation. These genes are an important source of variation which may be expressed in the phenotypes of later generations. Such variation is essential if genotypic change is to take place. For the same reason, sexual reproduction is of selective advantage – the different combinations of genotypes produced increase variation and allow for rapid adaptation by a population to a changing environment. Eukaryotes inherit one set of chromosomes from each parent, creating new genotypes that are translated into different phenotypes. Also, having a paired set of chromosomes doubles the number of genes and the potential variation at any particular locus. In addition, during meiosis (the process by which a single set of chromosomes is formed for the gametes)

recombination can shuffle alleles between chromosomes, creating further variation. This is not a strategy for the prokaryotes (the bacteria and blue-green algae) with their single, unpaired chromosome, although they have their own form of sexual reproduction and recombination.

Within a population new genetic variation can be introduced by immigration, effectively increasing the size of its gene pool. At the species level, **mutation**, a corruption of the genetic code, is the only source of wholly new variation in most higher organisms. This is particularly important for microorganisms, because their clonal, asexual reproduction means a mutation may be readily conserved within a population and their rapid population growth allows an advantageous mutation to become established quickly. These organisms can also transfer non-chromosomal genetic material between cells and this is perhaps their most important source of variation, accelerating their adaptation to new conditions. Additionally, using the mechanism called transformation, some bacteria are able to assimilate DNA directly from solution. This gives these organisms a very rapid adaptive capacity, despite their lack of true sexual reproduction. The details of these processes are outlined in Figure 3.1, and these form the basis of many of the technologies described later in this chapter.

The speed of adaptation depends upon the total genetic variation in a population. The greater the variety of individual phenotypes, the more rapidly a population will adapt to a changed environment. This was formulated in mathematical terms by Fisher as his Fundamental Theorem of Natural Selection: the rate of increase of fitness in a population is proportional to the amount of genetic variability available. A greater variety of genotypes produces more options for the population to adapt to its environment and its response to a stress will be faster. By the same token, a species lacking genetic variability is more likely to become extinct when the environment changes (section 4.7).

3.1.1 Adaptation and stress

Selective pressures determine which phenotypes survive to reproduce and which fail. The organism has either to withstand and accomodate these stresses, or die, failing to reproduce. Parsons (1987a,b) has studied these responses in the fruit fly *Drosophila* stressed by heat shock. As with any environmental variable, an organism will tend to have an optimum temperature range, beyond which some physiological functions become impaired (Figures 2.1, 2.12). Above 30°C *Drosophila* stop normal protein synthesis and instead begin producing special heat-shock proteins. These offer some protection against denaturation of the cellular proteins, probably by supporting their structure. This form of acclimation is found in all groups of organisms and the proteins are very similar between very different groups. They are thus comparable to the metallothioneins (section 2.8) – both have a highly conservative structure and are inducible at times of stress.

They also each have a metabolic cost: producing these proteins deflects resources from other processes and incurs the energetic costs of their synthesis. Interestingly, organisms that can tolerate high temperatures appear to have relatively low metabolic costs under stress (Hoffman and Parsons, 1989), an important advantage if the response itself is not to exacerbate the stress. A metabolic rate that does not rise in severe conditions is a characteristic of tolerant strains of *Drosophila* and mussels (*Mytilus edulis*). Mussels with high rates of protein synthesis suffer more under temperature stress: what is important here is not the rate of protein production, but rather the energetic efficiency of the process and its metabolic costs (Koehn and Bayne, 1989).

Acclimation allows an individual to survive longer in a stressful environment and reproduce under marginal conditions. A population may thus be able to sustain itself and, if the stress persists over a number of generations, go on to adapt genotypically. This is most likely in populations that show the greatest genetic variation, providing more opportunities to produce a tolerant genotype. In addition, those individuals with the greatest variation are more likely to accommodate a stress – organisms with many heterozygous loci are commonly found to have a broader physiological range. Growth rates also tend to be higher with increased heterozygosity in a range of organisms, as does scope-for-growth in some molluscs (section 2.9). Why genetic variation should have this effect is not fully understood. It does not appear to be significant at all loci, but most especially those involved in energy storage (ATP production) and protein catabolism (protein turnover) (Koehn and Bayne, 1989).

Under stress, the differences between individuals become more marked and, over several generations, the rate of genotypic adaptation increases. The selection of particular characters causes the direction of genetic change to become more fixed. This will reduce the variability in the population and limit the capacity of later generations to adapt to new conditions and new stresses.

Over the long term we might therefore expect the amount of genetic variation to decrease in highly selective conditions. There are two opposite views here – one suggests that selection does indeed refine the genotype, favouring those individuals best fitted to the dominant selective pressures and reducing the overall variation in the population. The other view argues that the advantages of heterozygosity, such as the greater capacity to acclimate to changing conditions, will maintain their variation. The argument cannot be easily resolved as the survival and reproductive success of an individual depends on many factors and not simply the heterozygosity measured at a few loci. In addition, many traits are linked: some alleles do not segregrate independently during meiosis, so that one character may invariably follow another – then its appearance in the genotype may have no adaptive significance.

On the other hand, a stressful environment is itself known to increase genetic variation. Temperature and a number of chemical stressors increase recombination rates in both mammals and insects, and the mutation rates of all organisms. Additionally, the physiological state of an individual can predispose it toward a higher mutation rate. Factors such as age and genetic constitution increase the chances of mutation as do any external agents which disturb the regulation of the cell. For example the mutation rate of *Salmonella typhimurium* increases with heat stress when growing on glycerol rather than glucose. This reflects the differing physiological states of the microorganism on the two substrates (Mac-Phee, 1985).

A more direct effect of stress on genetic variation is via the insertion of **transposons**. These are mobile genetic elements that can move between cells and are able to insert code sequences into the host DNA (Figure 3.1c). Rates of these transpositions increase in both prokaryotes and eukaryotes at times of temperature stress. This not only increases variation, but transposons will also directly improve the fitness of some prokaryotes, such as *Escherichia coli* under stress (MacPhee, 1985). Another group of bacteria, the pseudomonads, have a large capacity to degrade a variety of compounds, using enzymes coded on extra-chromosomal DNA – structures called **plasmids**. The movement of plasmids between individuals can lead to the rapid development of a tolerant population. The plasmid may itself become part of the host chromosome: at least one of the plasmid genes that code for the degradation of toluene and xylene will integrate with the host chromosome of *Pseudomonas* (Hardman, 1987).

Overall, evolutionary change will be most rapid at times of greatest stress both because of this increase in variation and in the intensity of the selection pressure (Parsons, 1987b). Some authors believe that short and extreme stresses are the most significant source of evolutionary change, at least for some species. It only requires short periods of exposure to stressful conditions to define different climatic races of *Drosophila melanogaster* (Parsons, 1987a). For such species, where resources are rarely limiting, competition for space or food will not be an important selective pressure. Instead they will be selected according to their ability to withstand extreme environmental conditions. Using protein polymor-phisms as a measure of heterozygosity in *Drosophila* for example, Parsons (1987a) was able to correlate around 66% of this variation with just four ecological factors defining their habitat, whereas those factors determin-ing their competitive ability were much less significant.

3.2 SPECIALIZATION AND GENERALIZATION

The amount of genetic variation present in a species will determine the speed with which it can adapt to new conditions. But it may also be constrained by its evolutionary history: a highly specialized species, adapted to one particular set of conditions may be less able to adapt to a major change in its habitat, compared with a species with a broader tolerance range. Although there is a possibility that the new conditions may allow the highly adapted species to flourish, this becomes less likely as the scale of the change increases.

Stenotopic species have a biology that fits them closely to their habitat. They tend to have narrow and well-defined tolerance limits and are confined to a narrow part of an environmental gradient: when this includes their use of resources and food, they may also be called **specialist species**. Specialists are found in relatively constant environments and are typically long-lived, competitive species, with little capacity to adapt to a major change in their environment. In contrast, **eurytopic species** are more adaptable, able to live and reproduce over a much broader range of conditions. These are commonly short-lived opportunists, also able to exploit a variety of resources, when they may be termed **generalists**. A generalist on one gradient tends to be a generalist on most others (Futuyama and Moreno, 1988): this helps to explain why some species have a broad geographical range and others are more localized. These two strategies require different life history characteristics, most especially those governing the rate of population growth (section 5.1). The response of the two types to environmental stress also differs and to understand this, we need to consider the concept of ecological niche.

3.2.1 Niche theory

The abundance of a species along any environmental gradient reflects its adaptation to a particular range of conditions. For each variable there will be an optimum range where the productivity of the species or some other measure of its performance is maximized (Figure 2.1).

The distribution of a species is rarely the product of a single environmental gradient. More often, this will be defined by a number of variables and their interactions. For example, the abundance of a plant may be governed primarily by the availability of water and nutrients, and will be maximized where both are optimum. The totality of all significant abiotic and biotic factors, and their interactions, define the **ecological niche** of an organism. This represents the sum total of a species adaptations (Pianka, 1983). A niche can be defined for an individual, a population or a species, and represents the response to the physical and biological gradients in the environment (Bazzaz, 1987).

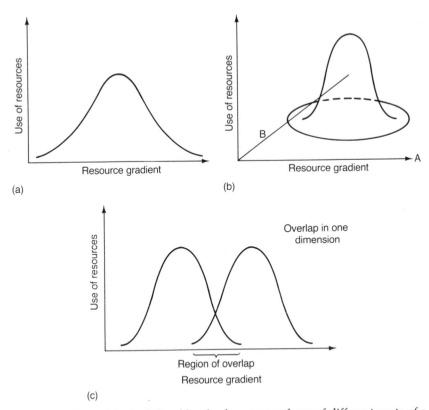

Figure 3.2 (a) A niche is defined by the frequency of use of different parts of a resource gradient. The distribution of a species reflects the extent to which each part of the gradient is used. Most individuals will be found within an optimum range (Figure 2.1) to which they are best adapted. (b) For two factors operating simultaneously, the pattern of use will describe their interaction. For example, food size and time of feeding. This produces a three-dimensional plot describing the organism's optimum range. With more than two dimensions, the niche exists only in hyperspace. (After Pianka, 1983.) (c) Overlap of resource use between two species along a single gradient. If the resource is in short supply in this region there will be competition between the two species. Adaptation by each species to one part of the gradient will eventually lead to each niche becoming more sharply defined, reducing the overlap.

To describe the niche for a species we could plot its distribution for each variable individually, but this gives no indication of how factors interact with each other (Figure 3.2). It is impossible to depict the organism's response to more than two factors graphically so that the niche can only be defined mathematically, in a 'space' of more than three dimensions (environmental factors). This hyperspace consists of n dimensions, where n is the number of gradients to which the organism responds. These factors operate simultaneously and the 'position' of the organism in this hyperspace represents its niche. Where conditions are optimum or within the zone of tolerance, the organism will reproduce successfully and its niche is defined by this position along each significant gradient.

Environments are continually changing so that there can be no external, fixed niche to which a species is fitted. Instead organisms respond to changing selective pressures, never being perfectly adapted to their habitat. However, there is a debate over whether natural selection fits an organism to a niche, some particular role within the community, according to a kind of community templet, or if the niche is actually a property of the organism itself. There are numerous examples of different species with very similar morphologies fulfilling equivalent ecological roles in different parts of the world (Colinvaux, 1986). This suggests that the demands of a niche are a product of the community of which it is a part (Giller, 1984).

Nowadays niche space is most often measured as the use of resources by a species, such as food or shelter, although its niche is defined by all significant factors including its relations with other organisms. The effect of these factors on growth and reproduction can be difficult to measure; even relatively simple parameters, such as temperature, can produce complicated responses through their interactions with other parameters. Variation in some factors may be more readily tolerated than others and one problem is weighting gradients according to their significance for the organism. These patterns of use may change with time and with the life cycle: for example, the niche required by a seedling may differ considerably from that of the adult plant, requiring more or less shade, differing soil depths and so on (Grubb, 1977).

3.2.2 Niche and competition

We should also distinguish between the niche that a species could occupy and that which it actually occupies. The former, the **fundamental niche**, refers to the range of a resource that a species would exploit in the absence of competing species. Competition and interactions with other organisms will make the actual or **realized niche** somewhat smaller than this. Competition for resources between species has been taken to be the principal cause of contraction of the fundamental niche, although other interactions, such as predation, are now recognized as also significant (Schoener, 1989). These interactions are often modelled along a single resource gradient, while in reality an organism may be responding to competition from several neighbours along different gradients.

The competition between two species can be visualized as their **niche overlap** (Figure 3.2c). For example, two predators consuming the same prey might be separated along this gradient because they take prey of different sizes. Where there is significant overlap some prey sizes are being consumed by both species. If there is a limited number of prey there will be competition between the two predators and then the most efficient will tend to occupy the largest part of the common resource, to the detriment of the other species. The efficiency of a predator can be measured simply as the rate at which its converts the resource, its prey, into offspring (section 5.3). With limited resources, niche overlap would thus be expected to reduce the population size of at least one of the competitors. Or, with considerable overlap, the loser might be excluded altogether.

Under some circumstances, however, niche overlap may be no evidence of competition at all: if two species are separated along on some other dimension, such as time of feeding, they may never be in direct competition for that resource. Also an abundant resource may be able to support two or more species and then high overlap is an indication that there is actually little competition. Niche overlap can only be a direct measure of competition when it includes some measure of resource availability (Pianka, 1983). In iguanid lizards, for example, high overlap can indicate both increased or decreased competition depending upon the abundance of resources (Schoener, 1989).

The range of a resource that a species can use is termed its **niche width**. Specialist species have a narrow niche width, concentrating on a small part of the resource gradient which they exploit with great efficiency. These organisms compete for resources over the long term, becoming highly adapted to that part of the resource gradient. Generalist species will fail in these competitions. They have a broad niche width instead, exploiting a wide range of resources, with their populations growing rapidly to use short-term opportunities. This is often the case in very changeable habitats, and is why generalists are characteristic of the early stages in the development of a community (section 7.1).

These different strategies are adaptations to the variability of the resource being exploited. When the supply of a resource is sustained over long periods, extended competitive battles can develop and the most efficient user selected – that is the species most effective in converting the resource into offspring. Over the long term these habitats are likely to contain highly specialized species with narrow niches, each adapted to a small part of the resource gradient. In contrast, the generalist can sustain itself in a highly changeable environment and is able to use non-optimum resources to survive shortages. Their broad niche width reflects their powers of acclimation. The differences in the adaptability of these two types explain why specialized species are more susceptible to environmental variability (section 4.6).

Even within a species, some individuals will be better able to exploit particular conditions. A species will consist of a number of different strains or races and a large variety of phenotypes, that together define its overall niche width. This total width can thus be partitioned between these types, each with its own optimum range. Bazzaz (1987) suggests that most competition for resources is between different races rather than between species. This **intraspecific competition** has been proposed as a force expanding niche width: if resources are limited, phenotypes able to exploit unused parts of the resource gradient will be favoured, extending the range of the species. **Interspecific competition** causes niche width to contract: here competition for limited resources between species will favour specialists able to use a resource more efficiently. Then the resource is partitioned into finer fractions, each occupied by a highly adapted species. The evidence for niche width being determined by the balance between these opposing forces is, however, somewhat equivocal (Giller, 1984).

Locally adapted populations of a species are called **ecotypes** and are typically found in extreme conditions, toward the tolerance limits of a species. Ecotypes appear in polluted habitats where few competitors can survive – conditions which favour opportunists. These can thrive in the absence of competition because their inefficient use of the resources carries few penalties. The metal-tolerant grasses described below are an example of this – although they flourish on contaminated mine spoils, these races lose out in competition with non-tolerant forms in non-polluted soils.

Niche theory is one of the central concepts in population and community ecology and a number of models and methodologies have been developed to measure competition and niche overlap (Rosenzweig, 1987). Yet a theory which depends solely upon competition as the mechanism for placing species along a resource gradient is something of a simplification. Parsons (1987a) argues that competitive effects may only be significant for larger vertebrates where resources are limiting. For many invertebrates resources are abundant, making competition for food or space relatively unimportant. Habitat choice in short-lived species, such as many insects, may not be determined by resources but by the chances of meeting a mate: a sexual rendezvous chosen according to whom you are likely to meet there. In this case, a resource is being used without reference to any form of interspecific competition (Futuyama and Moreno, 1988).

3.3 USING TOLERANCE AS A POLLUTION INDICATOR

Ecotypes arise when a species is unable to adapt phenotypically to some disturbance. If a stress persists over a number of generations and does not wipe out the entire population, it will be the least susceptible phenotypes which remain to reproduce. Their genotypes will dominate future generations and the direction of genetic change in a population will become more fixed. In this way genotypic adaptation occurs and tolerant races become dominant in stressed habitats. Indeed, Luoma (1977) has suggested using the presence of such ecotypes as an indicator of a persistent selective pressure, a direct measure of the biological impact of a pollutant over the long term.

One example is the work of Grant *et al.* (1989), who used zinc and copper tolerance in the marine polychaete *Nereis diversicolor* to map the zones of impact for these two metals in a contaminated estuary in Cornwall. Worms were collected from different parts of the estuary, and offspring from these worms were bred and reared under non-contaminated conditions. Thus any tolerance to the metals on their part could only be due to genotypic adaptation. This was measured as an LT_{50} for copper and zinc (Table 2.1). The young showed genotypic adaptations to both metals, though copper tolerance declined with the distance of the parental site from the source of the contamination. Zinc tolerance showed a quantal response (it was either present or absent) and was found at only two sites. This contrast suggests that copper and zinc tolerance are not linked, at least in *Nereis*. Further, the lack of ecotypes tolerant to zinc or copper at the non-contaminated sites indicates that here the ecotypes were at some competitive disadvantage with the normal population.

One possible reason for the gradation in copper tolerance might be some 'drift' of tolerant individuals or gametes into non-contaminated areas. Any such movement would have to be allowed for when mapping biologically significant copper pollution. Additionally, the presence of copper and zinc tolerance in *Nereis* may be little guide to the impact of these metals on other members of the community.

Some ecologists have looked for more direct evidence of adaptive change as a measure of pollution impact. One way an organism may adapt to a stress is to modify the structure of its most sensitive enzymes. Many enzymes occur in various structural forms differing by the sequence of particular amino acids. These **allozymes** are the expression of different alleles and thus represent a direct measure of genetic variation at specific loci. The number of allozymes in an individual can be used as a measure of its heterozygosity at those loci and also to identify different genotypes between individuals for each enzyme assayed. The various polymorphisms of these proteins are readily detected by their rates of migration under electrophoresis.

Much of this variation may have no adaptive significance (section 3.1) but, if one allozyme functions better than another under a stress, it will come to dominate in the population. Some allozymes show a generalized response to pollutants, whereas others are very specific. The mosquito-fish, *Gambusia affinis*, is a widespread freshwater species which has developed tolerance to a range of pesticides. Newman and his co-workers (Newman *et al.*, 1989) studied a population of *Gambusia* which had no history of exposure to mercury or arsenate, and found that individuals which were homozygous for one form of the enzyme glucosephosphate isomerase had the shortest survival time when exposed to both poisons. This was a rare genotype in this population. The same was found for two other rare homozygous genotypes (malate dehydrogenase and isocitrate dehydrogenase) in response to mercury alone. Most significantly, the survival time of the mosquitofish increased with heterozygosity at six of the eight loci studied (Diamond *et al.*, 1989), an indication that genetic variability alone carried some advantage.

Allozyme frequencies respond to a range of stresses, including organic and thermal pollution. Nevo and his co-workers (1986) related the occurrence of allozymes to a measure of niche width in three genera of marine molluscs (Figure 3.3) and compared this with their response to different stresses. For each genus they matched two species: one found predominantly in the upper littoral zone, classified as 'broad niche' (because of its capacity to withstand periodic exposure and desiccation), and a second species found only at the lower shore, deemed to be narrow niche. As expected, the broad-niche species in each genus had a greater genetic diversity, as indicated by its allozymes, and its survival was greater when exposed to each pollutant.

Variations in enzyme structure will have a selective advantage in changeable environments if a number of allozymes provides greater adaptability. This may be why heterozygosity appears to be a feature of broader-niche species, and why highly specialized species show a smaller amount of genetic variability (Lavie and Nevo, 1986). Heterozygosity at some loci can also be of selective disadvantage however – this appeared to be the case with glucosephosphate isomerase in all mollusc species studied here (Lavie and Nevo, 1986).

The frequency of certain allozymes in a population may be used as some indication of a past exposure to a stress, even though environmental levels may not currently be stressful. In effect, the genetic code

---

Figure 3.3 The survival of broad-niche species (*M. turbiformis*, *L. neritoides* and *C. scabridum*) and narrow-niche species (*M. turbinata*, *L. punctata* and *C. rupestre*) from three genera of littoral molluscs exposed to various pollutants. Species were classified according to their position on the shore: upper-shore species were assumed to be able to adapt to a greater range of conditions associated with periodic exposure, and termed broad niche (open boxes). In the same way, narrow-niche species (filled boxes) were from the lower shore. For the pollutants shown here, the broad-niche species always had longer survival times compared to unpolluted controls. (Redrawn from Nevo *et al.*, 1986.)

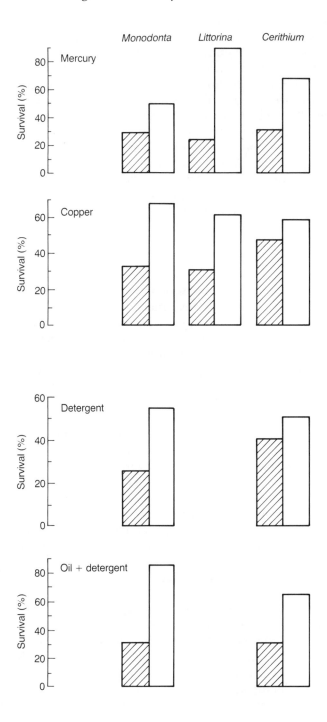

records a history of exposure, though for this information to be retained in the population in the absence of the stress, a tolerant allozyme would have to suffer little competitive disadvantage to non-tolerant forms.

Much of this work requires definitive studies to demonstrate how the allozyme confers tolerance to a pollutant. For an allozyme to be used as an indicator we need to describe why this form should have a selective advantage. It may be that its prevalence in the population is correlated with some other trait which is actually the subject of selection. Much of this variation may be neutral, due to genetic drift, and without any selective significance. Also the sensitivity of an allozyme of one species may be no guide to the impact of a pollutant on other members of a community. Given all this, allozymes do offer an insight into the nature of the toxic impact and on the interactions between pollutants (Ben-Shlomo and Nevo, 1988).

3.4 METAL TOLERANCE IN PLANTS

One of the best known examples of an evolved response to a changed habitat is the development of toxic metal tolerance in plants, especially the ecotypes of various grasses associated with metalliferous wastes. The spoil or tailings produced from extraction and smelting still contain considerable amounts of metals, sometimes at levels that make re-working economically viable (McNeilly, 1987). These concentrations are toxic to most plants, but many spoil heaps are vegetated by a few species of grass. These tolerant ecotypes are probably derived originally from the surrounding plant community, even though most of its present members show no such adaptation. We thus have the interesting question of how such adaptation has occurred and also how quickly.

Several grass species have produced varieties tolerant to one or more heavy metals: *Agrostis capillaris* has ecotypes tolerant to copper, lead and other metals. Although the tolerance is usually only to one metal, sometimes a low level of tolerance is conferred for a second metal: Wilson (1988) describes a correlation of copper and lead tolerance for *A. capillaris* collected from a number of sites in Wales. Although some plant genera never produce tolerant varieties, others, such as *Alyssum* or *Thlaspi* can accumulate several toxic metals to high levels and still go on to reproduce (Dickinson *et al.*, 1991).

The adaptation of grasses to metalliferous spoil heaps is a major source of evidence about evolution rates and gene flow in plant populations (Bradshaw and McNeilly 1981). Sowing ordinary grass seed on to a contaminated soil will often produce some tolerant individuals, albeit at a low frequency. Even relatively inbred commercial grass seed will produce one or two tolerant individuals from a large sowing (Wilson, 1988), although some species, such as *Poa trivialis*, never do (Macnair, 1987).

Many populations must therefore have some store of genetic variance at the crucial loci that determines their capacity to colonize contaminated soils. Again, species with low variability are less adaptable and less likely to produce tolerant ecotypes (McNeilly, 1987). Toxic metal tolerance is found rarely in plants from stable and unpolluted habitats, where stenotopic species would be favoured. Tolerance is also less common in species whose reproduction is mostly vegetative, as this offers less scope for new variation to be introduced into the population.

3.4.1 The evolution of metal tolerance in grasses

Three factors contribute to this variation – the size of the population (N), any selective pressure against tolerance (s) and the mutation rate (μ). A large population has more opportunities for different alleles at each loci, whereas a small and highly localized population will tend to be inbred, with little variation. Ecotypes tolerant of toxic metals are more often found in common species with large populations (Macnair, 1987) – these are typically broad-niche species and their genetic variation is evidence of their non-specialized nature.

A **selection coefficient**, (s), measures the relative selection pressure against a phenotype, calculated as the proportion of its progeny not surviving to reproduce compared with that of the most abundant phenotype. Where there is no pollution stress, a low selective pressure against tolerance will ensure that this trait remains in the population. The positive advantage of tolerance for life in a contaminated habitat will rapidly fix this character in a population subject to the stress.

The development of tolerance will also depend upon the number of genes contributing to the adaptation and their rate of mutation: tolerance will be less likely to occur if several genes have to mutate rather than just one or two. Zinc tolerance in *A. capillaris* is due to a number of genes and is also a dominant trait: together these suggest a strong selection pressure has driven this genetic change (McNeilly, 1987).

Tolerance can occur very quickly – significant differentiation of populations of grass is known from spoil heaps less than 100 years old. Tolerant ecotypes of *Agrostis capillaris* and *A. canina* appeared within three years of a zinc–cadmium smelter beginning operations in Germany (Ernst 1986), further evidence that evolution can be very rapid under extreme selective pressure.

Most spoils are marked by their low nutrient levels, extremes of pH, poor soil structure and texture, as well as semi-arid conditions. This will favour eurytopic species able to adapt phenotypically to the poor conditions and perhaps to go on to adapt genotypically. In the absence of competitors, the colonizing population will undergo rapid genetic change. This founding population will begin with a small number of

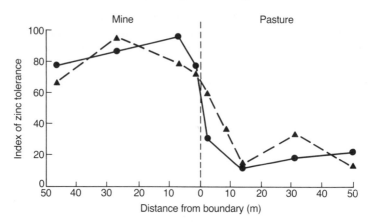

Figure 3.4 The rapid loss of zinc tolerance in two species of grass – *Agrostis capillaris* (solid line) and *Anthoxanthum odoratum* (broken line) – across the boundary between a lead mine and surrounding agricultural land at Trelogan, North Wales. This uses an index based on the ratio of root lengths of the grasses when grown in zinc-dosed and zinc-free solutions. These genetic differences are known to be maintained by different flowering times either side of the boundary which reduces gene flow between the two populations. (Values for *A. capillaris* are × 3.) (After Hickey and McNeilly, 1975.)

individuals and consequently a small gene pool. They will only be a 'sample' of the gene pool of the original population and, by chance, might be very uncharacteristic of this population. Further differentiation will occur if the genotype of the founding population is not 'diluted' by genes arriving from outside. This gene flow can be limited in plants by a combination of four mechanisms that encourage inbreeding (Macnair, 1987):

1. self-fertilization;
2. flowering at a different time from the parent population;
3. alteration of the receptivity of the stigma to foreign pollen;
4. the development of hybrid non-viability.

A. capillaris is one of several grasses which show increased self-fertility in its tolerant populations. In addition, the flowering times of tolerant and non-tolerant varieties of both *A. capillaris* and *Anthoxanthum odoratum* differ on either side of a boundary between contaminated and uncontaminated soils (Figure 3.4). Despite these differences, further away from this zone, either on the spoil or the natural soil, both forms of each species flower at more or less the same time. In effect, those tolerant individuals most exposed to the pollen from the normal populations are least likely to be fertilized by it. This helps to preserve their genetic differences in the offspring. There may also be some selective disadvantage of having genes contributed by a wild-type parent: hybrids of copper-tolerant and wild-type Monkey flower (*Mimulus guttatus*) are non-viable due to the presence of a single gene, even though this locus is not responsible for the tolerance (Macnair, 1987). Along with the strong selective advantage of tolerance itself, such factors ensure that the tolerance gene becomes rapidly homozygous throughout the population, a process termed **fixation**.

3.4.2 Phenotypic responses to metal-contaminated soils

Before any genotypic response is initiated, a number of mechanisms may protect the plant from the toxic effects of high metal concentrations. A variety of plants show some form of acclimation, a physiological or morphological response initiated by exposure to a pollutant. This becomes especially important in the survival of long-lived species, including some trees, where genetic change is less frequent (Dickinson *et al.*, 1991). Some plants grow limited root systems in high metal soils whereas others, such as birch trees (*Betula* spp.), may produce metallothioneins (section 2.8). Indeed, the extent to which genetic adaptation contributes to metal tolerance in plants is still disputed, as a sizeable proportion of the tolerance may be lost when clones are transplanted into non-contaminated soils (Dickinson *et al.*, 1991).

The physiological basis of metal tolerance in plants has yet to be fully described. There is evidence that plants produce compounds which bind the metal, effectively making it unavailable to the more sensitive sites within the cells. A metallothionein capable of binding copper has been isolated from a tolerant strain of *Agrostis gigantea* (Rauser and Curvetto, 1980). Although it shares some chemical characteristics with animal metallothionein, it is structurally closer to proteins isolated from yeasts. Similar proteins have now been isolated from both tolerant and non-tolerant plants (Dickinson *et al.*, 1991) suggesting that they may be part of the plant's essential metal regulation, perhaps for zinc. Tolerance often appears to be a change in the gene regulating the production of such proteins, at least as far as copper tolerance is concerned (Macnair, 1987). If this is the case, the adaptation is a modification of the homeostatic system regulating a physiological response – like the heat-shock proteins, an increase in the rate of protein production to accomodate the stress.

Physiological acclimation to a new habitat may also be accompanied by morphological changes. A copper-tolerant population of *Mimulus guttatus* grows on a spoil in California, even though its natural habitat is wet and waterlogged soils. Spoils are typically much drier habitats and plants have to withstand drought conditions. Although the wild population sports numerous seed capsules in its native habitat, *M. guttatus* growing on the waste has few, and shows much greater morphological variation (Macnair, 1987). Similar variation is found in grasses and other plants growing on spoils, and is evidence of the greater phenotypic variability that may occur in the absence of competitors, particularly in eurytopic species.

3.4.3 The costs of tolerance

Most of the evidence would suggest that metal tolerance in plants causes some loss of competitive ability, that the organism has to meet additional metabolic costs to survive in a contaminated soil. The grass *Agrostis capillaris* growing on old mine and spoil sites appears to suffer such costs. This species mainly reproduces by vegetative growth, so clones from a single genotype can dominate some areas. Wilson (1988) compared the relative growth rate of tillers from various contaminated and uncontaminated sites by weighing the dry matter of their above-ground production after 17 and 28 days' growth in soils without metal contamination. Wilson was able to show that the growth rate of a population was negatively correlated with its copper tolerance, although variations between individuals were often as great as those between populations. The nature of the trade-off between metal tolerance and growth rate is not known.

Some cost of tolerance would explain the competitive failure of ecotypes resistant to the herbicide atrazine in two weeds, *Senecio vulgaris* and *Amaranthus retroflexus*. Normal (susceptible) varieties of both species show more prolific production of seed and biomass than resistant ecotypes when grown together in competition in an uncontaminated soil (Figure 3.5). The competitive disadvantage of the tolerant varieties may account for their low incidence where atrazine is not used (Conrad and Radosevich, 1979). The same is likely to be true of metal-tolerant ecotypes of grasses – their failure to establish themselves in normal soils probably results from their competitive exclusion by non-tolerant varieties which bear none of these additional costs (Nicholls and McNeilly, 1985). If the selective pressure against tolerance was low, the gene(s) responsible would be selectively neutral on normal soils, but a major advantage on contaminated soils. The genes would then be be easily fixed in these plants. Instead we find their occurrence is low, even in adaptable species like *A. capillaris*, an indication that there is some cost associated with their expression (Wilson, 1988).

3.4.4 The use of tolerant grasses in reclamation

Some varieties of tolerant species are available as commercial seed mixtures for use on grossly contaminated soils. Metalliferous wastes are often highly friable and windblown dusts from a site can severely pollute surrounding ecosystems. Any vegetational cover helps to stabilize surfaces and tolerant plants may be the only species able to grow in the

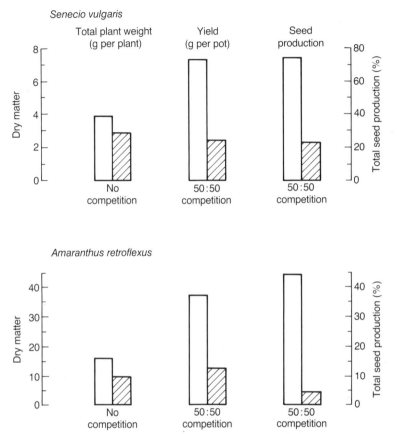

Figure 3.5 Biomass and seed production of atrazine-resistant (filled boxes) and non-resistant varieties of (open boxes) two flowering plants. In all cases, the herbicide-resistant varieties are less productive when sown in 50:50 combinations with the non-resistant variety in pot trials. These differences became more marked as the competition from the wild types was increased. (Redrawn from Conrad and Radosevich, 1979.)

early stages of a reclamation. Several genotypes can withstand the very dry or acidic conditions of spoil heaps, as well showing tolerance to one or more toxic metals.

Early colonizing plants help to ameliorate the toxic threat for later colonizers. The organic matter produced by the first plant growth improves the texture of the spoil and helps it to hold water, binding the soil particles together. It also binds metals, reducing their availability to subsequent colonizers (section 7.4). In addition, the plant debris is important as a carbon source for a community of tolerant microorganisms in the spoil (Klein *et al.*, 1985).

From these beginnings, the most important soil functions of decomposition and nutrient release become established. Spoils are typically deficient in the major plant nutrients and few nitrogen-fixing plants will survive: legumes are especially sensitive to toxic metals (Bradshaw and Chadwick, 1980). The first plants will need fertilizers to survive and adding manure or other organic fertilizers also serves to increase the organic matter in the spoil. The debris from the plant community adds to this organic component, further helping to hold nutrients in the soil. Even so, repeat applications of fertilizers are often necessary over a number of years.

Reclamation offers large scope for investigating rates of evolutionary change, gene flow and the mechanisms by which ecotypes differentiate into new species. The rapid acquisition of metal tolerance by some grasses has led McNeilly (1987) to suggest that importing tolerant ecotypes to vegetate these sites may be unnecessary. Instead, he suggests we use material collected from the local plant community, sowing species known to produce tolerant ecotypes. Besides reducing the costs of reclaiming such sites, it also limits the amount of non-native genetic variation brought into the area. As Cairns notes (1987) we have little understanding of the ecological significance of minor genetic variation and replacing the native flora by alien species effectively reduces the gene pool of the former. Also substituting one species with another means that the restored ecosystem rarely matches the undisturbed community. Nor should we assume that the gene flow is only one way, into the restored community.

3.5 THE USE OF TOLERANT MICROORGANISMS IN SOIL REMEDIATION

There are about 70 000 old industrial sites in West Germany with grossly polluted soils, many of which are contaminated with mineral oils (Puschnig *et al.*, 1991). This list will have grown considerably since German re-unification in 1990, just a small part of the area of dereliction and spoilt land that extends throughout Europe and most industrialized nations. These soils are contaminated by a formidable range of pollutants and many sites will require new techniques and much ingenuity to restore them at an affordable cost.

The soil is the major site of degradative processes in terrestrial ecosystems, with an impressive capacity to reduce relatively intractable substances to much simpler compounds. This capacity is derived from the catabolic enzymes of the bacteria, fungi, actinomycetes and invertebrates that comprise the decomposer community. Soil communities are highly complex, dominated by the bacteria and fungi which, in their degradation of organic compounds, provide resources and nutrients for higher organisms. This decomposer community is composed of **chemo-heterotrophs** – organisms that use organic compounds as their source of carbon and energy, relying ultimately on the photosynthetic activity of the plant community above ground.

Many organic pollutants or **xenobiotics**, are synthetic compounds to which the microorganisms in most soils have never been exposed before. However, given time, and their rapid adaptive capacity, most carbon compounds can be degraded and used as an energy source by the bacteria. We can exploit this adaptability in restoring a soil and accelerate the degradation of a range of xenobiotics including pesticides and oils. Similarly, we can use resistant species to establish a community in soils with large amounts of non-degradable wastes.

In most soils carbon and the other major nutrients are in short supply and the competition for these resources means there is little scope for an invasive species to establish itself. This is one reason why many of the previous attempts to inoculate soils, especially agricultural soils, with non-native microbes have failed (Focht, 1988). Even so, one prospect for treating highly contaminated soils and wastes is to use introduced species or whole decomposer communities. The advent of recombinant DNA technology provides us with the means to select or construct the genotype in a microorganism, adapting it to degrade certain organic compounds. With these techniques genetic code can be moved from one organism to another, allowing us to compile a genotype for a specific function. The organism then acquires an adaptation without ever being subject to a selective pressure.

3.5.1 Metal tolerance in soil microorganisms

Given time, natural selection will usually develop a population capable of re-colonizing the most contaminated of ecosystems. Indeed, grossly polluted soils and waters are often fruitful sites for finding genetic code for use in reclamation. Metal-contaminated soils are typically dominated by Gram-negative bacteria, such as *Flavobacterium*, *Alcaligenes* and *Pseudomonas*, although it is possible to isolate metal-resistant forms from uncontaminated soils (Duxbury, 1985).

Metal resistance is held on the extra-chromosomal DNA of the plasmid and is closely linked to drug resistance. Similarly, tolerance to mercury is located on other transposons (Figure 3.1c) in *Pseudomonas*, *Klebsiella* and *Citrobacter* (Duxbury, 1985), although its capacity to move between individuals, given the low population densities of the soil, is largely unknown. This may happen more readily in the crowded conditions of the gut of a consumer, especially decomposers such as woodlice or earthworms.

Cadmium and mercury resistance is plasmid-linked in *Staphylococcus aureus*, though the mechanism of tolerance appears to differ: cadmium resistance relies on limiting accumulation by the cell, whereas mercury is transformed to a more volatile or less toxic form (Gadd and Griffith, 1978). *Klebsiella* accumulates more cadmium in the presence of sulphur and will also change shape as it develops resistance to cadmium, probably as a physiological adaptation (Aiking *et al.*, 1982). Tolerance to mercury may involve the cation being reduced to elemental mercury to increase its volatility, or oxidation to a less toxic form. The methylation of elemental mercury will also increase its volatility, but this also increases its toxicity. A large number of different groups, including fungi, aerobic and anaerobic bacteria use this mechanism, producing methylmercury (CH_3Hg^+) or dimethylmercury (CH_3HgCH_3). This can be readily accumulated by higher organisms and has led to a significant number of human poisonings. Other metals may also be methylated, though with less dramatic effects.

Different detoxification mechanisms are used for other metals. Copper is precipitated in intracellular granules within the microbial cell, a strategy adopted by a range of higher plants and animals dealing with high metal burdens. Sulphur-reducing bacteria are protected from several toxic metals by producing metallic sulphides which precipitate the metal on the cell wall. Their extracellular sulphur can also protect other, more sensitive species growing at the same site and the production of citric acid will have a similar effect (Gadd and Griffith, 1978).

Where there is a source of organic material to supply their carbon needs, tolerant microorganisms may be found in the most contaminated of soils and spoils. The products of microbial metabolism can aid the leaching of the metals from the soil, and increase their availability to higher plants. Then the interactions between metals also become important (Francis, 1985): iron, particularly in the form of pyrites (FeS_2), can generate acidic conditions which increase the mobility and availability of toxic metals. During the reclamation process, it is important to mop up this excess acidity, to reduce metal availability. Again, applying abundant organic matter aids this process by buffering the acidity of the soil and by binding the toxic metals. It also provides a substrate on which the microbial community can grow (Table 7.1).

3.5.2 The use of tolerant microorganisms in degrading organic pollutants

Soil microbial communities around natural oil seepages have had a long time to become adapted to the complex hydrocarbons finding their way above ground. In more recent times, oil spills have become sufficiently frequent for the same effect to be found over a wider area.

Crude oil consists of four basic components that become more

Table 3.1 Major classes of petroleum hydrocarbons

Class	Example	Molecular weight	Biodegradability	Properties in water
Saturates	Alkanes e.g. pentane	Lowest	Highest	Volatile – readily lost
Aromatics	Arenes e.g. low MW – benzene high MW – naphthalene			Forms emulsions and mouses
Asphaltenes	Phenols, fatty acids, ketones, esters, porphyrins			
Resins	Pyridines, quinolines, etc.	Highest	Lowest	Forms tar balls

recalcitrant (less degradable) as their molecular weight increases (Table 3.1). Aromatic hydrocarbons with two to four rings degrade slowly, whereas those with five or more rings are largely intractable (Atlas, 1988). The rate of degradation is also governed by the concentration of a compound, its physical state, its water solubility and its sulphur content (Table 3.2). Thus the relative proportions of the four main classes of hydrocarbons determine the degradability of a crude oil, as do features of the environment into which it has been released. This latter includes the presence of any bacteria with a previous exposure to the pollutant: the damage caused by the *Amoco Cadiz* incident off the Brittany coast was ameliorated by its past history of frequent spills and the enhanced power of its microflora to attack hydrocarbons (Atlas, 1988).

The initial stages of microbial degradation involve oxidation by oxygenases, but the formation of 'mousses' or tar balls in water, or of oxygen-poor conditions in the soil, slow degradation considerably. Under these conditions, some degradation of oxidized aromatics occurs and microbial hydroxylation of the aromatic ring of toluene and benzene then uses water as a source of oxygen (Leahy and Colwell, 1990).

The bacteria use the hydrocarbons as a source of carbon and energy, but their growth can only be sustained by the presence of other essential nutrients, most especially nitrogen and phosphorus. One common technique to accelerate degradation is to apply nitrogen and phosphorus in excess to the spill, so that they are not limiting to microbial growth. Paraffinized urea, which mixes readily with the oils, has been used with success on contaminated soils where traditional mineral fertilizers are lost too readily by leaching.

Table 3.2 Factors affecting the degradation of petroleum hydrocarbons (after Atlas, 1988; Leahy and Colwell, 1990)

Factor	Effect slowing the rate of degradation
Chemistry of hydrocarbons	Proportion of high MW compounds
	Complexation with soil humic materials
	Presence of volatiles toxic to microorganisms
	High sulphur content and high proportion of aromatics
Physical properties of oil	Formation of a stable mousse in sea water, reducing oxygen supply
	Formation of tar balls from resins and partially oxidized asphaltenes limiting oxygen supply
	High loading producing anoxic conditions
	Low solubility and small surface area of hydrocarbons
	Increased viscosity and low rates of volatilization at low temperatures
Environmental conditions	
Oxygen	Lack of turbulence (in protected bays or inlets), small tidal range reduces oxygen supply
	Anoxic sediments poor in oxygen
Temperature	Low temperatures slow biological activity and also retain toxic short-chained alkanes
Reducing conditions	Some components (e.g. benzene, toluene) may be degraded during methanogenesis, albeit slowly
Availability of nutrients	Shortage of nitrogen, phosphorus and iron will limit microbial growth
pH	Extreme pHs limit microbial and especially bacterial activity

Any single species of bacteria or fungi may only be able to degrade a small number of the many hundreds of hydrocarbons present in a crude oil, so a diverse community of decomposers is needed to complete the process. The main bacteria found in both marine habitats and soils include *Achromobacter, Alcaligenes, Bacillus, Flavobacterium, Nocardia* and *Pseudomonas*. Fungal genera found in both environments include *Aspergillus* and *Pencillium* and, additionally, there are other species found only in each habitat. Although the bacteria are probably the more important degraders of these hydrocarbons, each group increases their numbers following a spill (Leahy and Colwell, 1990).

The extent to which this increase in size follows a genotypic adaptation by these populations is not known: much of the increased degradative capacity is due a predominance of certain genera within the microbial community and the inducibility of specific enzymes. However, the capacity to catabolize polycyclic aromatic hydrocarbons (PAHs) is almost

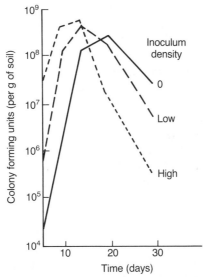

Figure 3.6 An increase and decline in *Acinetobactor* sp. able to degrade biphenyl in a soil laced with the hydrocarbon. The soil was inoculated with three densities of the bacteria. These were indigenous bacteria whose numbers responded to the availability of the xenobiotic as their carbon source. A large inoculum rapidly depletes the carbon source and the population of the bacteria then declines. (With permission, from Focht, 1988.)

certainly genetically based: microbial populations in sediments with this capacity have greater numbers of colonies showing plasmid-linked toluate and naphthalene oxidation (Sayler *et al.*, 1985). Similarly, the ability to oxidize a number of hydrocarbons has been found to be plasmid linked in strains of *Pseudomonas*, conferring a selective advantage on these strains during a spill (Leahy and Colwell, 1990).

Whether genotypic or phenotypic, these adaptations offer the possibility of seeding spills with strains of bacteria and fungi with an enhanced degradative capacity. **Bioremediation** is the deliberate use of the catabolic properties of living organisms to rid ecosystems of organic pollutants. These techniques have been used to reclaim heavily polluted soils, especially old chemical and gas works, spills from storage tanks or other forms of dereliction. Atlas (cited in Leahy and Colwell, 1990) suggested that seeding microorganisms for oil spills should have the following properties:

1. an ability to degrade most petroleum compounds;
2. genetic stability;
3. viability during storage;
4. a high level of enzymic activity in the environment;

5. a rapid rate of growth in the environment;
6. be non-pathogenic or have no capacity to produce toxic metabolites;
7. an ability to compete with the native microflora.

One way to meet these requirements is to seed with a community of species collected from a contaminated site: these will respond rapidly to an increase in their substrate (Figure 3.6), particularly if collected from the same area. Commercial cultures of hydrocarbon degraders are available, though of dubious value (Atlas, 1988).

Potentially useful strains can be isolated from polluted habitats or areas of natural seepage. At grossly contaminated sites, isolates may be cultured from the least polluted areas where the greatest variety of species are likely to be found. These are then grown up in the laboratory (Figure 3.7) and further selection of resistant strains is attempted by increasing levels of exposure to the hydrocarbon. At this stage, an organism is acclimated to the conditions of a particular site, such as moisture, pH and so on. With soils, beds can be prepared to receive the seeding culture: this involves increasing the aeration of the soil by turning, adding supplementary nutrients, such as nitrogen or phosphorus, and perhaps adding enzymes which may commence the degradation. Any substances that are likely to interfere with the process, such as plastics, wood or other rubbish are removed before seeding. Because of the low mobility of bacteria in soil, thorough mixing of the seed culture with the soil is essential. This can be improved if the soil is moved into embankments or confined areas for the degradation period. In some cases special enclosures may be used to optimize temperature and humidity, and a liner fitted to collect the leachate to prevent contamination of the groundwater. Sowing grasses on exposed sites can stabilize the surface and this also helps to establish a community of microinvertebrates early in the remediation of the soil (Puschnig *et al.*, 1991).

The single most important factor is the degree of oxygenation and this may require the soil to be turned several times during the process. This was found to be critical in the pilot-scale facility established by St John and Sikes (1988) to treat the soil from an old industrial site in Texas (Figure 3.8). In this study, microorganisms were collected from the least contaminated parts of the site and an inoculum prepared for the enclosure. They compared the efficiency of the bioremediation process (using levels of phenanthrene as an indicator of the rate of degradation) with single or multiple inoculations, and with and without nutrient additions. All treatments accelerated the loss of phenanthrene under the optimum moisture conditions of the enclosure, reducing its half-life to 33 days, but the degree of oxygenation was shown to be more important than either inoculation or nutrient addition.

Figure 3.7 A 150 litre pilot-scale fermenter. A bioreactor of this type allows precise control of the abiotic factors of the microbial culture to optimize the ecosystem before industrial-scale production. This would test the feasibility of using a bioreactor to harvest the microbial products or to grow up the culture before its release.

Exploiting variability

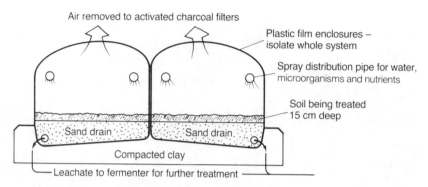

Air removed to activated charcoal filters

Plastic film enclosures – isolate whole system

Spray distribution pipe for water, microorganisms and nutrients

Soil being treated 15 cm deep

Sand drain

Sand drain

Compacted clay

Leachate to fermenter for further treatment

Figure 3.8 The use of a plastic film enclosure to control the bioremediation of a soil from an old crude oil field site in Texas. This allowed treatment of the leachate (via a fermenter) and vapours generated during the treatment period. Nutrients and microorganisms could be introduced via the spray system. The enclosures were used to evaluate various treatment regimes. This established that the single most important variable was the oxygenation of the soil and the degree of contact with the microorganisms. The whole facility was able to increase significantly the rate of phenanthrene degradation (after St John and Sikes, 1988.)

The success rate of seeding appears to depend upon the habitat being treated. It is generally more effective for oil spills in marine rather than fresh waters. On land it is thought that competition from large populations of native decomposers, adapted to the particular conditions within their soils, prevent it being used successfully at the larger scales (Leahy and Colwell, 1990).

3.6 THE USE OF GEMs IN BIOREMEDIATION

While much of this work on bioremediation has used locally adapted strains of microorganisms, increasing attention has been directed to the potential use of genetically engineered microorganisms (GEMs), in the hope of extending the range of xenobiotics that might be degraded. The first ever patented GEM was an organism able to degrade a number of low molecular weight hydrocarbons. However, the potential problems associated with the release of such organisms has meant that it has never been used on a real pollution incident (Atlas and Sayler, 1988) and may only be used to treat waste in containment facilities.

Recombinant technology enables us to equip bacteria with a capacity to degrade a range of hydrocarbons, hopefully without any pathogenic or toxic side-effects. This exploits the ability of genetic material to move between microorganisms (Figure 3.1), a natural phenomenon that occurs for many critical enzymes, especially when the microorganism is under

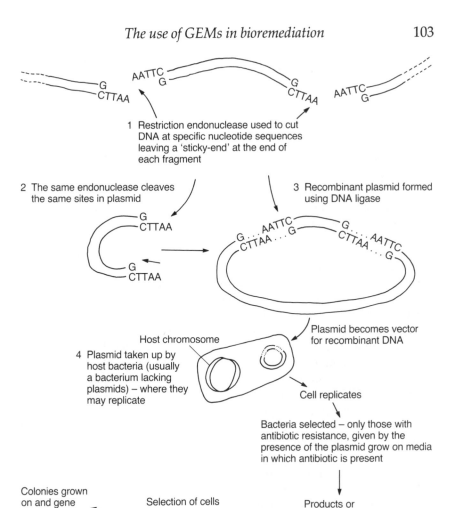

Figure 3.9 A simplified account of the gene cloning techniques used in recombinant DNA technology. Restriction endonuclease enzymes cut a length of DNA into a series of fragments, which are then inserted into plasmids. These act as vectors, introducing the genetic material into a host cell lacking the plasmid, using transformation. These are cultured and can be selected from other colonies by virtue of some other character carried on the plasmid, such as antibiotic resistance. Further selection for the desired gene from the original fragments will use a gene probe, a portion of radioactively labelled RNA or DNA which will hybridize with the desired gene, allowing it to be identifed. After replication, the plasmids can be isolated from these cells, and further treated to isolate the pure gene itself.

stress. Recombinant technology uses transposons or plasmids to incorporate particular genes into the genome of a host cell (Figure 3.9). Thus by selecting particular mutations and facilitating the transfer of genetic code

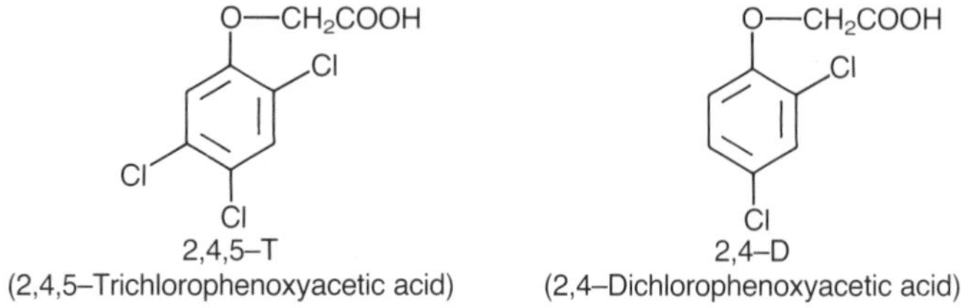

2,4,5–T
(2,4,5–Trichlorophenoxyacetic acid)

2,4–D
(2,4–Dichlorophenoxyacetic acid)

Figure 3.10 The herbicide 2,4,5-T differs from 2,4-D by the presence of the additional chlorine at the number 5 carbon in the ring. This increase in the degree of halogenation makes 2,4,5-T recalcitrant, resistant to biodegradation. In contrast 2,4-D is readily degraded by *Pseudomonas* sp.

between species we can change the complement of existing enzymes within a species and enhance its degradative capacity.

Some organic compounds owe their recalcitrance to the inability of microbial enzymes to 'recognize' the hydrocarbon, perhaps because of the attachment of a halogen to some part of the molecule (Figure 3.10). The presence of a chlorine, bromine or some other halogen prevents degradation as the compound has to be dehalogenated before it can be attacked by the conventional suite of enzymes. For example, the degree of chlorination determines the rate of degradation of polychlorinated biphenyls (PCBs) (Unterman *et al.*, 1988). This degradation can be achieved by 'relaxing' the specificity of the enzymes of a bacteria, or by adding new enzymes to its catabolic pathway. A less specific enzyme will recognize structurally similar compounds, such as a halogenated organic molecule. Pseudomonads able to attack benzoates can be induced to attack chlorobenzoates, by gradually increasing the concentration of the halogenated form in the substrate (Chapman, 1988). This has been termed a 'horizontal' extension of their degradative capacity.

All catabolic processes converge on the TCA (Krebs or citrate) cycle. A vertical extension of this catabolic capacity adds enzymes to the 'top-end' of the pathway, allowing a greater range of compounds to be degraded initially. For example, Timmis *et al.* (1988) describe equipping a strain of *Pseudomonas* unable to degrade salicylate with the capacity to attack both this substrate and chlorosalicylate, using a plasmid vector to add the code for a degradative enzyme. This extended the catabolic pathway for the bacterium allowing its products to pass on to the TCA cycle.

Halogenated organic compounds are not unknown in nature: they are found in a variety compounds of biological origin, including fungal metabolites. Some may be degraded as the result of **cometabolism** – that is, the halogenated compound itself is not the source of energy, but is degraded as a byproduct of the catabolism of some other compound. The pollutant thus represents a secondary substrate. In some cases, organisms adapted to the substrate use it as a primary energy source.

A dehalogenation reaction relies on dehalogenase enzymes that split the carbon–halogen bond hydrolytically. A culture derived from a contaminated soil can be enriched by gradually increasing the amount of the halogenated compound available as a substrate. This acts as a selective pressure, favouring those species which have the enzyme, and those with the most effective version. Some bacteria, most especially *Pseudomonas*, have developed the capacity to dehalogenate a number of hydrocarbons, opening up the possibility of treating soils with high levels of pesticides, PCBs and similar compounds. Increased rates of PCB degradation were achieved using a strain of *Pseudomonas putida* originally derived from a contaminated soil. After enrichment, a laboratory culture of the bacteria could lower PCB levels in a soil by 50% in just three days. Under field conditions however, the need for repeated inoculations of the soil meant that the performance was less spectacular (Unterman *et al.*, 1988).

The genetic information coding for a dehalogenase can be transferred between strains using plasmids or suitable plasmid vectors (Figure 3.9). By equipping one strain of *Pseudomonas* with several plasmids, a single species capable of oxidizing a range of hydrocarbons has been produced. It may also be possible to construct a complete catabolic pathway within a single bacterium. This would be particularly useful in hydrocarbon degradation, as complete mineralization of many compounds involves several steps and this normally requires a community of organisms (Hardman, 1987). Engineering a single species with this capacity would avoid some of the problems of interspecific competition in a diverse microbial community. It also simplifies the optimization of the bioremediation process in a containment facility.

The first organism to be engineered to degrade a specific compound was used to attack the herbicide 2,4,5-T (Figure 3.10). Degradation of 2,4-D was known to be plasmid-linked in *Alcaligenes paradoxus*. A number of species of bacteria were collected from waste dumps and mixed with other strains with known catabolic powers. The concentration of 2,4,5-T was gradually increased in the culture. One strain of *Pseudomonas cepacia* was derived with a capacity to degrade both 2,4-D and 2,4,5-T, with either herbicide being used as the sole source of carbon. This has been used successfully to decontaminate a soil. How the presence of the other bacteria, and other plasmids, assisted in the development of this capacity is not known (Hardman, 1987).

GEMs engineered for degrading halogenated compounds are likely to be confined to treating wastes before they enter the environment, rather than being used broadcast on a contaminated site. At this stage we know too little about the mobility of genetic material between species and the impact of engineered organisms on natural communities. By using anaerobic digestion (section 9.2) confined to a fermenter, we can accelerate the degradation of a waste before it leaves a site and lower the chances of its microbial community surviving outside the bioreactor. We may also be able to recover useful products, such as methane, in the process.

3.7 RECOMBINANT TECHNOLOGY AND THE RELEASE OF GEMs

Conventional techniques of plant and animal breeding are limited to the genetic library within a species. In most organisms, the small chances of producing fertile offspring from crosses betweeen species meant that novel genes could rarely be introduced into domesticated stock. Recombinant technology changes that, allowing DNA to be moved between species, especially in the bacteria. Many microorganisms and viruses have mechanisms for transferring and ensuring the survival of their genetic code, often by using a higher organism. Recombinant technology exploits this ability, allowing us to produce **transgenic organisms** which express characters acquired from foreign DNA.

Higher plants are especially amenable to manipulation because they can be readily grown in tissue culture. Tissue sections will take up DNA from solution when the permeability of the cell membrane is temporarily increased. The foreign DNA may then be incorporated into the host chromosomes. Similarly, DNA can also be inserted as viral or bacterial DNA where these produce lesions or galls in the host plant. For example, *Agrobacterium tumefaciens* carries a plasmid which induces crown galls in a host plant and, in the process, part of the plasmid may become inserted into the host DNA. This foreign DNA then becomes part of the genetic stock available to the plant breeder. These methods have been used to produce several transgenic strains of tomato plants. Amongst them are varieties with increased resistance to tobacco mosaic virus, resistance to the herbicide glyphosate (to allow spraying of the weeds surrounding a crop) and a capacity to express the insect poison, δ-endotoxin, which is produced naturally by *Bacillus thuringiensis* (section 5.7). These have now all been tested in field trials (Gaertner and Kim, 1988).

But there are inherent dangers here. Placing novel code into an organism which is then released to the wild could allow the DNA to find its way into other, unintended species. Herbicide resistance, for example, might possibly be transferred from a genetically engineered plant to a weed species, or a novel gene could become part of a pathway that leads to the production of a toxin. While there are barriers to this transfer, such movement is known to happen naturally in bacteria even if at a low rate (Table 3.3).

In wild populations, non-lethal mutations have naturally produced new and more virulent forms of particular species. With transgenic organisms however, it is the scale and pace of the change that make the

Table 3.3 Mobility of recombinant DNA between bacteria (after Miller, 1988)

Mechanism of movement	
Conjugation	Most probable mechanism of transfer in the environment. Known to occur in the bacteria of the vertebrate digestive tract, in sewage treatment plants, rivers and soils
Transduction	Range limited by host range of mediating bacteriophage. Generalized transduction most important as all genes are transferred with equal efficiency
Transformation	True role in transfer in environment largely unknown
Barriers to transfer	
Environmental	Association between organisms most likely only if they share similar habitat requirements. Conjugation occurs over a limited temperature range
Gene entry barriers	• The sex pilus of a donor cell needs to recognize the cell envelope of a recipient cell, and this can limit transfer between species
	• Transduction is also limited by the host range of the mediating bacteriophage
	• Transformation limited to competent cells.
Barriers to establishment	• Host endonuclease can cleave and inactivate foreign DNA to protect self DNA
	• Environmental stresses or phages can inhibit endonuclease activity and some foreign DNA commonly escapes cleavage
	• Presence of other foreign DNA which is incompatible prevents establishment and plasmid is lost
	• Transfer of chromosome DNA must have sequence homology with host DNA to be incorporated into host chromosome at crossover. Transposons do not require this for transfer
Barriers to gene expression	Phenotype may not express new gene due to failure to read properly or because product is not modified enzymatically to activate it

outcome more uncertain. Past experience with selective breeding of domesticated plants and animals or the release of modified viral forms, such as vaccines, suggests there may be relatively few problems. This might reflect the non-competitiveness of these inbred genotypes, which usually, but not always, lose out to the more vigorous heterozygotes of the wild type. Even so, a significant number of our domesticated plants and animals have become pests as feral populations (Williamson, 1988).

Because bacteria proliferate asexually, foreign DNA can be rapidly fixed within a population, especially if it confers some selective advantage. Although the novel code in the GEM may have little ecological impact, in the genotype of another species – perhaps in a different location in the bacterial chromosome – it might produce a combination which is more serious. This could be a particular problem with transposons which do not require a homologous sequence for transfer (Figure 3.1).

Models to predict the rate of spread of novel genetic code have begun to be developed. These incorporate measures of dispersal of the transgenic organism itself as well as the foreign DNA. Many are based on simple population models (section 4.2) and incorporate terms for changes in gene frequencies (Kim *et al.*, 1991; Manasse and Kareiva, 1991). Each of these terms will depend on the life history of the organism and the survival advantage the gene itself confers. To ensure that the GEM or its novel code does not persist, some have suggested that a modified bacterium should only be released if it is at a competitive disadvantage to other species in the receiving habitat: the danger with a competitively superior GEM is that we have little prospect of containing it within the target area. Even inferior GEMs will still carry a risk of persisting beyond their useful life – Kim *et al.* cite three possible mechanisms that might allow them to survive.

- They acquire an advantageous mutation and then go on to increase their population size.
- They are able to establish themselves in a highly structured habitat, with very discrete microhabitats from which it is difficult to dislodge them – for example, the soil.
- They eventually become competitively superior and then displace the native species.

Given its mutation rate and its rate of population increase, models can predict how quickly a competitive strain might be produced, or how quickly an inferior strain would be lost. The optimum solution is for the GEMs to have a low persistence in the environment and to use a strategy which requires large and repeated inoculations of organisms with low fitness (Kim *et al.*, 1991).

Many GEMs may be at a competitive disadvantage for reasons we have already discussed, such as having to meet the costs of synthesizing additional proteins. The induced production of an insecticidal protein δ-endotoxin by *Pseudomonas fluorescens* represents 30% of its total protein output (Gaertner and Kim, 1988). Plasmid-free strains of a bacteria will have lower energetic costs and might be expected to outcompete a GEM of the same species. The presence of plasmids increases the generation time of *E. coli* and there is also evidence that the products of the foreign gene's expression impair some aspect of the host's metabolism (Lenski and Nguyen, 1988). Not surprisingly therefore, when the selective pressure favouring the plasmid disappears the plasmid is often lost from the population.

Nevertheless, we should not assume that all genetically manipulated species will be at a competitive disadvantage. Where there is some selective benefit, the organism could become dominant, threatening the survival of its competitors and the stock of genetic variation which they represent.

We thus have to balance the risk of some deleterious effect against the benefits of the GEM. The risk is measured as the probability of the event occurring multiplied by its cost; the longer the GEM is in the environment the larger the risk. If the escape of the GEM has no environmental impact the risk will be deemed minimal or zero. Using this definition, the higher the utility of a GEM the greater the risk we would be prepared to take, but our problem is quantifying these possibilities, especially with the complicated interactions that occur between species in natural communities. For example, the introduction of a non-specific bacterial toxin into an agricultural plant could decimate both an insect pest and its natural enemies. Such risks can be minimized by making the agent – the recombinant DNA or its product – as specific as possible, a principle that holds for other forms of biological pest control.

Any risk assessment before a release can only provide a list of negative results, confirming that an organism does not have a particular effect (Simonsen and Levins, 1988). Consequently, any release has to be thoroughly monitored for its ecological impact and to map the movement of the recombinant DNA between species. If a specific code does cross between species or genera, some assessment of its risk to its new owner and the commmunity has to be made. At this point there is little that can be done anyway – the genetic material is in the environment with a proven capacity to spread and we have limited scope for containing it further.

Williamson (1988) suggests that the best guide to these potential problems is our experience of other invasive species. These are not always wild types, with little competitive advantage: weeds, for example, do not divide neatly between wild and domesticated plants. What seems to be more important is the habitat into which they are released and their scope for population growth. Eurytopic species, opportunists capable of rapid population growth, have frequently caused large-scale extinctions of indigenous species (section 6.5).

Devising safe practice for the release of GEMs will thus require relatively complete knowledge of the ecology of the released species and the community it is about to join. The soil, for example, is a highly structured habitat, which does not normally allow the transfer of microorganisms or their DNA over long distances. We would thus expect the dispersal of GEMs to be very slow in this environment, compared with aquatic environments (Stotzky *et al.*, 1991). Transduction in the soil, however, may be aided by bacterial colonies adhering to soil particles. The survival of the introduced DNA will also be important – high clay soils may hold and protect bacteriophages, increasing the persistence of these viruses (Stotzky *et al.*, 1991). Conjugation can be very efficient in liquid cultures in the laboratory, though these rates of transfer are unlikely to apply to natural habitats (Istock, 1991). Bacterial viruses may be a more significant route of transduction in waters where there are low bacterial numbers. Similarly, the motility of other DNA, for example as pollen grains from transgenic plants, will depend upon a larger scale ecology.

It may be possible to identify species that are potential recipients of the foreign DNA, based on their ecology and physiology. Various techniques have been developed to detect recombinant DNA in laboratory cultures and these are now used in the field to trace its movement. These methods include hybridization with a matching piece of DNA or RNA which has been labelled with a radioactive isotope (usually with ^{32}P in the molecule) to create **gene probes**. The detection of the radioactivity indicates that the DNA has found its way into a different organism. Antibodies, specific for the products of the recombinant DNA can also be used to map, indirectly, the distribution of the gene. Other methods use **gene markers** – particular genes associated with the recombinant DNA and which flag its presence by their activity. One example is the *lacY* and *lacZ* genes, which allow *Pseudomonas fluorescens* to use lactose and, in doing so, betray themselves by turning the bacterium blue in the presence of a dye. Other gene markers include antibiotic and toxic metal resistance, both of which are plasmid linked.

The first GEM released into the environment was *Pseudomonas syringae*, using a strain which lacked a specific gene. This was released to compete with the native leaf-colonizing species and, by displacing it, improved the frost resistance of a field of strawberries by 1°C. This scale of improvement may or may not be significant, but it demonstrates how the release of a GEM could be used to interfere with the composition of a community, in this case by competitive displacement. Here the release appears to have been controlled: the GEM could not be detected more than 20 m away from its point of inoculation, and it did not become established in the population (Gaertner and Kim, 1988).

By 1989 there had been six planned releases of genetically modified organisms in the UK and 59 in the USA, mainly of modified plants. Regulations in most countries require that planned releases of GEMs have to be notified to a review committee for vetting and for licensing. In the USA this is the BSCC (Biotechnological Sciences Coordinating Committee). The scale of the review depends upon the pathogenicity of the organism to be released and the size of the trial to be undertaken. In the UK similar work is undertaken by the Health and Safety Executive. These panels of experts are building up a body of experience to set standards for risk assessment and containment measures. As yet there are no internationally agreed regulations, despite the ease with which many of these techniques can be developed. International co-operation in both designating GEMs, establishing patent rights, in developing safe practices and control will be crucial as these become more widespread.

Summary

Organisms may adapt to changing conditions in their environment by either phenotypic or genotypic change. The capacity to adapt genetically is a function of the amount of genetic variation in a population. Stress can serve to both increase this variation and accelerate the pace of genotypic adaptation. According to one simple classification, organisms may be divided into specialists or generalists – depending upon the range of environmental conditions they can accomodate and the resources they use. Specialists tend to have narrow niches, adapted to a small range of conditions, and are typical of relatively constant habitats. Competition is probably the main force defining their niche width. Generalists are opportunists, found in more variable conditions and more likely to colonize stressful environments. Species able to adapt to polluted environments typically have more genetic variation within their populations and can be classified as broad-niche species. Such species are most likely to produce tolerant races.

These tolerant ecotypes have been used to monitor pollution, especially benthic invertebrates in marine environments. They are typically highly heterozygous organisms, as indicated by the protein polymorphisms of their enzymes. Metal-tolerant grasses are used to revegetate spoils and wastes. Again, these ecotypes are usually confined to contaminated soils because they are at some disadvantage in competing with non-tolerant varieties on normal soils. Very often resistance to pollution appears to incur costs for both higher organisms and genetically engineered microorganisms.

Microorganisms also adapt to high levels of metals and hydrocarbons, and their capacity to degrade crude oil is exploited in both aquatic and terrestrial ecosystems. Through a process of selection, often using locally adapted populations, these can be induced to degrade the more recalcitrant forms, including halogenated hydrocarbons such as pesticides. With recombinant DNA technology it is possible to add additional genetic code to some organisms and extend their catabolic powers. Not only can this accelerate their degradation of these compounds, it can, in bioreactors, avoid the competitive interactions which might inhibit the process. The question of releasing GEMs into the environment is still being debated, but the whole discussion is slowed by our lack of basic information. A number of models have been developed which attempt to measure the spread of GEMs or of novel genetic material through different ecosystems.

Further reading

Ginzburg, L.R. (ed.) (1991) *Assessing Ecological Risks of Biotechnology*, Butterworth-Heinemann, Boston.
Omenn, G.S. (ed.) (1988) *Environmental Biotechnology*, Plenum, New York.
Shaw, A.J. (ed.) (1990) *Heavy Metal Tolerance in Plants: Evolutionary Aspects*, CRC, Boca Raton, FL.

A humpback whale off Cape Cod, Eastern USA. (Photograph courtesy of Frank Clark.)

4

Managing populations

Some populations are exploited for their economic value, as a source of food or of raw materials. Others are seen to have some intrinsic value, perhaps as game, or some scientific or aesthetic interest, and are then worthy of conservation. In either case, man adopts the same strategy: encouraging the population to grow at its fastest rate within the resources available. The environmental conditions and resources of certain species can be controlled to increase their productivity – we cultivate them. Others are exploited in their 'wild' state. Then we can only manage our effort in catching a species and the size of individuals we harvest. This is true of open-sea fishing and some aspects of forestry. Here we aim to maintain the capacity of the population or stock to provide maximal yields from one season to the next and to avoid reducing a population to a level from which it cannot recover. In managing its exploitation, we thus seek to conserve both its numbers and the population's capacity for growth. For this we need to know both the size of the population and its rate of growth. In the first part of this chapter, the theoretical background to population growth is reviewed, together with the concept of a 'maximum sustainable yield', a central principle in the regulation of fisheries, forestry and game management.

The size of a population is only one indication of whether it is endangered: under the pressures of exploitation or the fragmentation of its habitat, a population may decline to levels from which it is unable to recover, and the threat will be determined by its rate of population growth. We need to review each of these sources of uncertainty in endangered species and, in the second part of this chapter, we use these models to predict the probability of extinction. Finally, the genetic implications of small population sizes are reviewed.

To be modelled, a population has to be defined in both space and time – for example, the number of fin whales (*Balaenoptera physalus*) in the north Pacific in 1900 – though well-defined boundaries usually exist only in the mind of the researcher. Natural populations are rarely confined by any such notional boundaries. Population models can be further simplified if it is assumed that there is no migration of individuals across these boundaries. The **population** then consists only of those individuals of the same species capable of breeding with each other, within the defined area at the specified time.

Most field studies are unable to measure total population size and instead use population density – the number per unit area. Although we shall refer to total number or abundance throughout, all that follows can be equally applied to populations measured in terms of densities.

4.1 POPULATION GROWTH

A univoltine species is one that completes its life cycle in a single year, with all adults dying at the end of the season. For example, many species of insect are univoltine, only sustaining their populations by the eggs they leave behind to overwinter. They are said to have discrete generations and their population size next year will depend on the number of adults (and more especially, the number of egg-laying females) alive this year.

Not all of the eggs laid will survive to become the adults in the next generation. The rate at which adults in one generation produce new adults in the next generation is called the **reproductive rate** (R_0):

$$R_0 = \frac{N_g}{N_0} \tag{4.1}$$

where N_0 is the number of adults in one generation and N_g is the number of offspring reaching breeding age in the next generation. R_0 is simply the average number of offspring reaching reproductive age produced per adult. The population size will only increase in the next generation if R_0 is greater than one.

Many plants and animals have life cycles longer than one season and their populations will include survivors from previous generations. Their generations overlap. Both survivors and offspring thus contribute to the size of the new population. A census taken before and after some interval of time will give the **net reproductive rate** (R_N):

$$R_N = \frac{N_t}{N_0} \tag{4.2}$$

where N_t = number alive (survivors and offspring) after the time interval t and N_0 = number alive before t.

R_N is a net reproductive rate because it combines both the birth rate and the survival rate per individual during the time interval t. Again, R_N represents the average rate of increase (or decrease) per individual, over the period of the census.

The time interval can be any length of time, greater than or shorter than a generation. The length of a generation (g) is measured as the mean time elapsed between the birth of the parents and the birth of the offspring. If the interval is set to one generation ($t = g$), R_N and R_0 measure the same thing – the average rate of change per individual over one generation.

These are termed finite rates of change as they measure incremental changes in population size between discrete time intervals. However, for many species reproduction and death occur continuously. Some organisms, such as bacteria, also have very short generation times and their rate of population increase is continually changing. We then need to know the increase over a small increment of time – an instantaneous rate of change.

This instantaneous rate is derived as the natural logarithm of the finite rate of increase:

$$\text{instantaneous rate} = \log_e \text{finite rate.}$$

So the instantaneous rate of change per individual in populations with overlapping generations is

$$r = \log_e R_N.$$

The variable r is termed the **intrinsic rate of population increase**. It represents the net effect of birth rates and death rates at a particular time.

If there are no deaths, this rate of increase is due to the additions from births alone and r is simply equal to the instantaneous birth rate. More usually, the rate of loss due to deaths has to be subtracted from the rate of addition:

$$r = (b - m)$$

that is, the rate of change per individual represents the difference between the birth rate per individual (b) and the mortality rate per individual (m). Again, these are instantaneous rates. Obviously, without immigration a population can grow only if the birth rate is greater than the death rate.

The rate of change in the population size (dN/dt) can now be written as

$$\frac{dN}{dt} = rN. \tag{4.3}$$

That is, the change in the population size over the time interval t is equal to the rate of change per individual multiplied by the number of individuals. This rate of change is thus proportional to the population size, with r being the constant of proportionality.

With a fixed value of r, the population at any time can be derived by integration of equation 4.3. This will give

$$N_t = N_0 e^{rt} \tag{4.4}$$

where e $= 2.71828$ (the base of the natural logarithm); t is the time interval; N_0 is the population at time zero; and N_t the population after time t.

Managing populations

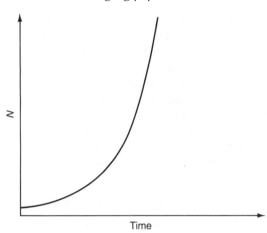

Figure 4.1 The exponential growth in a population with time, with no limitations on population size.

Thus, assuming that r does not change from one generation to the next, we need only know the size of the starting population (N_0) and the length of time over which the population has been growing to derive N_t. Plotting N_t for a number of time intervals produces a curve of exponential population growth (Figure 4.1). Any value of r greater than zero ($b > m$) will produce this shape, but the curve will rise more steeply if r is larger. If the death rate exceeds the birth rate, r will be negative and N_t will then decline.

It may seem unrealistic to plot the decline or growth of a population over an extended period using a fixed value of r. The rate of increase of a population will vary with its age structure: if there is a high proportion of old individuals, the birth rate is likely to be low and the death rate high. Similarly, a rapidly growing population with its high birth rate may be thought to have a value of r which is continually increasing. In fact populations growing in an exponential manner can be shown to achieve a 'stable age distribution', where the fraction of the population in each age group tends to become fixed. Here there is a high proportion of individuals in the youngest age groups and the balance between birth and death rates for each age group becomes constant. Under the conditions of exponential growth, the assumed constancy of r for the population is therefore justified.

4.2 THE LIMITS TO POPULATION GROWTH

No population can continue to grow indefinitely. Food and other resources in an ecosystem are finite and this imposes a limit on the number of individuals it can support. This is called the **carrying capacity** (*K*), the maximum number (or density) of individuals of a species that an ecosystem will sustain. As the carrying capacity is approached, so more individuals are competing with each other for the dwindling resources which remain. The intensity of this intraspecific competition for space, food or other resources increases as the number of competitors grows.

The availability of limited resources is thus dependent on the population density. These **density-dependent** factors will tend to stabilize a population around the carrying capacity: when *N* is greater than *K* the population must decline as demand for resources exceeds supply. Any unused resources will support population growth up to *K*. This can be important in population management because it will determine the rate at which new individuals are recruited to the stock. In general, fishery managers assume that juvenile mortality is density-dependent, increasing as the population density increases, whilst adult mortality is taken to be density-independent, typical of species with a high reproductive rate (Sinclair, 1989). These are often highly variable populations, sometimes termed *r*-selected species because of their reproductive rate (section 5.2).

The limits to the growth of the population can be incorporated into the model. A proportion of the carrying capacity will already be occupied by the existing population, that is *N/K*. The remaining capacity for population growth is therefore

$$1 - \frac{N}{K}.$$

As *N* increases this capacity is reduced and this acts on *rN* to reduce the rate of population growth:

$$\frac{dN}{dt} = rN\left(1 - \frac{N}{K}\right). \tag{4.5}$$

Plotting this gives a logistic curve of a population growing in a limited environment (Figure 4.2). The effective rate of population growth is increasingly slowed as the habitat fills. With no population growth possible at *K*, the birth rate must here be balanced by the death rate. In this simple model, the population is thus regulated by density-dependent factors.

The carrying capacity represents a stable equilibrium, established by this density-dependent regulation of population growth. The size of the equilibrium population is determined by the capacity of the ecosystem, but the speed of approach to *K*, and also the time taken to return to it after disturbance, depends on *r* alone (May, 1981b). Thus *r* is of crucial importance when populations are pushed from their equilibrium density by exploitation. Usually, the rate of population increase (and its variability) is much more important to the survival of populations facing

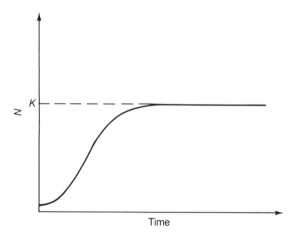

Figure 4.2 The sigmoidal growth of a population in an environment limiting the population to a maximum size, *K* (the carrying capacity).

extinction and *K* only becomes significant when space is limited. For this reason, most conservation efforts are directed at species with low values of *r*, typically larger animals with long generation times and a low reproductive potential or **fecundity**. These are commonly species with naturally constant population sizes, often termed *K*-selected species (section 5.2).

Notice that the most rapid increase in the population is where the line is closest to the vertical, at some value of *N* below *K*, in a region where competition for resources is not yet slowing population growth substantially. This can be seen more clearly by plotting the rate of increase against time (Figure. 4.3). The line forms a parabola: the rate of increase rises because *N* gets larger (*r* is constant). The subsequent decline in the rate of increase follows as growth is slowed by the limiting environment. Eventually these density-dependent effects halt all population growth as the carrying capacity is reached.

With increasing competition for resources, individuals not only reproduce at a slower rate but their growth rate is also slowed and they may reach a smaller size as adults. As a result, increases in numbers or weight (biomass) will follow the same sort of pattern to that in Figure 4.3. On the other hand, the rate of growth in an exploited population may increase as harvesting leaves more food or other resources available for the remaining stock. Exploited fin whale populations appear to grow faster and achieve sexual maturity earlier (Meredith and Campbell, 1988). Indeed, ovulation and the frequency of pregnancy may also be determined by the available food resources.

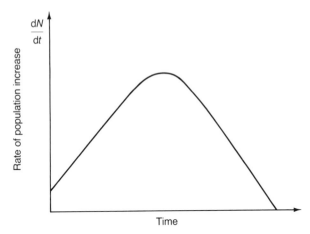

Figure 4.3 The rate of increase in population size with time. This rises as N increases but begins to decline as density-dependent factors slow population growth when N approaches K.

Such density-dependent factors become important in determining how many individuals are added to a catchable stock. When properly managed, this allows the productivity of a population to rise to match the losses to the harvest. Ideally, the stock would then remain constant from one year to the next.

4.3 DETERMINANTS OF STABILITY

In reality K is unlikely to be fixed over any extended period of time: fluctuations in environmental conditions will continually alter the size of population the ecosystem can support. Density-dependent effects are also unlikely to operate instantaneously on the population, and there are likely to be time lags in its response. For example, food shortages in one generation may determine the number of offspring in the following generation. Such delays can have consequences for the stability of the population. If the time lag is longer than the time taken for the population to recover its equilibrium after disturbance, the return of the population to K will be relatively unchecked and, as a result, more rapid. The population will then overshoot K and oscillate around its carrying capacity before settling back down to that equilibrium. With a larger time lag relative to the recovery time the oscillations persist as a series of cycles (stable limit cycles) around K. Even chaotic population movements can be predicted from such simple models without needing to blame environmental fluctuations (May, 1981b). These models may help to explain the variability associated with real populations, especially in exploited and endangered species. They also highlight the dangers of basing fishery management decisions on limited data in a highly variable world.

Since the return time to K depends on r, so will the response of the population to environmental fluctuations: populations with a large r will respond more quickly to change. Similarly, in a variable environment the overall carrying capacity will depend on the size of the fluctuations in K and large-scale fluctuations can hasten the loss of small populations. The fragmentation of many natural habitats effectively reduces their carrying capacity, representing one of the principal threats to many endangered species. Exploited populations that are held below their carrying capacity by harvesting have a reduced capacity (a smaller N) to recover from such environmental variability.

4.4 EXPLOITING A POPULATION

Harvesting the maximum number of individuals would mean catching the whole population and thereby destroying it. Although this can make short-term economic sense, a more enlightened approach would harvest only that which can be replaced by growth in the population. This is the idea of a **maximum sustainable yield** (MSY), a central concept in the management of forests, fisheries and game. MSY represents the maximum size of harvest which a population is able to restore by growth in its numbers, biomass or both over a given period. Ideally, the population remains the same size: the size that gives its maximum capacity for growth. The aim is to avoid overfishing which will endanger the stock. There are two ways in which this may happen:

1. growth overfishing: where too many small fish are taken so that too few reach their full weight. This is common in many fisheries;
2. recruitment overfishing: where the number of spawning females is significantly reduced, thereby lowering recruitment.

The latter is relatively uncommon, but is disastrous for the long-term survival of the stock.

What constitutes the population or stock is often defined by the fishing activity (including such details as the mesh size) as well as the details of its geography (Gulland, 1983). Individuals only become available for catching when they are of sufficient size – when they have been 'recruited' to the **catchable stock** (N_c). We therefore have to distinguish between births which add to the whole population and **recruitment** which adds only to the catchable stock. The biomass or weight of the catchable stock is increased both by recruitment of new individuals and the growth of new tissues. The model that was previously used to describe a population's growth can be applied to growth in the biomass of the catchable stock: we need only make N equivalent to the biomass in the catchable stock.

Many of the principles to be discussed below apply to forest management and to a lesser extent to game management (where the emphasis is upon numbers and not biomass). However, the ideas have originated in and retain the nomenclature of fisheries management. Two different models of management are considered here, the surplus yield model and the dynamic pool model, both derived from the logistic equation previously described. However, we first need to review how an estimate of the catchable stock can be made.

4.4.1 Estimating the catchable stock

In a fishery, the death rate consists of two components – natural mortality (M) and the fish taken in the catch (F). If the population is not to decline, these losses have to be met by recruitment of new fish (R). If the biomass of the stock is to be maintained, tissue growth (G) and recruitment have to match these losses:

$$F + M = R + G. \tag{4.6}$$

F, the instantaneous fishing mortality rate, is the product of two components:

$$F = qE \tag{4.7}$$

where q = constant measuring the catchability of the fish using a particular set of gear (coefficient of catchability) and E = fishing effort (for example boat-hours per unit area).

In effect, q is the proportion of the catchable stock caught per unit effort, or more simply, the efficiency of the fishing operation. So qE is the fishing mortality rate (F) for a given level of effort (E). The **yield** (Y) to the fishery is therefore the fishing mortality rate multiplied by the size of the catchable stock:

$$Y = FN_c. \tag{4.8}$$

A yield is at equilibrium when the population can maintain itself by replacing the harvest through G and R (equation 4.6).

The catch per unit effort (U) is simply the yield divided the effort:

$$U = \frac{Y}{E} \tag{4.9}$$

or from equations 4.7 and 4.8, $U = \dfrac{FN_c}{E} = \dfrac{qEN_c}{E}$.

As q is a constant proportion of N_c, if N_c increases, we expect more fish to be caught for each unit of effort. We can therefore derive N_c by simply dividing the catch per unit effort by q:

$$\frac{U}{q} = N_c. \tag{4.10}$$

Thus the catch per unit effort (U) is an indication of the stock size (N_c). Any decline in catch size will indicate a decline in the catchable stock. However, this can only be an accurate indication of N_c when the harvesting is at equilibrium – when the yield to the fishery is matched by stock replacement. If Y is greater than the rate of replacement, the stock will have been overfished and the estimate of stock size based on U will be too high.

The assumption that the catchability of the fish (q) remains unchanged is critical. We need to measure the variability associated with q to check whether a given amount of effort always results in the same yield. One method is to check whether a percentage increase in effort realizes the same proportional increases in the size of the catch. A range of factors can alter the relationship between q and N_c, including the previous fishing activity, changes in the experience of the fishing crews or the way in which the fish are distributed in the sea, especially those species which shoal. For these reasons, q is usually estimated for a fishing season, with a particular set of fishing gear (Gulland, 1983).

To manage the fishery, some other estimate of fishing mortality, independent of catch size, is needed to measure its effect on the catchable stock. By tagging and releasing fish, the rates of mortality for fish of different ages can be determined, and their appearance in the nets gives some estimate of both F and M. Tagging can also allow estimation of the rate of recruitment. On its own, catch per unit effort provides fishery managers with some indication of N_c based on the size of catches landed at the quayside, but there are numerous sources of uncertainty and this is an unreliable guide to the size of the harvestable stock (Cushing, 1981).

4.4.2 Surplus yield models

Any growth in a population above that required to maintain a constant size can be said to be surplus. An equilibrium yield will harvest that surplus and no more. Surplus yield models aim to predict the value of F that will permit the maximum equilibrium yield. Given an estimate of the catchable stock, the fishing effort needed to harvest that weight of fish can be determined.

These methods are based on the logistic model of population growth, though this now only applies to the catchable stock (N_c), measured as numbers or as biomass. Incorporating estimates of catchable stock and of fishing mortality into the model allows for the rate of growth to be reduced by the losses to the fishery. This is called Schaefer's model:

$$\frac{dN_c}{dt} = rN_c\left(1 - \frac{N_c}{K}\right) - FN_c \qquad (4.11)$$

where FN_c represents the yield to the fishery, as in equation 4.8 above. The effective rate of increase in the catchable stock is thus slowed due to harvesting. An equilibrium yield is achieved when the rate of harvesting is balanced by the rate of replacement from both recruitment of new individuals to the stock and of new tissues. This can occur at any population size:

$$rN_c\left(1 - \frac{N_c}{K}\right) - FN_c = 0.$$

This does not, however, indicate the maximum possible yield that can be replaced.

As we saw in Figure 4.3, with the logistic model the rate of increase in the population or its biomass describes a parabola with time. As N rises with time, this rate of growth follows a similar curve when plotted against N (Figure 4.4). This growth represents the surplus that can be taken by the fishery at equilibrium. The graph shows the rate at which new individuals are added to the population and can be harvested without the population declining. These are thus equilibrium yields for different population

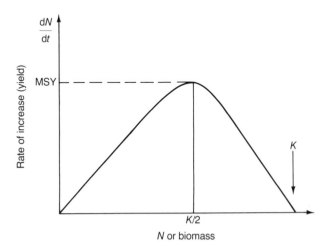

Figure 4.4 The rate of increase in numbers or biomass also rises and falls as N (or biomass) approaches the carrying capacity. The maximum rate of increase represents the maximum sustainable yield (MSY), and this is equivalent to a half of K.

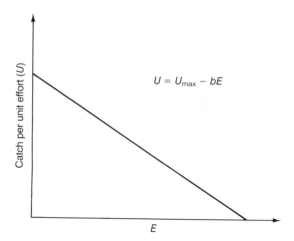

Figure 4.5 A line of best fit for catch per unit effort against effort E. This would be calculated using linear regression to derive the equation for the line, using data from fishing records using the same fishing gear. U_{max} is the maximum catch per unit effort, the intercept on the y axis, and b is the slope of the line.

sizes. The maximum sustainable yield (MSY) is achieved where the rate of replacement is at its greatest: with the logistic model this is when N_c is half the carrying capacity.

Unfortunately, K is rarely known for a fishery and managers have to decide on MSY using the evidence of their catch sizes. Using past records for seasons in which the same fishing gear has been used, the observed relationship between catch per unit effort U and fishing effort E is established using statistical analysis (Figure 4.5). From this the manager can predict the yield as a function of fishing effort and then derive the appropriate effort for the MSY (Figure 4.6).

There are several dangers here. The assumption that the fishery is in equilibrium, with yields balanced by replacement, cannot be tested using the catch data alone. Also, the logistic model is an incomplete description of a population's dynamics because it does not allow for different age distributions, and assuming that MSY and E are fixed in relation to K is unrealistic in populations with variable age structures. It is equally unrealistic to assume K remains constant for any length of time. Altogether, this makes it difficult to regulate the fishing effort close to MSY over an extended period, especially when managers are relying on incomplete or inadequate data sets for their estimates of F, K or MSY. Nevertheless, this has been the principal method of determining fishing activity in most fisheries around the world. Pitcher and Hart (1982) give a clear description of the collapse of the Peruvian anchovy fishery because

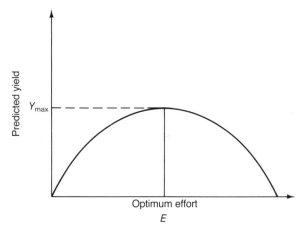

Figure 4.6 The prediction of maximum yield (Y_{max}) from fishing effort E. This can be shown to be:

$$Y = U_{max}E - bE^2$$

where U_{max} is the maximum catch per unit effort.

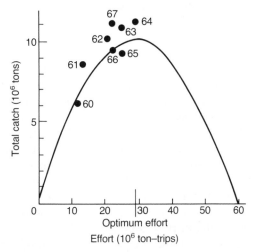

Figure 4.7 The prediction of the yield from the Peruvian anchovy fishery (*Engraulis ringens*) before its collapse in 1972. The yields achieved in different seasons are also shown with the effort needed to achieve them. (From Boerema and Gulland, 1973; with permission of the editor, *J. Fish Res. Board Can.*)

of a simple reliance on MSYs (Figure 4.7), without due allowance for the inherent variability of crucial ecological processes.

There are also economic reasons for not fishing too close to the MSY. If the unit price of the yield is fixed and the price of the fish remains the

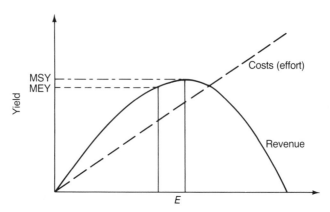

Figure 4.8 As the yield rises and falls so does the revenue which it generates. Costs will increase with fishing effort (*E*). The difference between revenue and costs is the profit. The greatest difference occurs somewhat below MSY. This is maximum economic yield (MEY).

same whatever the supply, the revenue from the catch will simply follow the yield curve. Additional fishing effort will incur additional costs: in this case, costs are assumed to rise in a simple manner with fishing effort (Figure 4.8).

The profits from the catch are the difference between the revenue received and the costs. Here, the greatest profits are made by fishing at some level before the MSY is reached. Thus the maximum sustainable yield does not necessarily produce the maximum economic yield.

There are other, more realistic variants of the Schaefer model. Some adaptations of the principle do not rely on the logistic model and fit different population growth curves. Others introduce a time lag so that recruitment is derived from a previous population size. An alternative strategy, sometimes used in managing game, particularly large mammals, sets a constant effort fixed close to a population size that will give the MSY. Yields then rise and fall in response to *N* alone: if *N* should fall, so will the yield per unit effort. If, on the other hand, effort was allowed to increase when yields are falling, the population would be driven to lower levels and eventual extinction. In a fishery when estimates of N_c are necessarily tentative, it makes more sense to relax the effort when yields fall: this allows the catchable stock to recover more quickly and also reduces the costs (Figure 4.8). One alternative is to harvest a constant catch, although such fixed yields will endanger stocks when populations fall very low. A constant escapement policy, allowing a fixed proportion of the yield to escape, can mean closing the fishery down during years with low numbers. This problem is avoided with the constant effort strategy – it has the advantage of stabilizing the investment in the fishery from one year to the next and, as before, yields vary with the stock size. The evidence from a number of fisheries is that it also promotes greater stability in the size of the stock compared with a constant catch approach (Getz and Haight, 1989).

One reason for the precarious balance between yield and recruitment at MSY in a fishery is the amount and quality of the data used to derive it. Generally, some estimate of the relationship between the size of a stock and the size of its recruitment is needed. Recruitment appears to be density-dependent in many fisheries – rising as the stock population declines. This appears to result from greater larval mortality, with the intensity of the intraspecific competition greatest in those fish with the highest fecundity (Sinclair, 1989). In the Schaefer model the capacity of the population to replace losses to the fishery assumes that yields are at equilibrium, and this requires validation from measurements other than catches landed on the quayside. Additionally, a plot of yield against fishing effort may not show a sharp maximum in a particular fishery and then an MSY will be difficult to define. Fixing an MSY for several years is unlikely to be realistic when stock size is varying from one year to the next and we may then use the alternative, MSAY, the **maximum sustainable average yield**, derived from several years' data.

4.4.3 Dynamic pool models

One major difficulty with the Schaefer model is that it takes no account of the distribution of age or size within a population. Yet natural mortality depends on age, and recruitment on age and size. Different age-classes (or cohorts) will have different rates of reproduction and different average weights. Beverton and Holt, or dynamic pool, models allow for these differences between cohorts within the fishing stock (or pool). Estimates of the yield from individual cohorts are added to derive the total harvest to the fishery.

The total yield is derived as a series of simple steps: the number caught in each age-class is the fishing mortality rate for that age-class multiplied by the number of fish in it (F_tN_t). This is an age-specific version of the Schaefer model. Multiplying this catch size by the average weight of a fish in that cohort gives the biomass it yields to the fishery. The total yield is the sum of these values for all cohorts in the catchable stock:

$$Y = \sum_{t=t_c}^{t_m} F_tN_tW_t \qquad (4.12)$$

where Y = the total yield to the fishery; N_t = the number of fish in age-class t; W_t = the average weight of a fish in age-class t; F_t = the instantaneous fishing mortality rate for age-class t; t_c is the age at which an individual is recruited to the fishery; and t_m is the maximum age of fish in the stock.

Managing populations

These are again instantaneous rates and the total yield to the fishery may thus be calculated for any time interval, integrating for all ages. However, modern versions of this model treat the cohorts as discrete increments and use their average value for the summation of total yield.

Thus, to derive the total yield, estimates of N, F and W are needed for each age-class. The number in a particular cohort (N_t) will depend upon the number of recruits entering the catchable stock and their subsequent survival. The exponential model is used to derive this as:

$$N_t = R\,e^{-(F+M)t} \tag{4.13}$$

where N_t = number of fish still alive after t years in the catchable stock and R = the number of recruits that originally entered the pool in this cohort.

This is an equation we met earlier when integrating the exponential population growth curve (equation 4.4). Now it measures the number of recruits remaining after losses due to fishing mortality (F) and natural mortality (M).

Although fishing mortality is often taken to be a constant for the whole stock, the model will allow for different values of F for each cohort. Again, natural mortality (M) is often taken to rise simply with age, but better estimates using empirical data (based on mark–release–recapture measurements) may be made. In the same way, predictions of rates of growth in biomass can be derived from real data or from theoretical models. Many commercial fisheries run submodels to derive each of these terms and then feed these predictions into the Beverton and Holt model (Gulland, 1983). However, allowance has to made for the uncertainty of these estimates. One reason is the by-catch problem – fisherman landing fish which are not wanted or which are protected. These are usually returned to the sea dead and never appear in the statistics used to derive F.

Recruitment is invariably difficult to measure accurately. The dynamic pool model assumes R to be constant from one year to the next and simply records the proportion of a cohort surviving with time. This allows easy calculation of the probability of a recruit surviving to a particular age.

Figure 4.9 The effect of two different annual fishing rates on the biomass caught. This is a hypothetical example, based on cod, and taken from Hardy (1959). In each case, 1000 fish are recruited to the stock in year one. The shaded bar shows the total catchable stock, and the open bar, the number caught over successive seasons. In (a) they are fished at an annual F of 80% and are fished out in 6 years. (b) At an F of 50% the cohort lasts for 11 years. Older fish have a higher biomass, and by multiplying the number of fish caught by the average weight for that age we get the biomass that they yield to the fishery, (c and d). At a low F more fish survive for longer and increase the biomass harvested.

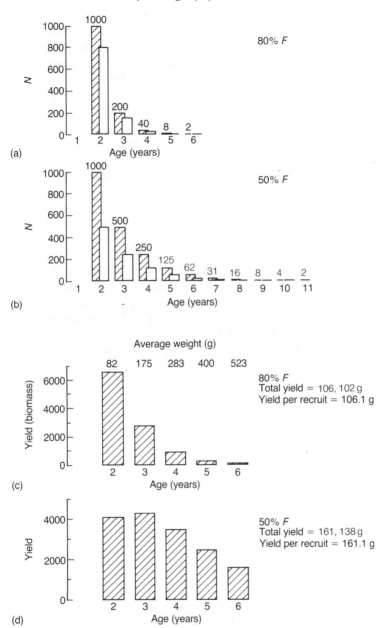

(a)

(b)

(c)

(d)

Additionally, given the proportion of survivors in a cohort and the biomass of its yield, we can derive the average yield per recruit at a given rate of fishing mortality. Adding these values together for all cohorts allows us to calculate the yield per recruit for the whole fishery at that *F*.

These values can be highly informative and, indeed, they are the basic measurement used in the management of a fishery with this model. Figure 4.9 demonstrates how the growth rate of the fish combines with the fishing mortality rate to determine the yield per recruit. This is a hypothetical population devised by Hardy (1959), based on cod. Annual fishing rates of 50% or 80% will each catch the same number of individuals (1000), but these are caught over a longer period at the lower rate (11 rather than 6 seasons). This allows the fish to grow to a larger size. The later catches of heavier fish result in a larger average biomass per recruit.

The size of the fishing effort is thus a major factor determining how long a fish remains in the stock, and consequently, how large it will grow. The choice of fishing effort governs the yield to the fishery, exactly as it did in the Schaefer model. In dynamic pool models the yield per recruit for different values of F is calculated and a decision then taken on the optimum fishing effort (Figure 4.10).

If the recruitment of fish is unaffected by F, the amount of fishing that gives the maximum yield per recruit will be simply that needed to harvest the MSY. The model indicates the optimum age at which fish should be caught, and thereby the fishing gear and mesh size needed to catch fish of this size. Every individual above this age of recruitment can be caught, whereas those below this age are allowed to escape.

The optimum fishing mortality will vary with the growth rate of the fish. A slow-growing species will need longer to put on weight, and a

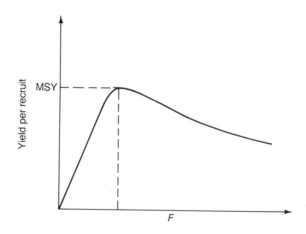

Figure 4.10 Derivation of the optimum F for the maximum yield per recruit. Yield per recruit rises initially with fishing effort but declines as we begin to harvest younger and smaller fish. The optimum F occurs where the yield per recruit is maximized.

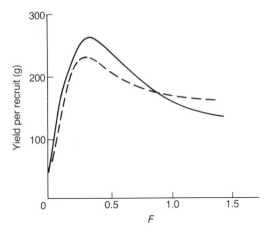

Figure 4.11 Yield curves for two populations of the same species, growing at different rates. One is subject to density-dependent growth (dotted line) and produces a lower yield per recruit compared with the population where the growth rate is constant (unbroken line). (After Beverton and Holt, cited by Cushing, 1981.)

large number will thus have to escape capture to achieve maximum growth. The optimum yield of this species will be at a low rate of fishing mortality. The alternative is a lower yield per recruit at higher values of F. For example in Figure 4.11 a lower yield per recruit is produced in a population whose growth is slowed by density-dependence. Modern dynamic pool models incorporate complex growth curves rates to allow more accurate estimates of optimum fishing mortality.

4.4.4 The problem of recruitment in dynamic pool models

Generally, it is a lack of information on all the parameters, and especially annual recruitment, that limits predictions using dynamic pool models. Recruitment is thought to decrease as stock increases (Sinclair, 1989), but the relationship between the size of the stock and the size of the recruited population is still not fully understood (Hall, 1988). In their simplest form, dynamic pool models assume a constant recruitment over a range of stock sizes and are only effective for populations in equilibrium (Getz and Haight, 1989). Additionally, many species are fished from a mixture of cohorts, and no fishing gear can deliver perfect selectivity of the stock. Over the longer term, the relationship between recruitment and stock size needs to be established to estimate accurately the productivity of the fishery.

Recruitment will be reduced with very low populations and a small stock of adults will be less able to maintain recruitment under poor

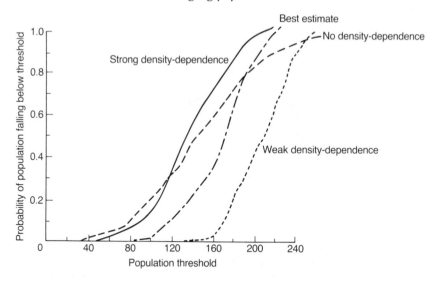

Figure 4.12 The effect of density-dependence on the probability of cod (*Gadus morhua*) falling below a range of population thresholds over a 40-year period, based on records of fishing stocks. This is equivalent to the fish becoming extinct. The density-dependence is shown operating at four levels on recruitment, with one being the best estimate for the wild. Applying weak density-dependence to the population models lowers the probability of extinction compared with a population where it is absent, but as this dependence increases so does the probability of extinction. (After Ginzburg *et al.*, 1990.)

environmental conditions. Large-scale physical processes, such as average wind speed at the time of spawning, can also generate large variations in recruitment rates (Peterman and Bradford, 1987). Fishing quotas based on single estimates of an MSY are highly susceptible to fluctuations in fish stocks resulting from such environmental variability. Pitcher and Hart (1982) note that pre-recruit surveys are required to derive models of recruitment for all commercial species and to estimate their capacity to replenish themselves under a range of conditions.

Recruitment is probably density-dependent in many fish stocks by virtue of its effect on larval survival (Sinclair, 1989). More recent models using empirical data based on cod (*Gadus morhua*) suggest that the effect of density-dependent recruitment on survival of the population varies with the intensity of intraspecific competition (Figure 4.12).

Indeed, the long-term survival of some cod stocks may depend on the intensity of this competition. However, in the case of the northeast Arctic stock, the loss due to fishing outweighs all other sources of mortality put together (Law, 1991). In the past, the fishing mortality increased as harvesting moved to the feeding grounds, taking smaller fish. In doing so, it may have imposed a selective pressure on the stock, favouring more rapid development, and this probably accounts for the early sexual maturation of these stocks, reduced from 9–11 years to around 7–8 years (Law, 1991). The same appears to be true of a number of other fisheries around the globe, and possibly also the fin whale. However, with the whale population at least, this early puberty might also be due to the increased abundance of food available to the reduced density of an exploited stock (Meredith and Campbell, 1988). Thus any effect of selection has to be distinguished from the lower intensity of intraspecific competition. Nevertheless, Law (1991) raises the interesting possibility of managing the resource by using the selective pressure of the fishing strategy to promote high yields.

4.4.5 Variations on the dynamic pool model

In species where there are detailed and accurate measurements for F and M for each age-class, the progress of cohorts with time can be followed using life tables. The effects of killing individuals from different age-classes on the overall survival of the population may then be accurately modelled. Indeed, the population can be managed to achieve the stable age distribution associated with rapid population growth. Although this is rarely possible with a commercial fishery, Goodman (1980) describes such an example for the Pribiloff fur seal (*Callorhinus ursinus*) of the north Pacific, once an endangered species, and now managed for an MSY by taking young males and older females.

Dynamic pool models have several major advantages over the surplus yield models, not least their capacity to accomodate more elaborate models for age-specific F, M and G. This adaptability also allows them to manage multispecies fisheries, which, if yields are partitioned, can run models for each species side by side. The principal weakness of the simple dynamic pool model is its reliance on a population at equilibrium, an assumption that may only be valid if a very large recruitment has taken place. Its main value lies in the relationship it establishes between fishing effort and yield, and the merits of a selective harvesting policy based on size-classes.

4.5 MANAGING FORESTS FOR MAXIMUM SUSTAINABLE YIELD

Harvesting a forest has much in common with fisheries management. Both aim to take the maximum sustainable yield which, as the dynamic pool model demonstrates, means calculating the age at which the maximum rate of growth has been achieved. But, in contrast to the fishery, the forest stand is stationary, easily counted and measured, and can be harvested as selectively as we choose.

The forest manager also has control over a number of variables which can determine the size of the harvest collected. These include:

1. the density of the trees;
2. the species of the tree harvested;
3. the size of the tree harvested;
4. the composition of the tree community;
5. their nutrient supply (by the use of artificial fertilizers);
6. the population structure of the stock and of the harvest;
7. the age (size) of recruitment to the stock;
8. the quality of the data collected.

Obviously, here the forest manager has more in common with the farmer cultivating a crop, although, as in fisheries management, the harvest is determined by the size and age of the stock. There are also a number of factors over which the forest manager may have no control:

1. the size or carrying capacity of the habitat;
2. the climate and the species that will flourish;
3. any aesthetic or social value placed on the forest;
4. the broader ecological implications of forest management – the impact on wildlife of forest clearance, or on soil erosion within the watershed, for example.

These are not exhaustive lists, but they do demonstrate the greater information and greater control available to the forestry manager. If the prime aim is an economic return on the timber being grown, then a monoculture using even-aged stands will be favoured, at least in the short term. Over the longer term, a more in-depth review of the ecology of the forest and the larger environment may be needed.

The species grown will depend on the end-use of the timber – faster growing varieties are used largely for paper production; the better quality timber needed for structural and other uses tends to be slower growing. On the other hand, a site with many species may have an enhanced ecological value, particularly if these form an uneven-aged stand.

The harvesting strategy will also reflect these differences. Even-aged monocultures are most easily harvested using a system of clear-cutting, where a stand within the forest is felled at one time. Recruitment might then rely on seeds arriving from the surrounding canopy, though more often, seedlings are planted directly (Packham and Harding, 1982). In some cases, a number of trees are left after the main felling operation to provide seeds and cover for the regeneration of the stand, chosen according to their size and growing vigour, so it is their seeds that form the next generation (Packham and Harding, 1982). The thinning of the stand to lower the population density will also remove the poorer quality trees.

The value of these stands follows an asymptotic curve – broadly following the increase in the volume of the wood. The maximum return is achieved when the volume or value accumulated in unit time is maximized (Getz and Haight, 1989). However, the value of a clear-cut will also depend on the distribution of size classes of the trees. Predictive models for the size and number of trees based on the current stand age and density are relatively effective, and indeed, can even be extended to an individual tree (Getz and Haight, 1989). This data is collected from permanent plots where the growth of the stand is monitored very closely, or by using the tree-ring data from harvested stands.

Uneven-aged management uses selective harvesting of specific age classes which, at the same time, can control the spacing and growth of the remaining trees. This is very often the case with stands containing several species. Models for such stands operate on size-class analysis, predicting the progress of a tree from one class to another to decide the optimum time to harvest (Getz and Haight, 1989). As with the fishery models, recruitment to the stock commences when a tree is large enough to merit harvesting, and a maximum sustainable yield is based on removing trees of a particular size. A number of models are able to include the effects of stand density and competition, on tree growth and the rate of recruitment of seedlings to the stock.

Although much fisheries and forest management has been based on simple population models, the lack of allowance for variability in the environment, or in r or K, mean that these are relatively crude management tools unless some measure of recruitment is included. Indeed, some have argued that we could learn much about the effects of density on recruitment by deliberately overfishing some stocks under controlled conditions (Cushing, 1987). Certainly, there is much scope for refining these assumptions further (Hall, 1988).

Many of these models have been extended and developed to incorporate greater detail on the rates of growth and recruitment, as well as features of the harvesting method (Getz and Haight, 1989). Nevertheless, many fisheries have suffered major declines in stock due to the inadequacy of the data and our understanding of these processes. Real populations rarely show a stable equilibrium, while even simple deterministic models can predict a range of possible states as population parameters change. Many ecologists question whether concepts of stability can be applied to populations, either in models or in the real world. Even so, it is certain that the capacity of an exploited population to avoid extinction will depend upon its rate of population growth. Since the world-wide ban on the taking of the fin whale in 1976 (Figure 4.13), its population has begun to recover and is projected to reach around 90% of its unexploited population size in the north Pacific by the early part of the 21st century.

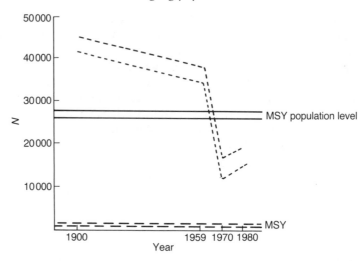

Figure 4.13 The decline and recovery of the fin whale (*Balaenoptera physalus*) in the north Pacific. The dotted lines show the range of estimated population sizes on four occasions since whaling for this species commenced at the beginning of the century. The range of population sizes necessary to support a MSY is shown by the solid lines, and this would produce a yield of around 1200 whales (broken lines). At current rates of recovery this population could support such a MSY in around 8–16 years time. Deep-sea whaling for this species began in 1956 and was banned in the north Pacific under international agreement in 1976. This population represented about 10% of the total world population before exploitation, and without further exploitation could achieve 90% of its original size in 25–30 years. (After Meredith and Campbell, 1988.)

4.6 CONSERVING POPULATIONS

The fossil record suggests that, on average, most species survive for between 1 and 10 million years. Although species will become extinct through natural processes, the changes that industrialization has made to the environment in the past 200 years have caused a major increase in this extinction rate. Very often these losses have been an incidental consequence of habitat loss, others have resulted from overexploitation. In some cases, populations have declined through a combination of both factors (Figure 4.14); elsewhere, we have made more deliberate attempts to wipe out a species – as with the American bison in the 19th century.

Particularly endangered are those species whose numbers are reduced by the fragmentation of their habitat, constrained by the smaller carrying capacity of their environment. As with exploited species, we attempt to conserve a sustainable population, though now these numbers become translated into the minimum areas of habitat needed to support a viable population. For the same reason, the conservation of a species needs to

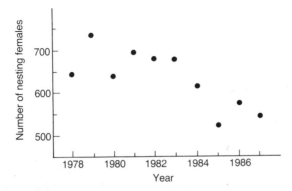

Figure 4.14 The estimated number of nesting female Kemp's ridley turtles (*Lepidochelys kempi*) at their primary nesting site in the Gulf of Mexico, 1978–87. This is declining by about 3% per annum. The numbers are based on the assumption that each female produces 1.3 nests. In 1947, there were an estimated 47 000 breeding females, and the decline is largely due to the harvesting of eggs and of females on this beach. (After Thompson, 1988.)

consider the design of nature reserves, ideas which are developed in section 6.4. Our aim here is to decide what features of a species' population ecology make it prone to extinction and how vulnerable populations can be managed to conserve them.

The logistic model of population growth is a deterministic model – for a given input of population growth rate per individual (r) and carrying capacity (K), only one value of N is derived for a particular time (equation 4.4). The outcome is 'determined' and no other result is possible. In the real world, r is likely to vary with a number of population parameters, such as population density and also as environmental conditions change, such as the availability of nesting sites. There is thus some uncertainty, some variability, associated both with demographic and environmental parameters.

The logistic model predicts that the population will eventually settle to a stable density around K, regulated by density-dependence. Real populations occupy heterogeneous ecosystems which change in time and space, and hence have a variable carrying capacity. Nor do populations respond rapidly to changes in environmental conditions, or have a constant reproductive rate. Indeed, density-dependent regulation of population growth will only be important at the highest densities, when death rates start to match birth rates. In the middle ranges of a population, away from K and away from a low N, birth and death rates are less responsive to population density (Strong, 1986). Additionally, the history of a population will also determine its capacity to grow: past events will have determined its current sex ratio and age distribution – say if a cohort has been lost in a bad year and a stable age distribution is not achieved. These variable prospects for survival and reproduction mean that r can then assume a range of values. Overall, there are now a large number of possible outcomes to the model. The closest approximation to an equilibrium density is likely to be a population that fluctuates within a range, showing 'range stability'.

Populations which vary over a large range are more difficult to manage. These are the populations most likely to become extinct, falling below a threshold simply by chance, from which they cannot recover. Modelling extinction has to incorporate such variability – if the growth rate of a population varies widely then we might expect that population to have a higher probability of expiring. Consequently, in trying to predict the survival prospects of a species, ecologists have concentrated on r and its variation, as well as the actual number of individuals alive. Leigh (1981) believes that the amplitude of a population's fluctuations is the best measure of its chances of extinction.

In effect, given a sufficiently long time, chance alone will lead to all current species becoming extinct. However, some species obviously have better survival prospects than others and would be expected to survive longer. We thus have to decide what constitutes a secure future for a species over a given period of time, and which species should be considered endangered. There is no agreed level of security, but some acceptable, if a rather arbitrary, level of risk has to be specified. It might be decided that a 95% chance of surviving the next 100 years is acceptable while 99% security would be too expensive in terms of reserve area or other costs. Such decisions do not just involve ecologists – a broader planning process is usually involved if only because sufficient habitat also needs to be conserved (Chapter 6).

4.6.1 Minimum viable population

A minimum size of population will be needed to conserve a species within the level of security that has been specified. Two types of **minimum viable population** (MVP) can be defined – one based on population and environmental parameters alone (demographic models) and a second estimating the MVP for a viable genetic population. These will nearly always represent two different values. For the most part these have yet to be combined into a single coherent model, as both are at an early stage of refinement and are yet to be tested empirically. The genetic implications of small population sizes are reviewed in a later section.

We begin by considering the effect of r, N and their associated variability on the **persistence time** of a species – the predicted length of its survival. Here we use the predictions of one theoretical model, the birth and death process model (Goodman, 1987) which incorporates variability in r arising from both demographic and environmental sources. The model is not described here: there are several possible approaches to measuring the average time to extinction, depending on the amount of demographic information available (Leigh, 1981; Goodman, 1987). From these estimates for a species, MVP can be derived. In what follows, extinctions due to catastrophes (earthquakes, floods and so on) are not considered, although there have been attempts at modelling even these (Ewens *et al.*, 1987).

4.6.2 Persistence time and variability in r

When it happens, we all have a good reason for dying, but the excuse varies from individual to individual. In very small populations close to extinction, a knowledge of these causes of mortality is essential to protect the species and aid its recovery. With this information we may then intervene directly, perhaps using a captive breeding programme, or by managing specific features of the species habitat.

For populations which are not reduced to this size a more general approach is adopted, by examining the relationship between population size, growth rate and its probable survival time. In the absence of major catastrophes or some form of genetic impoverishment, Goodman (1987) recognizes two sources of variability in the rate of population growth per individual (r):

1. **Demographic variability** (V_1) – the variance associated with the chances of meeting a partner, of giving birth, of dying, of giving birth to female offspring and so on. If we take the average variance in r for an individual as V_1 (where this is independent between individuals) then the total effect for the population is

$$V = \frac{V_1}{N}.$$

   In a constant environment this would be the only source of variation in r. Since V_1 is fixed, the average variation in r declines with N. For this reason demographic variability is likely to be most important at small population sizes.
2. **Environmental variability** (V_0). When all the changes in the birth and death rates are due to environmental factors, we can simply state

$$V = V_e$$

   where V_e is a constant for the whole population measuring the effect of environmental variability on r. As this does not change with N, V_e will be important at all population sizes. Consequently V_e is much more important than V_1 for large populations.

This is something of a simplification because, as a population expands into new areas, it will encounter different degrees of environmental heterogeneity, and these would be reflected in V_e. However, such effects are thought to be minor and are ignored in the model.

The total variation in r is simply

$$V = \frac{V_1}{N} + V_e. \tag{4.14}$$

Large variations in r will produce wide fluctuations in N, increasing the likelihood of extinction by chance. Given a sufficiently long time, a highly variable population will go extinct by chance alone: the probability of extinction can thus be expressed as the mean persistence time of a population (Leigh, 1981; Goodman, 1987). These models are based on the maximum population size (N_m or the 'population ceiling') that a species can achieve. N_m can be somewhat larger than K because a population with large fluctuations may overshoot its carrying capacity and achieve higher populations for short periods.

How do these variations affect survival times for populations of different sizes? Demographic variability is only important for very small populations, and its effect on survival times quickly become insignificant as the population grows. The models suggest a geometric increase in persistence times as N increases (Figure 4.15). This justifies our attention to the details of the reproductive biology of a species when it is reduced to very small numbers, a strategy followed in many large-mammal breeding programmes (Allendorf and Servheen, 1986).

Environmental change is more random, and its effects are generally independent of population size. The time to extinction rises more or less linearly with N (Figure. 4.15) and there is no threshold population beyond which the population can be said to be secure. This sort of variability will be the most significant for most species. Because of this linear relationship, a much greater increase in population size is needed to extend the persistence time of the species. The effect of environmental variation on the rate of population growth is difficult to quantify (Goodman, 1987) and ecologists cannot always agree on what size of population gives adequate protection (Wilcox, 1986).

The total variance (demographic and environmental variability combined) will produce a rapid decline in the population size as it approaches its expected persistence time – half the population will have been lost before 70% of the persistence time will have elapsed (Shaffer, 1987). This implies that simple estimates of a mean persistence time will give no indication of the increased vulnerability of the population as time

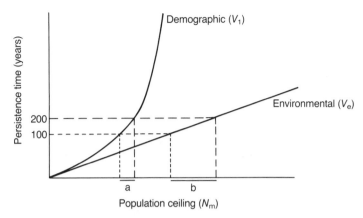

Figure 4.15 The increase in predicted persistence time of a population with increasing population size. Under demographic variation in r, a small increase in N_m will double the persistence time (a), but a much larger increase is needed when environmental variation operates (b). (After Shaffer, 1987.)

proceeds. To guarantee its survival for this period, Shaffer suggests that the population would have to be much larger than that derived from the model.

4.6.3 Persistence time and the rate of population growth

What is the relationship of persistence time with the size of r, the rate of population growth? In the absence of environmental variation, time to extinction increases with r, even for small populations (Figure 4.16a). Obviously, as long as the birth rate is greater than the death rate, the population will grow, and the major threat can only be from environmental change, such as loss of habitat or pollution. From Goodman's model it appears that if r is density-dependent the pattern remains the same, although the rate of increase in persistence time is much smaller (Figure 4.16a): evidence for this effect was observed in small populations of the cod seen earlier (Figure 4.12), where an increase in the intensity of competition raises the probability of extinction (Ginzburg *et al.*, 1990). This effect will be relatively unimportant in a constant environment as the variation in r is expected to decline as the population increases (equation 4.12) (Goodman, 1987).

When environmental variation in r dominates, the size of the population growth rate has a minor effect on persistence time (Figure 4.16b) and density-dependence now has little significance for the population's survival. For this reason, environmentally induced variation in the mean population growth rate is likely to be a major cause of extinction in many

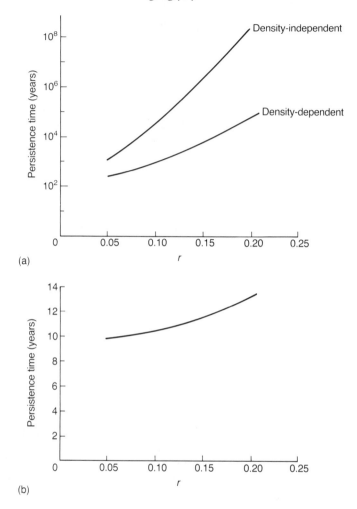

Figure 4.16 The effect of the rate of population growth per individual (r) on the predicted persistence time of a population, using Goodman's (1987) model. In (a) the variation is attributable to demographic factors alone ($V_1 = 1$), and persistence times increase rapidly with r. A density-dependent r shows a smaller increase, but follows the same pattern. In (b), $V_e = 1$, the variation in r is attributable to the environment alone. Now, persistence times become much shorter (note the change of scale on the y axis) and the effect of r is marginal, whether it is density-independent (as above) or density-dependent.

endangered species. Very large populations are then the only assurance against extinction.

These predictions are depressing as only large populations will give a reasonable level of security in a fluctuating environment and this model suggests that most efforts to date have underestimated the size of population needed to ensure survival. The model is, however, simplistic in making no allowance for the effects of genetics or differences in age or sex structure of the populations. Their predictions will be refined using empirical data, especially estimates that partition the variability in r. Shaffer (1987) believes such data will lead to reductions in the necessary size of the MVP or conversely, increases in the mean persistence time.

4.6.4 The persistence of real populations

Collecting data in the field to test these models will require a large effort, but there has already been some confirmation of the relationship between N and persistence time. Using historical records Berger (1990) concluded that the survival of isolated populations of bighorn sheep (*Ovis canadensis*) was determined by small population size (Figure 4.17), probably due to its implications for the genetic viability of the species.

Others have attempted to overcome the lack of field data by looking for general rules which relate the demographic characteristics of a species to other features of its biology, such as body size. Larger animals may be more susceptible to extinction because of their longer generation times and greater demands upon the environment. Although their recovery might be slow, a long-lived animal is, however, more resistant to short-term environmental changes. On the other hand, small animals show greater adjustment stability – their rapid rate of population growth allows a fast recovery from low numbers.

Belovsky (1987) has examined the relationship of body size in mammals to N_m, r and V, and its effect on persistence to generate data to test Goodman's model. Within most groups of vertebrates there is a simple relationship between body weight and reproductive rate (Southwood, 1981):

$$r_m = \propto W^\beta$$

where α and β are constants for a group of animals; W is the body weight; and r_m is the maximal rate of population growth per individual.

This is exactly the same relationship that we encountered earlier when considering the effect of body weight on pollutant uptake (section 2.4). Although this is only an approximate relationship for any particular group, it does allow an estimate of the rate of population growth for a species from its average weight.

In fact, a large number of characteristics of a species can be related to body size, including longevity and litter size (Wood, 1983a). Observed K values for populations also appear to show a relation to body weight, again as empirical correlations. In this way, Belovsky uses the body

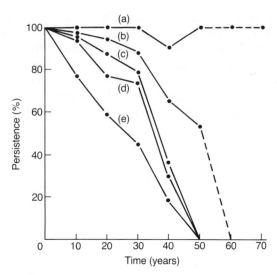

Figure 4.17 The decline in the proportion of populations of different population sizes of bighorn sheep (*Ovis canadensis*) from the western USA. (a) = 101+; (b) = 51–100; (c) = 31–50; (d) = 16–30; (e) = 1–15. Dotted lines indicate estimates based on sample sizes of less than four. Inbreeding in small populations is probably responsible for a reduced resistance to disease. (After Berger, 1990.)

weight of various mammals to indicate their maximum population sizes in a given area with known food resources.

To test the birth and death process model, estimates are also needed of the variance in the rate of population growth. Belovksy measures V_e in r as the variation in food availability – for herbivores this can be measured directly, as the standing crop of its diet, or some primary determinant of that, such as rainfall. This is a highly indirect measure of variance in r, and has to be scaled into units of r using statistical analysis. These estimates of N_m and r allow Belovsky to predict the persistence time of a mammal with a given body weight. This can only be a rough estimate, as two animals in the same habitat with the same body weight can have very different population growth rates and population sizes. Nevertheless, it provides some idea of which species are most likely to be susceptible to extinction.

Belovsky has used such estimates to test Goodman's model for herbivorous mammals from isolated montane forest in the western USA and he obtained a good correlation between observed and expected persistence rates (Figure 4.18). This is some confirmation that Goodman's model provides an adequate description of persistence times for single isolated populations.

The theoretical models, and these first attempts to validate them, suggest that relatively large populations will be needed for survival in

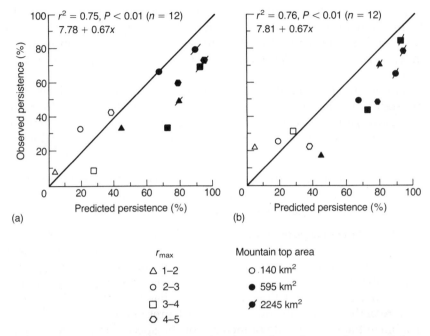

$r^2 = 0.75, P < 0.01$ ($n = 12$)
$7.78 + 0.67x$

$r^2 = 0.76, P < 0.01$ ($n = 12$)
$7.81 + 0.67x$

(a)

(b)

r_{max}	Mountain top area
△ 1–2	○ 140 km²
○ 2–3	● 595 km²
□ 3–4	⬤ 2245 km²
◇ 4–5	

Figure 4.18 The correlation between the observed persistence times and those predicted by Goodman's model for herbivorous mammals in the montane areas of the western USA ((a) = Great Basin; (b) = southern Rocky Mountains). In both locations, the mammals were divided into four classes according to the size of their r_{max} (the maximum value of r) and three classes by their K value (the area of their habitat), giving 12 possible combinations. The observed persistence was derived from data covering the last 8000 years. Each region shows a significant correlation and a similar regression equation for these mammals. This suggests that Goodman's model is a useful approximation to the effect of r_{max} and K on persistence time, at least for these mammals. (With permission, from Belovsky, 1987.)

fluctuating environments. In reality our efforts are often constrained by the size of habitat needed to conserve a viable population; many of the existing reserves for endangered species are too small to ensure long persistence times (Wilcox, 1986).

4.7 LOSS OF GENETIC VARIATION

A living organism adapted to a particular habitat must be capable of further adaptation: being able to change in a changing world is essential for long-term survival. Genotypic adaptation results from the selection of phenotypic traits produced from the variety of genetic information in the population. Small populations have less genetic variation available to them. They also have a greater chance of breeding with near relatives with similar genotypes. Together, these reduce the variation between individuals in the population, essential for genotypic adaptation (section 3.1).

Limiting the number of potential partners also increases the likelihood of producing offspring with deleterious genotypes. They show **inbreeding depression**, a phenomenon well known from the breeding of domesticated plants and animals, and consisting of three principal effects:

1. the production of deleterious, homozygous recessive combinations;
2. a reduction in variation for subsequent generations;
3. a lack of heterozygote vigour (a reduction in heterosis).

The decline of small populations may be hastened by the genetic effects of inbreeding depression, which often express themselves as increased juvenile mortality, lower growth rates and, in some cases, greater susceptibility to disease. The conservation of a species cannot simply maintain a given population size, but also has to attend to the genetic welfare of the species. By minimizing the degree of inbreeding, we aim to maximize the intrinsic rate of increase of the species and its capacity to adapt to future environments.

Heterozygotes not only maintain variation in the population, but as individuals they tend to show improved growth and reproductive vigour. An increasing body of evidence suggests that being heterozygous confers some advantage in a number of biological functions: increased growth rates and scope-for-growth in molluscs (section 2.7), greater disease resistance in avian eggs and more effective metabolic enzymes in butterflies (Allendorf and Leary, 1986). Together such effects mean that being heterozygous carries some selective advantage (**heterosis**) over homozygous members of the population.

Larger populations will tend to accumulate and retain more variation than smaller populations, but all can lose variation through a process called called **random genetic drift**. Consider a simple population – a group of diploid organisms that show random segregation of genes into their gametes (and with no mutation or migration). The proportion of alleles that are A_1 in the next generation should be exactly the same as the proportion in the parent generation. This is the law of the conservation of allele frequencies. However, not all gametes will be used, and only a proportion will give rise to offspring. The genetic information in the gamete pool is 'sampled'. Chance may mean that the allele frequencies in the offspring differ slightly from the parents and this drift in their frequency then becomes fixed in the population. Given a sufficiently long time, this chance sampling of the gene pool at each generation might allow A_1 to dominate completely or alternatively, be lost entirely. The population will then be homozygous for A_1 (or alternatively A_2) and all variation at that locus is lost.

Because it relies on chance, such fixation usually takes a very long time, but these times shorten in small populations (Christiansen and Feldman, 1986). A large population may balance losses through drift by its mutation rate, but small populations have fewer opportunities for mutational change. Breeding with a relative sharing a similar genotype will also increase the rate of fixation.

Thus a small population faces two threats to its genetic variation – an acceleration in random genetic drift and a higher propensity to inbreeding. As this lowers the capacity of the population to produce viable offspring, this interaction between the demographic and genetic factors has been called the 'extinction vortex' driving the population lower (Gilpin and Soulé, 1986).

4.7.1 Effective population size

The size of the population is therefore important for conserving the genetic viability of the species. But a simple head count of individuals is no indication of the number contributing to the gene pool – the size of the breeding population is usually somewhat smaller because:

1. there will be individuals not of breeding age;
2. not all adults will produce offspring;
3. reproduction may be dominated by a small number of adults (say a male with a harem).

Nor will the total genetic variation in this breeding population be fully represented in their offspring: this will be somewhat smaller because of inequalities in the 'sampling' of the gene pool, as, for example, when some adults produce more offspring than others. In terms of its genetic properties, this makes the effective size of the population smaller still. The **effective population size** (N_e) has a precise meaning to geneticists, who define it as the size of the ideal population that would undergo the same amount of random genetic drift as the actual population (Lande and Barrowclough, 1987). The essential point is that N_e is invariably smaller than N because the total variation in a population is rarely represented in its offspring. The determination of N_e depends upon whether simple mendelian genetics can be assumed and whether generations overlap (Lande and Barrowclough, 1987). These equations are not developed here, but we can demonstrate the effect of two population parameters, the number of offspring and the number of partners, on N_e for a simple univoltine species:

1. Some alleles will be under-represented if there is great variance (σ^2) in progeny number. That is, if some parents produce large numbers and others very few:

$$N_e = \frac{4N}{2 + \sigma^2}.$$

So as σ^2 gets larger, N_e decreases.

2. Where there is an unequal sex ratio between partners, say one male fertilizes several females, one set of genes will be over-represented in the offspring:

$$N_e = \frac{1}{(\dfrac{1}{4N_m} + \dfrac{1}{4N_f})}$$

where N_m is the number of males and N_f the number of females. If a breeding population of 1000 consists of only 50 males, the effective population size is just 190.

In this way, the details of its reproductive biology can lower the effective population size of a species, and the amount of genetic variation available for each generation. Fluctuations in population number (Franklin, 1980), or subdivision of populations into groups which are unlikely to inter-breed will further depress the effective population size. Lande and Barrowclough (1987) give details of these and alternative derivations of effective population size for overlapping generations.

It is the effective population size which determines the long-term future for a species and its capacity to adapt. A poor genotype may not only worsen the long-term prospects for adaptation and survival, it can also have more immediate consequences, perhaps hastening the demise of the species through reduced population growth. Based on the experience of animal breeders, Franklin (1980) suggested that an N_e of 50 would provide sufficient protection against inbreeding depression in the short term, though an N_e of 500 is necessary for long-term protection against random genetic drift. Lande and Barrowclough concluded that the N_e depended on the most significant type of genetic variation for a species, extending the idea beyond the single-locus approach adopted here. By also specifying the type of selection that may be acting on this variation, their calculations suggest the need for a large N_e. This will translate into much larger actual population sizes. An N_e of 500 and an actual population of 2500 has been suggested as necessary for saving each of the five species of rhinocerus in Africa and Asia, though the area of reserve needed to support this MVP differs between species. The same N_e has been used in the conservation management of the red-cockaded woodpecker (*Picoides borealis*) in North America, but Reed *et al.* (1988) calculate this to be 509 breeding pairs and an actual N of 1323. Few of the existing populations surveyed reach this size, nor are their reserves (based on the size of the territories needed to raise a successful brood), likely to be large enough to support the calculated MVP.

The first priority is to save whatever remains of the genetic information of a species, no matter how few individuals this resides in. There are no golden rules or magic numbers which apply to all species under all circumstances, and it may often require captive breeding in artificial environments. Where natural populations are fragmented, they may exhibit rather more genetic variance than a single population of the same size (Lande and Barrowclough, 1987). This can be an important additional source of variation, and is considered next.

4.7.2 Populations and metapopulations

Both the demographic and genetic parameters of a population are, in part, determined by its geography. The populations of a species are frequently fragmented and divided amongst a series of patches. This population of populations is termed a **metapopulation** (Lincoln *et al.*, 1982). The size of each constituent population, their distance apart and their geometric configuration determines the frequency of interchange between each. Ideally, we would know the rates of flow of genes between populations, of colonization and of extinction to produce a coherent management plan. This integration of the genetic, demographic and geographical considerations has been termed **population vulnerability analysis** (PVA) by Gilpin and Soulé (1986).

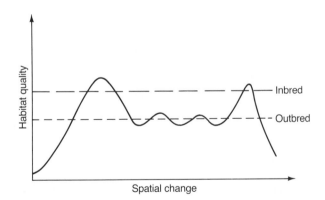

Figure 4.19 Possible habitats for colonization in inbred and outbred populations. Most can support an outbred population with its attributes of adaptability and lower susceptibility to adverse conditions. Habitats have to be less demanding for the inbred population to survive. Here it is confined to smaller areas which are further apart, neither of which bodes well for its future. (After Gilpin, 1987a.)

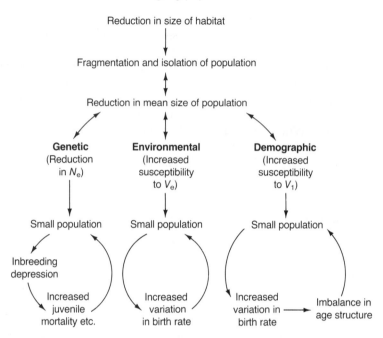

Figure 4.20 The interaction of demographic, environmental and genetic factors in the population decline of a species. Each loop will add further impetus to the other two, driving the population closer to extinction.

Each patch will have its own carrying capacity, and its degree of inbreeding and genetic drift will depend on its population size. Overall, if there is no interchange between the patches, the effect of fragmentation will be to lower the N_e for the metapopulation. Any exchange of individuals and their genes will depend on the geometry of the patches and the mobility of the species. Within a metapopulation, some populations will be stable and never suffer extinction whereas others might suffer extinctions fairly regularly. Their patch then has to be recolonized. Colonization and extinction in a patch will depend on its habitat quality and its proximity to adjacent populations (section 6.4). The genetic constitution of the colonizers will also determine their success rate (Figure 4.19).

The loss of habitat and fragmentation of the population can start a series of processes leading to the extinction of a species, Gilpin and Soulé's 'extinction vortices' (Figure 4.20). These are a series of positive feedback loops that drive a population lower by increasing V_1 and V_e or lowering N_e in each patch. These processes are all interlinked and the proper conservation of a species has to attend to each of these factors.

Moving individuals between patches may help to reduce the effects of inbreeding in many species. Perversely, there is another danger here, of **outbreeding depression**. This is where populations show a high degree of adaptation to local conditions, and hybrids between these ecotypes show reduced vigour. This may be a particular problem with trees (Ledig, 1986) and appears to be more prevalent in organisms with a low degree of developmental stability (and a greater capacity to adapt to local conditions). This danger has been highlighted in attempts to conserve the grizzly bear (*Ursus arctos*)in the USA. Plans to move individuals or semen between the six small inbred subpopulations (with low N_e) have to contend with the possibility of outbreeding depression (Allendorf and Servheen, 1986).

However, we should not assume that habitat loss is the only cause of extinction. Some species are more susceptible than others, perhaps because of their history. Heavy poaching reduced the Great Indian rhino in the Chitwan National Park in Nepal to an N_e of around 20–30, and an actual population of around 80 animals. It went through this 'bottleneck' about 30 years ago and is now has around 400 animals, a respectable rate of recovery, probably due to significant variation remaining in the breeding population. The cheetah may well be recovering from a bottleneck in its past when its population was severely reduced (Figure 4.21).

Species are naturally going extinct without human intervention, but the fragmentation of habitats through our activity has greatly accelerated the pace of species loss. Population ecologists and conservation managers do not always agree on the value of these models. There has, for example, been considerable argument about the best strategy for saving the Java rhino, following a PVA and the suggestion that a captive breeding programme was needed. Others argue that some of the assumptions and estimates need to be supported by better data and perhaps the models themselves need refining. Instead, they suggest that more effort needs to be directed towards conserving their habitats. One crucial question is whether the surviving populations should be collected together in one large reserve, a point we take up again in Chapter 6.

Summary

Managing populations to exploit or to conserve them relies on the quantification of several population parameters. Starting with a simple logistic model, ecologists distinguish the rate of growth (r) and the carrying capacity (K) to model the dynamics of a population. In these simple deterministic models the population will settle around a stable equilibrium size (or density) as long as the environment remains constant and its parameters stay within a small range.

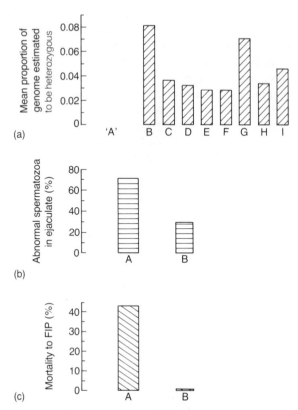

Figure 4.21 The lack of genetic variation in the cheetah (*Acinonyx jubatus*) and its implications for its survival. The cheetah is the only surviving member of its genus and is markedly divergent from the rest of the Felidae. Largely a solitary species, it has a low density throughout its range, and its overall population size is low. There is some evidence that it has been through a population bottleneck in its recent past, when it suffered a high degree of inbreeding. Now the species shows all the signs of low genetic variation. Amongst these (a) note the low proportion of its genome estimated to be heterozygous and that its spermatozoa show a high degree of abnormality (b). This lack of variability is thought to be responsible for a high rate of juvenile mortality and an increased susceptibility to disease (c), such as feline infectious peritonitus (FIP). (A = cheetah; B = domestic cat; C = lion; D = serval; E = leopard; F = caracal; G = ocelot; H = tiger; I = margay). (From O'Brien *et al.*, 1985.)

Two simple fisheries management models are described here based on the concept of a maximum sustainable yield. The surplus yield model suggests that the maximum harvest may be taken at about half the carrying capacity of the population. This is when the population is growing most rapidly and most able to replace the loss to the fishery.

However, this is critically dependent upon an accurate measure of the population size and its rate of growth. The dynamic pool model is a derivation of this, partitioning the population into size- or age-classes. This is appropriate for selective harvesting, removing only individuals for a particular size from the stock. Using this method, the fishing effort can be tailored to harvest the maximum sustainable yield. This model has been extensively refined to allow improvements in the estimates of other sources of mortality, and growth rates. Similar principles can be applied to forest management.

The capacity of an exploited population to grow depends upon the rate of recruitment. Small populations enter a dangerous area where they are increasingly susceptible to both demographic and environmental variability, and to the genetic implications of a small number of partners. The fragmentation of habitats leads to small isolated populations with a reduced genetic pool from which to produce subsequent generations. Estimates of the probability of an endangered species surviving in the long term depend upon the variability in their reproductive rate, and the size of the habitat needed to maintain an effective breeding population.

Further reading

Gulland, J.A. (1983) *Fish Stock Assessment*, John Wiley, Chichester.
Pitcher, A.J. and Hart, P.J.B. (1982) *Fisheries Ecology*, Croom Helm, London.
Soulé, M.E. (ed.) (1987) *Viable Populations for Conservation*, Cambridge University Press.

Bracken is one of the major global weeds and is the dominant ground cover over much of Richmond Park, an area of heathland in southwest London.

5

Managing pests

Not only does agriculture create habitats for the plants and animals we cultivate, it also provides resources for others from which we gain no yield. Just as weeds sprout from a ploughed field, disturbance to any ecosystem provides space for a range of organisms to invade. Predators and parasites are attracted to the concentrations of energy and nutrients in our crops, our cattle and ourselves. We compete with other species because of the niches we create as well as the niche we ourselves represent. The economic, social and health costs of this competition are immense.

Whereas most species have to rely on their genetic variation to compete, we have used our intellectual capital to try to win these battles. Following the early successes and later problems of synthetic pesticides, it is ironic that we are now starting to use the genetic variation of other species to control pests, applying recombinant technology to biological control.

Much of the improvement in the productivity of western agriculture in the 20th century has followed from the widespread use of chemical treatments. Cheap, easily used fertilizers and pesticides were the first applications of chemical technology to agriculture. Yet while total yields have risen, so has the proportion lost to pests. Since 1945, pre-harvest losses of food crops to insects in the USA rose from 7% to 13% and to weeds from 8% to 12%, yet during this time the use of insecticides increased 10-fold and of herbicides 100-fold (Pimentel *et al.*, 1984). This is due both to changes in the methods of cultivation and to the high, 'cosmetic' standards of food presented to western consumers.

The ill-considered use of pesticides carries additional and unseen costs. Too often our practices destroy or simplify the interactions within a community, interactions that are vital to its functioning and regulation. An agricultural soil needs a plant, animal and microbial community to decompose organic matter, to hold moisture and nutrients, and to resist erosion. Similarly, competition, predation, parasitism and other interactions serve to limit the population growth of the members of the above-ground community. Any form of agriculture will simplify these communities, reducing the number of species and weakening the links between them, but we often make the problem worse by feeding weeds, or by poisoning non-target organisms, including the natural enemies of a pest. For this latter reason, new pesticides are now developed which are highly specific for their target pest (Pickett, 1988).

In many cases, the pest is an alien, introduced species, separated from its natural enemies and escaping the limitations imposed by its native habitat. It assumes pest status when its economic damage justifies expenditure on its control, or when its impact on a natural community merits action. Many introduced pests flourish because their new hosts have not evolved specific defences against them. Of the major agricultural pests in the USA, 40% are introduced (around 200 species) although 65% of these are not considered pests in their native habitats (Horn, 1988).

In this chapter we consider the use of biological methods to control pests and weeds, most particularly insect pests. In its simplest sense, biological control aims to reunite a pest with its natural enemy, in the hope that the predator or parasite can significantly depress the pest population. These methods still represent a tiny proportion of the overall pest control effort (Jutsum, 1988), but this is an expanding area of applied ecology, with a long commercial future. Up to 1980, around 150 species of insect and 30 species of weed had been controlled by natural enemy releases (Samways, 1981).

Practical experience can count for much: Brazilian farmers have used the remains of dead and infected moth caterpillars for some time as a potent insecticide against other moths attacking their crops (Samways, 1981), simply aiding the spread of a virus through the pest population. However, many attempts have failed, mainly because the programmes have relied too heavily on trial and error introductions – perhaps only 16% of these efforts have led to a permanent solution (Horn, 1988).

Many population ecologists acknowledge that the contribution of theory to the practice of biological control has been small (Waage and Hassell, 1982; Murdoch *et al.*, 1985). This is not entirely the fault of the theoretical ecologist: the results of many biological control programmes have been poorly documented and few records allow comparison of the pest population before and after treatment. More detailed theoretical models are needed to aid the selection of control agents and to determine the conditions for long-term coexistence of the pest and its natural enemy (Ehler, 1991). It is these two applications of the theory that are considered in this chapter, although it will also become apparent that both theory and practice have much to learn from each other.

After defining a pest we go on to look at their population characteristics and the conditions under which a pest might achieve epidemic population growth. A natural enemy might increase both its consumption and its numbers as the pest population grows and the logistic model (section 4.1) is developed to describe the population growth of a pest and its predator. Here 'predator' and 'prey' are used in the broadest sense, to include any natural enemy and the pest it attacks. The model also allows us to define stable conditions when the two populations might coexist to give long-term regulation of the pest. Finally, we consider a number of methods which are commonly grouped under the broad heading of 'biological control' but which manipulate other features of a pests' biology and ecology to control it.

5.1 DEFINING THE PEST

'**Pest**' is a status that we give to an organism by virtue of its effect on human health, comfort, convenience or profits (Horn, 1988), although an invasive species may be deemed a pest by virtue of its effect on native communities. A pest is normally defined in economic terms, when the damage it causes justifies expenditure on a control programme. The population density of the pest must pass a threshold, beyond which the cost it incurs is greater than the cost of its control – its **economic injury level** (EIL) (Figure 5.1). The size of the population at EIL may be small if the pest is attacking a very valuable resource, such as the timbers of your house. Others, such as infectious and dangerous pathogens have no EIL, and control begins as soon as it is detected.

As the EIL is passed, the most common action is to apply a chemical pesticide. There is usually little time lag between this treatment and the pest succumbing, so that further damage is prevented. With biological

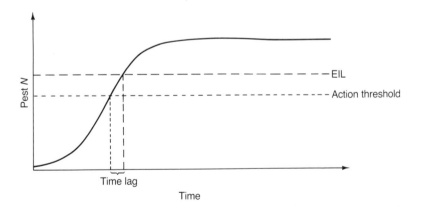

Figure 5.1 The economic injury level (EIL) is the size of the pest population that justifies expenditure on some form of control. If there is some time lag between a control action being taken and the pest population responding, an action threshold is specified to prevent the population reaching its EIL.

Managing pests

control it may be several days or weeks before there is a detectable impact on the pest. Anticipating this delay requires an 'action threshold', to initiate the programme before the economic injury level is reached (Figure 5.1). This highlights one important difference between the two methods: although pesticides offer rapid and emergency control of a pest, the depression of its population is rarely long lasting; in contrast, biological control methods may take several seasons before the pest population is significantly reduced, but it can provide control without the need for further action.

As the density of the pest and the damage it inflicts increase, so the value of a crop reduces (Figure 5.2). Set against this are the costs of controlling the pest. Profits are maximized when there is the greatest difference between the return on the crop and the cost of the control. This corresponds to the EIL: lowering the pest population below this level will incur additional costs which lower the profit. In this simple model the costs of control are the direct costs of a treatment and its application, but this does not allow for costs that are hidden or delayed. With pesticides, these may include the impact on non-target organisms, the need to monitor harmful residues or any health risks associated with their application.

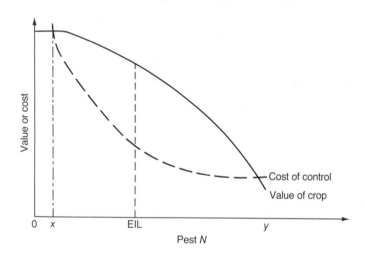

Figure 5.2 The EIL represents the maximum return on the cost of the control measure. As the number of pests increase, a point is reached (*x*) when the value of the crop starts to fall and at this point control commences. The costs of limiting the pest to a small population will be high, but these costs decline if a larger pest population can be tolerated (broken line). The EIL represents the maximum return on the cost of a control measure, when there is the greatest difference between the value of a crop and these costs. Beyond *y* the crop has no value and the control effort would cease.

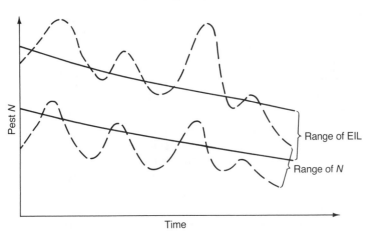

Figure 5.3 The value of a crop may not be fixed and will vary within a range. In the above example, the overall trend is for a decline in value with time. The pest population will also vary with time and the frequency of it crossing its EIL will depend on the value of the crop at any point in time.

The value of a crop is rarely fixed for extended periods, and the threshold of the EIL will change with the return on the crop. Additionally, there will be a large measure of variability in the size of the pest population: rather than there being a single, fixed threshold, a range of values is possible (Figure 5.3). Not only does this variability make pest outbreaks difficult to predict, it also makes an EIL difficult to specify.

In the real world, the ease of following a simple schedule means that chemical treatments tend to be applied not on demand, but according to convenience. When the costs are low, there is a temptation to use 'insurance measures' at the beginning of each season, anticipating a pest outbreak. Cheap pesticides are frequently used in this way, even though it may be relatively ineffective (Horn, 1988). Biological control is rarely used in response to a species passing its economic threshold, but rather as a long-term strategy.

5.1.1 Population characteristics of pests

The growth of a population may be limited by a lack of resources or by its interactions with other species. A pest population grows past its EIL because an abundance of resources removes or relaxes intraspecific competition. The same effect may result from the relaxation of interspecific competition or predation. A native species may pass its EIL with some change in its habitat, but the greatest proportion of pests are introduced or alien species – species which have escaped the regulatory mechanisms operating in their native habitat. This explains why the most impressive successes in biological control follow the importation of an control agent, when a pest is reunited with its natural enemy.

In the previous chapter, two parameters were shown to determine the rate of population growth in the simple logistic models: r, the intrinsic rate of population increase, and K, the carrying capacity of the ecosystem (equation 4.5). With limited resources, growth is checked and the intensity of intraspecific competition increases as the density of the population rises. The scope for population growth is then density-dependent. Where resources are unlimited the rate of population growth is independent of its density and is determined solely by the reproductive potential of the species (equation 4.3).

Rapid population growth will be advantageous in fluctuating environments when resources and the carrying capacity are continually changing. In contrast, an unchanging habitat will be filled with species close to their maximum population size and with few opportunities for growth. These habitats will favour individuals able to compete for the limited resources. Two different strategies will then be appropriate for each circumstance – a changeable habitat is likely to favour opportunist species with a high reproductive potential (termed **r-selected**), while competitive (**K-selected**) species will dominate constant environments. Species that are r-selected are subject to selective pressures which are non-competitive and largely independent of population density. K-selected species are the product of density-dependent selective pressures where efficient use of resources is of prime importance. This is a broad classification and most plants and animals will not fit neatly into either category. It is more useful to think of an r–K spectrum, with most species lying between these two extremes.

5.1.2 r-selected species

Where resources are transient and temporary, the population that grows most rapidly will prevail. The intrinsic rate of population increase, r, can be maximized in two ways – by increasing the birth rate or, more effectively, by reducing the generation time. Small organisms require a shorter time to reach maturity and reproduce, so smaller animals with short generation times have a higher r (Southwood, 1981).

This potential rate of increase is only achieved when there are few limits to population growth. Density-dependence rarely applies in these transient conditions and even if the population should overshoot its carrying capacity, a short-lived organism with discrete generations is unlikely to affect the resources of the next generation. Various features of the biology of r-selected species reflect their adaptation to these highly variable conditions (Table 5.1). Many of these characteristics are typical of animal pests, most particularly the insects.

Table 5.1 Characters attributed to *r*-selected and *K*-selected species

	Character	r-selected	K-selected
Habitat feature	Durational stability	Low	High
	Successional stage	Early	Late
Population features	Generation time	Short	Long
	Body size	Small	Large
	No. of reproductive events	One	Many
	Fecundity	High	Low
	Sensitivity of birth rate to population density	Low	High
	Mortality rate	High	Low
	Population stability	Low	High
	Investment in each offspring	Small	Large
Ecological features	Competitive ability	Low	High
	Efficiency in use of resources	Low	High
	Dispersal ability	High	Low
	Investment in defence mechanisms	Small	Large

Rapidly growing populations in transient environments will suffer high levels of mortality, especially amongst juveniles stages. A high rate of dispersal will be favoured because it can lower mortality by distributing the population amongst patches where there are no predators. Each individual invests a large proportion of its resources into a single reproductive event, often synchronized within the population at the end of a short lifespan. In this way they can outbreed their natural enemies: while they provide short periods of plenty for a predator, these are separated by long periods of scarcity. As a result, the population of natural enemies can rarely grow rapidly enough to limit extreme *r*-selected species. This is typical of many herbivorous insect pests such as aphids, where the populations of its predators appear to simply follow the fluctuations in aphid numbers (van Emden, 1988).

Many weeds do not fit neatly into this category, although a number illustrate some of these properties. Thompson (1988) surveyed the weeds invading reclaimed agricultural land in Guyana. The two most abundant weeds were *Echinochloa colonum* (Poaceae) and *Macroptilium lathyroides* (Fabaceae). Both are able to germinate with little seasonal control and mature rapidly. This enabled them to maintain a presence throughout the year, even though one of them (*M. lathyroides*) had a high death rate.

Additionally, they were both able to colonize bare soil more rapidly than their competitors. However, *M. lathyroides* also has a significant competitive ability, retarding the establishment of other species.

Many invasive weeds are annual species, able to flourish in the absence of competitors or natural enemies: 50% of the colonizers of waste ground in the UK are alien plants, whereas in established woodland these species comprise just 5% of the flora (Crawley, 1987). Plants may be more flexible in their strategies to a changing habitat – dandelion (*Taraxacum* spp.) is a weed capable of switching more resources to reproduction in shorter-lived habitats, and the same is true of different populations and species of goldenrod (*Solidago* spp.) (Southwood, 1981). However, the compliment of weeds in a region may change over the long term due to the competitive interactions between them, analogous to a secondary succession (section 7.2). These communities then become dominated by more competitive *K*-selected weeds (Forcella and Harvey, 1983).

5.1.3 *K*-selected species

A habitat that is stable over long periods will be populated by species close to their carrying capacity, with little scope for further population growth. A large body size may confer competitive advantage, but will in turn require a long generation time and thus lower *r*. Very often, this low fecundity of *K*-selected organisms is due to a long pre-reproductive period as well as the small number of offspring produced (Southwood, 1981). Their low fecundity is partially offset by a low rate of juvenile mortality. These species make a greater investment of energy and time in each offspring, extending in some cases to parental care. Their birth rate may also increase at low population densities as physiological factors (such as diet) or behavioural factors (such as territoriality) become less limiting. Most importantly, these are long-lived species which reproduce on more than one occasion: literally, their eggs are not all in one basket (Table 5.1).

Because of their large body size and long lifespan, the temporal and spatial scale over which these species range is larger than for *r*-selected organisms. This means their reproductive success has to be measured over a lifetime lasting several years. While a mosquito can complete its life cycle within a few months, the lifespan of some trees is several thousand years, during which time it can reproduce many times. The selection for competitive advantage make *K*-selected species highly specialized and consequently more sensitive to changes in their environment. They will also be less able to recover from low population densities and more prone to extinction (section 4.6).

These organisms are less likely to become pests: their powers of dispersal are often limited, and their rate of population increase too slow.

Nevertheless some can spread rapidly by modular growth of branches or polyps (Harper, 1981) and achieve pest status. Alien perennial plants may become weeds by their vigorous vegetative growth and competitive nature: *Rhododendron ponticum* is an introduced shrub that may, in the wild, outcompete the native British flora due to the depth of shade it casts.

The relationships between a species' life history traits and the durational stability of its habitat are derived from empirical correlations rather than any established cause-and-effect (Southwood, 1988). In so far as this classification groups together a variety of physiological and ecological characteristics, it helps to explain why some species are more likely to become pests. But these ideas are not without their critics, not least because of the difficulty of testing them. Most species are neither distinctly *r*- or *K*-selected and some may change between the two strategies according to environmental conditions. Additionally, the features of a species' genetic constitution, its size or physiology may constrain it to a particular strategy, rather than it being a simple response to environmental variability (Southwood, 1988). The various features of these two life history strategies are all attributed to the differing selective pressures of crowded or uncrowded conditions and it is probably unrealistic to assume that these factors dominate others which are not directly density-dependent, such as climate (Parry, 1981) (section 3.2).

A number of authors have suggested an additional strategy for those species adapted to life in very poor habitats, able to survive extended periods of stress. This has been given various names but was termed **adversity selection** by Southwood (1988), and it introduces a third dimension – the quality of a habitat. Stressful conditions may be accommodated by an organism at some metabolic cost that impairs growth or reproduction. Tolerating adverse conditions may lower the competitive ability of an organism, although this may not be disadvantageous if fewer competitors survive in the changed environment (section 3.4). Such species are of particular importance in the colonization of degraded or stressed habitats, and are considered more fully in Chapter 7.

5.1.4 Epidemic population growth

A pest population need not reach epidemic proportions to pass its economic injury level. However, major pest outbreaks are typically associated with rapid population growth, and we need to consider how these outbreaks might occur. Southwood and Comins (1976) describe a simple population model which examines graphically the conditions that allows a species (the prey) to escape regulation by a predator.

Competition for resources rises as the population density increases (equation 4.5). Thus, the proportion of mortality due to competition will

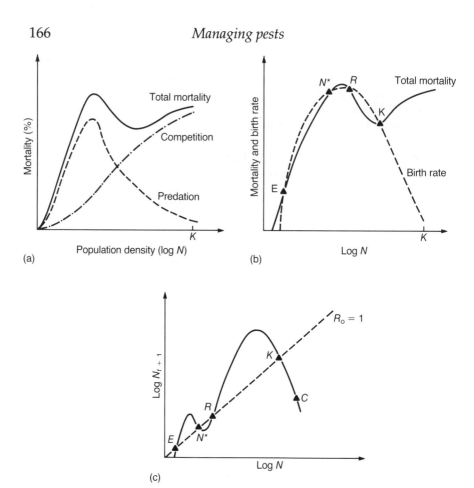

increase as the population approaches its carrying capacity (Figure 5.4a). Predation will also rise as the prey becomes more abundant, initially being density-dependent. However, as predators become satiated and their rate of population increase fails to match that of the prey, the proportion of mortality due to predation declines (Figure 5.4a). Above a certain population density, the prey effectively outbreeds its predator. The third plot shows the combined total mortality against population density.

The birth rate will also be density-dependent – initially increasing rapidly but declining as resources become limiting (Figure 5.4b). At the carrying capacity (K) the birth rate is matched by the death rate. The population will only grow when the total mortality rate is below the birth rate ($b > m$), when the reproductive rate R_0 (equation 4.1) is greater than one.

This is shown with an alternative plot in Figure 5.4c. Here the size of the one generation (N_{t+1}) is plotted against the preceeding generation (N_0): no net increase from one generation to the next ($N_{t+1} = N_0$) is indicated by the diagonal $R_0 = 1$. The continuous line maps the combined effect of the birth rate and death rate on R_0. Where this line dips below 45°, R_0 is negative and the population is in decline. Above the line the population is growing. At the intersections the birth rate and death rate are balanced and the population is at equilibrium, corresponding to the same points in Figure 5.4b.

The upper point K is the carrying capacity for the prey species beyond which no sustained increase in population size is possible. This limit is imposed by the resources of the environment and, being density-dependent, is therefore stable. A population moving too far above this point inevitably crashes (C). N^* represents the density at which predation is important in balancing the birth rate and death rate to give a lower equilibrium population size. This will also be stable – the population will tend to move to this position within its vicinity. At E and R the equilibria are unstable: if the birth rate drops below the mortality rate at E the population can only decline to extinction. Above R, the **release point**, the population escapes the control of the predator and the pest becomes epidemic. There are thus two possible stable states in this model, at N^* and K, determined by the presence or absence of a predator. Southwood

---

Figure 5.4 A synoptic model of epidemic population growth. (a) The proportion of mortality attributable to competition and to predation against population density (log values). In this simple model, competition causes perfectly density-dependent mortality, rising and then tailing off as population density increases. Predation is initially density-dependent, but the satiation of a predator causes the proportion of mortality due to predation to decline as the pest population continues to increase. Combining these curves produces the total mortality plot shown. (b) The birth rate grows rapidly as the population density increases but declines as the habitat becomes more crowded. By plotting the rate of mortality on the same graph, we can determine at what points the population is growing (when $b > m$) or declining ($b < m$). At four intersections the birth rate is balanced by the death rate. (c) This graph plots the size of one generation ($\log N$) against the size of the following generation ($\log N_{t+1}$). When there is no difference $R_0 = 1$, shown by the diagonal line. A composite of the the mortality and natality curve from 5.4b is shown by the solid line, which effectively maps the distribution of R_0 with different population densities. Where this line is above the diagonal the population is growing; the four intersections show the positions where there is no net population growth. E represents the extinction point, below which the population must be lost. N^* is a stable equilibrium that the population can return to – this is the population size determined by the action of a natural enemy. Above, R, the release point, the population will expand away from its natural enemy ravine and will achieve epidemic population growth. K is the carrying capacity for the pest in the absence of a natural enemy. At C, the density is too high and the population must crash. (After Southwood, 1975.)

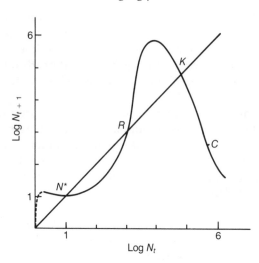

Figure 5.5 The population growth curve for *Cardiaspina albitextura*, a hemipteran pest of eucalyptus. The various points appear to match well the estimates derived from the original observations in the field – a release point of around 10–15 eggs per shoot and a carrying capacity of about 100 eggs per shoot. (With permission, from Southwood and Comins, 1976.)

and Comins (1976) demonstrate this pattern for an hemipteran insect that attacks eucalyptus trees (Figure 5.5).

The gap between N^* and R is the **'natural enemy ravine'**, the region in which the prey population is depressed by a predator. A control programme using a natural enemy must aim to confine a pest species to this density (or below it). Long-term stability of pest numbers thus requires sufficient predators to maintain this suppression.

Southwood and Comins (1976) have demonstrated the effect of extreme r- and K-strategies on these curves. The rapid reproductive capacity of an extreme r-selected species may be able to outbreed its natural enemy and effectively have no natural enemy ravine. Such a pest would be beyond the control of a predator and will have frequent epidemic outbreaks before crashing back. Its reproductive powers are such that it can recover from low densities and will be unlikely to go extinct. Aphids are r-strategists with some of these characteristics: various species of aphid have effectively no natural enemy ravine under most conditions (van Emden, 1988). For different reasons, an extreme K-strategist will also have no natural enemy ravine. This organism will have highly evolved mechanisms to avoid excessive predation or parasitism, and mortality due to predators is insignificant. These are populations regulated by intraspecific competition, and their long reproductive life and high rate of survival mean that they are maintained at a high density, close to their carrying capacity. As a consequence, such species enjoy much greater population stability.

The model highlights the importance of the population thresholds which govern extinction and epidemic, although such breakpoints and ravines can also be generated by other factors, including migration. It also demonstrates how the depression imposed by a natural enemy can be most important for those species which are intermediate between the two strategies: biological control methods are only likely to be effective against pests that are neither extreme r- or K-strategists. Indeed, Southwood and others have suggested that pesticides or some other techniques may be the only viable control strategy against distinctly r-selected species such as aphids (Horn, 1988).

5.1.5 Pest ecology

Beyond these strategies, other features of a species' ecology can contribute to it becoming a nuisance, including their response to changes in their habitat. For example, the invasion of bracken (*Pteridium aquilinum*) on upland farms in Britain can be traced back to recent changes in farm practices, and particularly the reduction in trampling as hill farmers have switched from cattle to sheep in recent years (Lawton, 1988).

Agricultural practices may in other ways encourage pest outbreaks. Many commercial plant varieties have a reduced resistance to insect attack and this, along with the large-scale monocultures of even-aged plants, produce a uniform habitat offering massive scope for epidemic pest outbreaks.

Equally, in introducing a natural enemy, we may provide it with the same opportunities, when it could itself become a pest by switching its attention to a non-target species. The introduction of the lacebug *Teleonemia scrupulosa* to control the introduced weed *Lantana camara* in east Africa led to later problems when it attacked sesame crops, even though this was not a problem in its native Mexico (Pimentel *et al.*, 1984). Although the agent was unable to complete its life cycle on sesame, the nymphs could produce viable young. Even so, this may have been only a short-term phenomenon, following the lacebug's population explosion on *Lantana* (Ehler, 1991). For these reasons, we need to understand both the ecology of the control agent and the pest before any release is contemplated. Lawton (1988) describes the search for a herbivorous insect to use against bracken in the United Kingdom, and the steps needed to ensure that any introduced control agent has no deleterious interactions with the native fauna and flora (Table 5.2).

Table 5.2 The necessary characteristics of potential agents for the biological control of bracken (*Pteridium aquilinum*) in the British Isles and the features of two possible control agents (after Lawton, 1988)

1. An insect of a seasonal, cool temperate climate similar to that of Britain
2. The insect must exploit the same subspecies and variety found in Britain (*Pteridium aquilinum* subsp. *aquilinum*, var. *aquilinum*)
3. The insect should be ecologically distinct from native British insects feeding on bracken to exploit a vacant niche, avoiding problems of severe competition
4. The insect should be taxonomically distinct to avoid parasitism or predation from the natural enemies of endemic insects on bracken
5. It must be specific to bracken
6. A species able to attack the rhizome, the principal method by which bracken spreads, would be an advantage

1 and 2 caused the search to be concentrated on the mountains of Cape Province in South Africa. Two possibe lepidopterans were identified whose caterpillars caused extensive damage to the same subspecies of bracken:

Conservula cinisigna (Noctuidae)
Although there are several noctuids feeding on bracken in Britain, this species tends to feed earlier in the year. It may therefore avoid attack by the parasitoid associated with the native species. Also, it attacks the young plant when it is most vulnerable, and appears to be monophagous on bracken, causing heavy damage to the fronds. It has been difficult to rear in the laboratory due to an unknown infectious disease.

Panotima sp. (Pyralidae)
No pyralids feed on bracken in Britain and *Panotima* has a life cycle different from any native species. The younger visitors defoliate the fronds and later mince the 'stem' (rachis). Again, attacks occur early in the season when the plant is most susceptible. *Panotima* appears to be monophagous on bracken, but is found over a wide climatic range in South Africa, possibly an indication of its adapability. Again, it has proved difficult to rear in the laboratory.

5.2 THE FUNCTIONAL RESPONSE OF A PREDATOR

The introduction of a natural enemy is an attempt to reduce the size of the pest population and to stabilize it around a new, lower equilibrium density (Waage and Greathead, 1988). Candidates for biological control agents have to suppress the pest's population to this density over the long term, or be able to wipe it out completely. We thus need to consider the ways in which a natural enemy might react to an increase in pest numbers. There are two possible responses – a **functional response,** when it simply eats more prey, and a **numerical response** when the predator increases its own population size. We examine how growth in the population of the prey and the predator might be linked in the next section.

Given a fixed number of predators, the capacity of a natural enemy to depress the prey population will depend upon the rate at which the prey can be consumed. The functional response of a predator reflects not only on its capacity to eat prey but also the time taken to find, catch, kill and consume an individual prey. This searching and handling time limits the number of prey that can be consumed in any period, even if the predator never becomes satiated.

Three different types of functional response were described by Holling (Figure 5.6). The simplest (type I) has a constant rate of prey consumption up to a limit: an animal filtering a constant volume of water per unit time will receive more prey if prey density in the water increases. However, beyond a certain point the predator is unable to consume the prey any faster (Figure 5.6a). Such a predator would thus have relatively little effect on a fast-rising prey population. This type of response appears to be rare in the animal kingdom.

A type II response is perhaps more common. The rate of consumption rises rapidly with prey density, tailing off at some saturation level (Figure 5.6b). This level is determined by the handling time for the prey: abundant prey are easily found by the predator, but the time taken to catch and consume the prey limits the functional reponse. A predator that is proficient in finding its prey will spend less time searching for its food and will show a faster functional response to any increase in prey numbers, at least up to the saturation level. Again this response is not density-dependent beyond this prey density.

The type III response differs from the type II curve by the presence of a brief lag in the rise of consumption (Figure 5.6c), but which then rises rapidly. This is often described as the sigmoidal response. The initial lag can be due to a predator learning a prey image, or simply switching to a new prey as that species becomes more abundant. Again, an increase in searching efficiency or a decrease in the handling time leads to the rapid rise in the rate of predation, up to the limits imposed by the handling time needed for each prey. The proportion caught is also density-dependent and stabilizing at low prey densities, before the predator becomes satiated. Above this density the prey again escapes the control of its predator.

With a fixed number of predators, all three responses indicate a limit to the rate of predation. More importantly, density-dependent mortality is only found to be significant at relatively low prey densities. Close to the saturation level, individual prey have less chance of being consumed, and rapid population growth will thus have a selective advantage – natural selection will favour species able to outgrow its predator.

In reality, the functional response of the predator is rarely so simple: larger, older and wiser predators will have a greater ability to catch prey and have a larger appetite. The size of prey consumed may also change

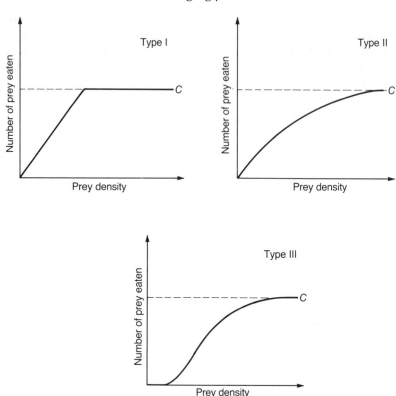

Figure 5.6 Holling's three types of functional response by a predator to an increase in prey numbers. All three are limited by the rate at which prey can be consumed – in I there is no handling time, and the limit is imposed by the ingestion rate. II and III are both limited by the time taken to find and consume the prey. They differ because III shows a time lag, either as the predator needs to learn a new prey image or because it has shifted from an alternative prey species.

with the age of the predator (Hassell, 1981); the age distribution of the predator population is therefore likely to be important to its functional response. The same is true of the prey population: mortality at different stages of the life cycle (particularly before reproduction) can have a greater impact on the fecundity of the pest. The parasite used to control cassava mealy bug, for example, selects instars of different ages in which to lay male or female eggs (Neuenschwander and Herren, 1988). Understanding its implications for both the pest and the predator requires partitioning their populations into age-classes, followed by a life-table analysis to quantify the proportion of mortality in each class. A full description of this method is given in van den Bosch *et al.* (1982). This method has formed the basis of some pest control programmes, although there is some contention over its capacity to detect regulation of the pest (Hassell, 1981).

5.3 THE NUMERICAL RESPONSE OF A PREDATOR

Classic biological control releases a natural enemy on a single occasion, thereafter to maintain the pest population at lower equilibrium within its natural enemy ravine (Table 5.4). Ideally, both populations persist – the pest to provide a food source for the predator, which may then survive to prevent future pest outbreaks. A stable interaction between the predator and its food source is essential if re-introduction is not to be a necessity. A number of mathematical models have been used to identify the attributes which can allow the two species to persist together. Their aim is to capture those features which govern the dynamics of the relationship, and in doing so, aid our choice of control agent.

To understand their interaction, we need to describe the numerical response of a predator as prey density changes. Here we shall use the simple deterministic models of population growth developed in Chapter 4, running two models side-by-side and allowing each population to respond to the density of the other.

5.3.1 Predator–prey relationships

The prey (N) would obviously flourish without predators (P) consuming them. Assuming no limits to their growth, we can show this as (equation 4.3):

$$\frac{dN}{dt} = rN.$$

However, a predator will kill and consume a proportion of the prey it encounters. This is measured as a, the **searching efficiency** or attack rate and represents the proportion of contacts between the predator and prey which lead to the death of a prey. At a fixed attack rate, more prey will be killed and consumed if there are either more prey or more predators and therefore more encounters. The rate of predation is thus given by aNP. This is the functional response of a fixed number of predators, so that as prey numbers rise, more are attacked. This assumes that all prey are available for consumption by the predators. Here we place no limit on the functional response, although this can be easily achieved in the model (Whittaker, 1975).

The rate of loss due to predation must now be deducted from the growth of the prey population:

$$\frac{dN}{dt} = rN - aNP. \tag{5.1}$$

The predator population can only decline in the absence of prey:

$$\frac{dP}{dt} = -mP$$

where m is the rate of mortality of the predator. This population will grow if prey are consumed to support its production of offspring. This can be measured as β, a coefficient quantifying the rate at which consumed prey become predator offspring. If $\beta = a$, there is a one-to-one correspondence between the consumption of a prey and the production of a new predator. This is the case with many insect parasitoids, where a single parasitic larva consumes and eventually kills one host.

The numerical response of the predator, its increase in population size, is given by the rate of predation (aNP) multiplied by its efficiency in converting these into new predators (β):

$$\frac{dP}{dt} = \beta aNP - mP. \tag{5.2}$$

These are the simple Lotka-Volterra models of predator–prey relationships. Running the two models together produces continuous cycling of both populations, with the cycle of the each population coupled to the other (Figure 5.7). A large prey population allows the predator numbers to grow, but this leads to a decline in prey abundance. Predator numbers have to follow this decline as food is increasingly scarce. Eventually predator numbers are so low that the prey poulation is able to recover and start the cycle again.

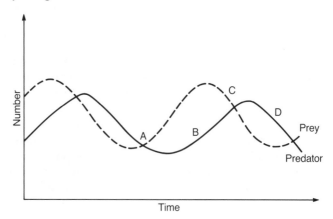

Figure 5.7 The population dynamics of a predator and its prey, derived from the simple Lotka-Volterra models. At A the prey population can increase because of low predator numbers. The predator population will then grow as its food source increases (B). However, predator numbers eventually cause a decline in the prey population (C), which in turn leads to a fall in the predator population (D).

This cycling will continue with a period and amplitude set by the terms of the equation. This is maintained unless the system is disturbed, when a new amplitude is set by the size of the disturbance. Although simplistic and decidedly unreal, it does demonstrate the close relationship that can exist between a monophagous predator and its prey, and their capacity to oscillate within a range (May, 1981c).

Adding greater realism can add greater stability to the model. Giving the prey a carrying capacity (K) limits its population growth in the absence of the predator or when predator numbers are small:

$$\frac{dN}{dt} = rN(1 - \frac{N}{K}) - aNP. \tag{5.3}$$

Now both populations assume a population size that remains constant as long as their environment does not change. These are stable equilibria to which they will return if the system is perturbed. It is the density-dependence of the prey that confers this stability, although this can be relatively weak if the prey population is depressed greatly below its carrying capacity (May, 1981c). Stability will also result from limiting the predator population, even if the prey has no limit.

There are additional modifications which can add realism to the model. The simplest is to place a ceiling on the number of prey the predator can

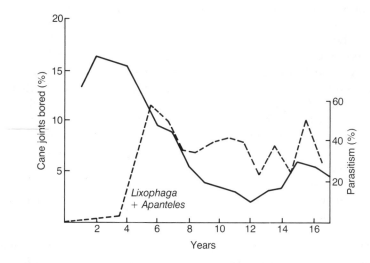

Figure 5.8 The decline in the incidence of sugar cane stem borers (*Diatraea* spp.)(solid line) in the presence of two parasitoids (*Lixophaga diaetraeae* (Tachinidae) and *Apanteles flavipes* (Braconidae)). The total percentage parasitism of the population is shown by the broken line. Notice that the pests and their parasites coexist for at least 10 years from these records. (With permission, from Waage and Hassell, 1982.)

consume, by assigning to it one of the three types of functional response. Because the density-dependent effects on prey mortality are confined to low prey densities, stability is lost when the prey density rises beyond the saturation level. The proportion of mortality due to predation is then greater at low prey densities, creating the instability (Hassell, 1981). Density-dependent predation, before the predator is satiated, will allow both populations to persist, confining the pest to its natural enemy ravine (Figure 5.8).

5.3.2 Stability and the distribution of attacks

The assumption that all prey are available for consumption by the predator is unrealistic. Prey are not uniformly distributed in their habitat, and some are more easily found than others. This will affect the stability of the predator–prey interaction. Hassell and his co-workers have studied this in insect parasitoids, using models with discrete generations (May and Hassell, 1988; Hassell and Anderson, 1989).

One in 10 of all metazoan animal species is a **parasitoid**. Most of these are wasps (Hymenoptera) and flies (Diptera) (Waage and Hassell, 1982). They are specialist parasites of a range of arthropods in which the larva grows inside the host, killing it as it matures. Parasitoids are thus somewhere between a true predator and a true parasite. This simplifies the model by specifying a one-to-one relationship between parasite and host, and by the assumption of discrete generations (which is true of many univoltine insects in the temperate regions).

The distribution of any prey will be patchy, with some patches richer in prey than others. Searching for prey costs time and energy, but a predator or parasite can reduce these costs by remaining in patches with a high prey density. The predator will thus tend to stay where food is most abundant. We might thus expect attacks to show a clumped distribution, aggregated where prey numbers are greatest. Patches with low densities will suffer reduced levels of attack and some will escape attack altogether. These partial refuges will ensure that the prey population is never exploited to extinction – important if both populations are to persist. As the degree of clumping or contagion in the attacks increases so does the stability of the two populations.

This stability results from the searching behaviour of the natural enemy but it can also derive from patchiness in the environment, providing partial refuges where the pest cannot be located. A similar effect will occur if there is a time lag between the life cycles of the pest and the natural enemy or if different patches are colonized at different times by the natural enemies (as long as both species tend to stay in a patch, once colonized). In fact, any mechanism which allows differential chances of being attacked serve to increase stability and allow the two populations to persist together (May and Hassell, 1988). There is also evidence that limiting the population growth of the parasitoid with hyperparasitism can add stability to this system. A number of hyperparasitoids attack *Trioxys pallidus*, a highly efficient control agent introduced against the walnut aphid (*Chromaphis juglandicola*) in California, part of an elaborate control programme that prevents overexploitation of the pest (Messenger, 1975).

The stability of this interaction is important if the parasitoid population is to persist, and respond to future pest outbreaks. These models are based on discrete generations and tend to have different stability properties to those with continuous growth: models with overlapping generations show greater adjustment stability, with both populations being able to return to their stable state over a much greater range of disturbances.

5.3.3 The depression of the pest population

The effectiveness of any biological control agent can be measured simply by comparing the pest population at its equilibrium density under predation to that in the absence of predation:

$$q = \frac{K}{N^*} \tag{5.4}$$

where q = degree of depression; K = carrying capacity in the absence of the predator; and N^* = equilibrium population size in the presence of the predator.

May and Hassell (1988) concluded that the success of a parasitoid in maximizing q (depressing N^*) depends on the balance between:

1. the net rate of increase of the prey;
2. features of the parasitoid efficiency in attacking the prey, including its search efficiency (a), the spatial distribution of attacks compared to that of the host, its sex ratio and the rate of mortality of its offspring.

As we have seen before, the search efficiency governs the speed of the functional response and sets the new pest population equilibrium of the pest (Figure 5.9). The search efficiency is some measure of the capacity of the natural enemy to find patches of prey when local extinctions and colonizations are occurring continuously. In an established environment without frequent disturbance, a specialist predator or parasitoid will be well adapted for finding its prey even at low densities, and thus have a higher search efficiency than some generalist predator. Beyond this, the size of the parasitoid population (and more especially the number of females) governs the equilibrium level of the pest.

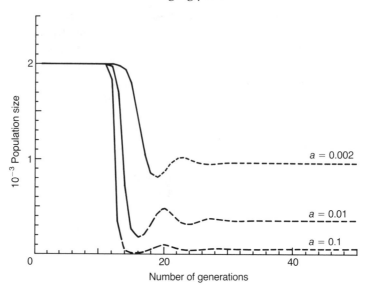

Figure 5.9 A greater reduction in the pest population is achieved by *a*, a higher search efficiency in the natural enemy. These curves, derived from mathematical models, show the depression from the pest's carrying capacity after the natural enemy is introduced at generation 10. Not only is this only more rapid, but it is also more pronounced in natural enemies with a high *a*. (With permission, from May and Hassell, 1988.)

Similar conclusions emerge from modelling the use of a pathogen (viruses, bacteria or protozoa) to induce disease in a pest species (May and Hassall, 1988). The rate of transmission of the disease depends on the rate at which it kills the pest and the size of the pest population. Most pathogens will be able to wipe out the pest completely if there is no form of partial refuge. A pathogen which is too virulent may actually need a large pest population to sustain itself in the environment. This will also be determined by any free-living stage of the pathogen and the length of time this is viable in the environment. For this reason, pathogens with an intermediate pathogenicity are likely to be most suitable in a biological control programme, requiring a low prey population to sustain them in the wild.

5.3.4 Multiple species models

It is unlikely that a biological control agent released against a pest will be the only species consuming it. In some cases, a biological control programme may release several agents at the same time, hoping that one or two will persist to regulate the prey. Models are therefore needed to describe how a control agent might establish itself in the presence of other consumers of the pest.

Such models have been developed by May and Hassell (1988) and confirm the value of common sense. Put briefly, two agents are most likely to coexist if they both contribute to the stability of their interaction with the prey. A control agent introduced into an existing predator–prey relationship will establish itself only if its search efficiency is higher than that of the resident predator. If it is very much more efficient it may displace its competitor altogether. Some separation of their feeding behaviour, perhaps by attacking different stages of the life cycle will allow coexistence.

Whereas an additional natural enemy can further maximize q, the effect of two predators acting together may not be additive (May and Hassell, 1988). However, they can work in concert: for example, the olive scale (*Parlatoria oleae*) in California is controlled by two parasitoids – one most effective in the hot, dry conditions of the summer (*Coccophagoides utilis*) and a second (*Aphytis maculicornis*) which is dominant during the rest of the year. The competition between the two appears to be mediated by the effect of the weather on their activity levels. Separated in this way, these introduced agents maintain the pest below its EIL (Messenger, 1975).

5.4 THE USE OF MODELS IN BIOLOGICAL CONTROL

This collection of models attempts to distil some general principles about the nature of the interaction between a pest and its natural enemies. These may then be used to indicate the necessary features of a successful natural enemy. Such deterministic models suggest that stability is more likely with a specialist parasitoid able to locate and concentrate its attacks in high pest densities, and also that their coexistence depends upon some of the pests escaping.

In a variable world these models can only be an abbreviated version of reality. The prediction that a patchy habitat is important for coexistence appears to be valid (Waage and Greathead, 1988). However, Murdoch *et al.* (1985) have questioned whether stable coexistence of pest and natural enemy is necessary for successful biological control: there appeared to be little empirical evidence of stable prey and parasitoid populations in a number of cases they reviewed. Both populations may show large fluctuations, including local extinctions, but if the EIL is never exceeded the pest is under 'control'. Also, in some cases it may be better to choose a polyphagous natural enemy which could survive a local extinction of the pest by shifting to alternate prey species. Further outbreaks would also produce a faster numerical response by this agent because its numerical response might begin from a higher population density.

Generalized models serve to remove the detail of particular predator–prey relationships, detail that is often crucial to the practice of biological control. A large amount of information may be needed to fully describe the particular conditions of one problem. Some practitioners prefer retrospective modelling using life-table analysis and multivariate methods. This is modelling after the event, using empirical data to find significant relationships between variables. The models are applicable to one problem only, but their predictive power is a valuable aid in managing the control programme. The *ad hoc* approach to releases in the past has meant that few programmes have collected data from the outset to enable such modelling.

One example of analysing a problem in this way has been the study of cassava and its pests. Cassava (*Manihot esculenta*) was introduced into Africa over 300 years ago and quickly became an important subsistence crop in the wetter tropics. During the 1970s two pests were accidentally introduced, cassava mealybug (*Phenacoccus manihoti*) and the cassava green mite (*Mononychellus tanajoa*), together capable of reducing tuber yields by up to 50%. Gutierrez *et al.* (1988a,b,c) describe a detailed model for the growth of cassava and of the pests, including such factors as the water and nitrogen status of the crop. These are known to govern the growth of the pest populations. The capacity of an introduced parasitoid (*Epidinocarsis lopezi*) and the native predators to depress the *Phenacoccus* population was measured using complex multivariate models. Simulations (holding some variables constant) were run to distinguish the relationship between the pest and its natural enemy from the stochastic 'noise' in the environment. The parasitoid was shown to follow a type II functional response to pest density when the mortality due to rainfall was held constant. Generalist predators (native coccinellid beetles) were unable to control the mealybug because their functional response was too small. Indeed, the introduction of *Epidinocarsis* caused a significant reduction in the coccinellid populations since the control agent was more efficient in its numerical response (Gutierrez *et al.*, 1988b). *Epidinocarsis* has been remarkably succcesful in controlling the mealybug (Neuenschwander and Herren, 1988).

Such detailed simulations of the ecology of a pest and its natural enemy have to follow the story as it develops, and after the agent has been introduced. In contrast, the analysis provided by deterministic models suggests general guidelines for the selection of control agents and makes predictions about the conditions that will allow effective long-term control. The use of specialist predators and parasites, efficient in finding their prey, are generally favoured for sustained control in stable environments, although some mechanism allowing a proportion of the pests to escape is required. Specialists are also less likely to have an impact on non-target species.

Several authors have questioned whether such models are the best guide to selecting control agents, arguing that they fail to provide sufficient predictive power (Ehler, 1991). Few models allow for indirect impacts of an agent upon its new community, perhaps by providing a food source for a native predator, or some other interaction. The mosquitofish, *Gambusia*, has been successfully used as a specialist predator of mosquitoes but has led to significant restructuring of the commmunities where it has been introduced (Ehler, 1991). Wapshere (1985) offers one protocol for selecting insects for attacking introduced weeds, according to a range of characters, including host specificity and the agent's population traits. This protocol also considers the match between the agent's original and new habitat, and the effect of native predators and parasites. The scale of the damage inflicted upon the weed is obviously important, although this will depend upon the density of infestation and the stage of development of the weed's population. As a general guide, Wapshere concludes that an agent effective against a weed in one location is the best candidate for other locations.

5.5 FINDING NATURAL ENEMIES

The search for an agent against an introduced pest will usually begin in the country of origin. This may not be easy to identify and, in some cases, the genetic variance of a population has been used to define its home range (Horn, 1988). Such work can be important in reducing later effort: 38 species of natural enemies were imported from different parts of the world and released against the black scale (*Saissetia oleae*) in California, but the only parasitoid that was effective (*Metaphycus helvolus*) came from South Africa, the home range of the pest (van den Bosch *et al.*, 1982).

Some have argued that the association between a pest and a natural enemy ought not to be too close (Pimentel *et al.*, 1984). After a long period of evolution together, a pest species might be so well adapted to a natural enemy that there is little depression when the two are reunited. A more effective control agent should then be derived from an area other than where the pest originated. However, most of the evidence suggests that highly adapted natural enemies are preferable. For example, some strains of the weed chondrilla (*Chondrilla juncea*) introduced into Australia can only be effectively controlled by host-specific strains of a pathogenic rust, the fungus *Puccinia chondrillina* (Cullen and Hasan, 1988). Harris (1986) also concluded that specialist natural enemies had a greater rate of establishment on weeds over one year (1984) in Canada. There may be merit in collecting from a number of populations of a natural enemy, from a broad area of the home range to increase the genetic pool from which the new population is established (Horn, 1988). A release will then include individuals able to survive a variety of conditions.

Other features of the ecology of an introduced species need to be considered, including potential antagonists in the host community which might interfere with its efficacy, or predators that might consume it (Table 5.2). To establish itself, a control agent may have to displace or survive competition with another consumer of the pest. It also has to survive predation itself. Predators feeding on the agent, or bad weather depleting its numbers, are the most common causes of failure (Crawley, 1987).

5.5.1 Selecting natural enemies

For animal pests, we can distinguish two basic strategies for biological control, effectively derived from the durational stability of the habitat. Specialist predators or parasitoids are more likely to provide long-term control, say in a citrus orchard, where the habitat is relatively constant. In annual crops a generalist predator will be more responsive (Samways, 1981). We discuss the range of cultural techniques that can be used to maintain generalist predators within an area later on.

In attempting to establish long-term control, several natural enemies might be found against a pest, but the specialist predator or parasite should have a number of characteristics (Table 5.3).

Table 5.3 Attributes of a predator or parasitoid as an effective biological control agent (after Samways, 1981 and Horn, 1988)

---

1. Rapid functional and numerical response to a rise in pest population density
2. High searching efficiency relative to the pest's rate of increase
3. Able to maintain a low pest population and sustain itself in the long term
4. Able to survive competition with native predators and parasitoids
5. High prey specificity with minimum impact on non-target biota
6. Synchronous activity with pest species, especially if attacking a particular stage of the life cycle
7. Easy to culture in the laboratory
8. Easy to release in the field
9. Cheap to use with rapid results to inspire confidence
10. No social nuisance

Specificity is important if the introduced agent is not to become a pest by attacking non-target species. This requires expert taxonomic skills and a thorough understanding of the ecology of a species. Some notable disasters have followed thoughtless releases: the introduction of the mongoose into Hawaii to control rats led to the extinction of a number of species of ground-nesting birds (van den Bosch, 1981), and similar problems were encountered on several Caribbean islands.

A specialist is likely to have a greater ability to find the pest, and the life cycle of pest and predator should be synchronized. Aphids reach epidemic proportions because the populations of their existing natural enemies are out of step: van Emden (1988) suggests that these predators and parasites will only be effective if their populations can be increased early in the season, to limit the later growth of the aphids.

The predatory behaviour of a natural enemy may be confined to one stage of the life cycle but due attention has to be paid to the ecology of the other stages. Only the larvae of syrphid flies attack aphids, while the adults feed on nectar. An abundance of wild flowers will feed the adult flies and encourage other polyphagous insects, which themselves can significantly lower aphid numbers (van Emden, 1988). The spraying of crops with yeast and sucrose mixtures is used to encourage adult feeding in the vicinity of aphids.

Overall, a rapid functional response appears less important than a rapid numerical reponse for successful control (Crawley, 1987). Also, a rapid rate of dispersal is essential if the pest population is not to expand beyond the control of the natural enemy. This has contributed to the success of *Epidinocarsis lopezi* against the cassava mealybug – the parasitoid was able to increase its range by 100 km in each dry season after it was established in west and central Africa (Neuenschwander and Herren, 1988).

5.5.2 Natural enemy release

The costs of finding a suitable candidate can be high, but often a small fraction of the costs of developing a modern pesticide (Horn, 1988). Having found an agent, there are then the additional costs of producing enough individuals for a release. The rearing of insects in the laboratory will require growing them on the pest itself or some substitute substrate. Field cages with high densities of the pest are commonly used to increase natural enemy numbers (and monitor performance) before release. Pathogens, including bacteria and viruses, are more easily raised in liquid culture.

The conditions for release of the natural enemy are frequently critical for their success, especially for non-persistent pathogens. Unless protected from ultraviolet radiation many pathogens have limited viability outside their host. Various techniques are now being used in commercial preparations to improve their long-term viability, including attempts to mask them from UV light (Payne, 1988). Similarly the release of insect control agents has to be timed for the appropriate time of year, location and so on. This has to match the abundance of the pest, particularly if the attack is confined to a particular stage in the life cycle of either organism.

Pest outbreaks of short duration or of rapidly growing weeds are most

Table 5.4 Four methods of release of biological control agents

Classical	Single introduction allowing long-term control from an established natural enemy population
Inundation	One release of a large number of natural enemies to control a single pest generation, with no anticipated follow-up action
Augmentation	Supplemental releases of natural enemies to raise their population
Inoculation	Periodic releases of a natural enemy which cannot sustain itself beyond a single season

readily controlled by chemical methods (Samways, 1981). Where the pest remains, long-term regulation has to establish a specialist natural enemy population which itself can persist in the habitat: unlike pesticides, a single treatment may be sufficient, even though it will take some time to be effective. Horn (1988) suggests that three years is a reasonable length of time in which to expect a control programme to work.

The four principal strategies for release are given in Table 5.4. The classic method – attempting to establish a natural enemy population – is largely confined to established ecosystems or those agricultural strategies that allow for strips of untilled land, patches in which both pests and the control agent can survive. The classic method has been most effective against introduced, alien species. About one-third of all introductions against weeds have been successful, although in Canada, establishment success is as high as two-thirds, reducing some weed densities by up to 95% (Harris, 1986).

Exclusion of a control agent by competition from the native fauna is a significant cause of failure in insects introduced as natural enemies (Ehler, 1991). Augmentation may be necessary if a sufficiently large population cannot be established. Occasionally, a second agent is needed if control is not achieved throughout the season. For example, the control of the weed *Lantana camara* on Hawaii required two insect herbivores to defoliate throughout the year (Harris, 1986). Similarly inoculation may be required where an agent cannot survive a complete season or where its life cycle is asynchronous with that of the pest.

Inundation operates by providing a rapid response, aiming to eradicate the pest by the sheer number of control agents released. This method is most often adopted with the release of pathogens as control agents.

The difficulty of growing the natural enemy in the laboratory may limit the size of a release, although Crawley (1987) suggests there is little evidence that the number released governs success. Exclusion experiments, in which the pest is protected from the natural enemy, can be used to assess q in the release areas, although many programmes rely on trial and error releases. This acknowledges that we have rarely described all the variables or chance events that we might later need to call bad luck.

5.6 THE ECOLOGY OF THE PEST

The proliferation of a pest and the damage it causes will depend on a range of biotic and abiotic factors in the habitat. In some cases, it may be possible to exploit some weakness in the pest's adaptation to its new environment to limit its proliferation.

The control of St John's wort or Klamath weed (*Hypericum perforatum*) is a classic example of its kind, although the success of its control agents depends upon the part of the world where they are applied. The weed is a temperate perennial plant, originally from Europe and Asia, which has invaded semi-arid regions of Australasia and the Americas (van den Bosch *et al.*, 1982). Being noxious to sheep and cattle, its rapid spread quickly ruins rangelands used for grazing.

A programme to control the weed in Australia began with a search for an insect herbivore in England and France. Two species of the beetle *Chrysolina* were found which were monophagous, feeding on the leaves of *Hypericum*. They were introduced with good results into Australia in the early 1940s. The weed also became a major problem in California, and *Chrysolina hyperici* and *C. quadrigemina* were introduced here from Australia. In 1944 the weed occupied two million acres of rangeland in California, yet just 12 years later the beetles had confined it to roadsides and shady canyons.

Chrysolina prefers to lay its eggs in open sun, accounting for the present distribution of the weed. In reducing its foliage, *Chrysolina* causes the plant to reduce its root growth. This lowers its resistance to drought and reduces the competitive advantage it enjoys over native species (Harris, 1986). It is now declining in the shaded areas because of the attention of second introduced pest, the root-boring beetle *Agrilus hyperici*. Neither *Chrysolina* species has become particularly abundant after introduction into British Columbia, but their lack of success in controlling the weed here may result from the moister conditions, where a reduced root system is less of a handicap. In more arid conditions, this temperate plant succumbs when its reduced foliage is unable to support the extensive root system needed.

The interaction between a pest and native competitors can also be exploited in a control programme. Some non-pathogenic bacteria provide protection against disease for many higher organisms. Mycorrhizal infections of the roots of higher plants prevent other infestations (section 7.2). The fungus *Dactylella oviparasitica* will protect peach orchards in California from the attentions of the root-knot nematode *Meloidogyne* (van den Bosch *et al.*, 1982), killing its larvae before they hatch. Competitive exclusion has been used to control crown gall in fruit trees and other plants, caused by *Agrobacterium tumefaciens*, by introducing a non-pathogenic strain of *Agrobacterium* that competes for binding sites on the roots (Campbell, 1990). This agent produces a bacteriocin that improves its competitive advantage against some races of the pathogenic bacteria. A genetically engineered strain has been cleared for release because it is unable to pass this plasmid-linked character on to the pathogen (section 3.6).

Establishing these interactions can be important for the stability of the community. Native coccinellid beetles quickly exploited the arrival of the cassava mealybug, but these in turn attracted native hyperparasities to the cassava crop. Along with other insects feeding on cassava, a food web was quickly established. The introduction of the control agent *Epidinocarsis lopezi* caused 10 species of hyperparasitoids to switch to this new host. Mathematical models suggest this hyperparasitism will stabilize the *E. lopezi* population, and thus be of long-term benefit in controlling cassava mealybug (Neuenschwander and Herren, 1988).

5.7 OTHER FORMS OF BIOLOGICAL CONTROL

Biological control uses a much broader range of methods than simply introducing a natural enemy to control a pest. There are other features of the biology of a pest or of a natural enemy that can be used in a control programme, including genetic and behavioural techniques. Genetic engineering offers great scope for further developments in these areas, should the questions about the release of genetically modified organisms be answered satisfactorily (section 3.7). We review some of the major techniques below.

5.7.1 Inundation using pathogens

A number of pathogens have been used against weeds, insects and vertebrates: perhaps the most notorious is the myxomatosis virus, derived from Brazil, and used with limited success to control the rabbit in Australia, New Zealand and the UK. Table 5.5 summarizes the range of microbial pathogens which have been used against insect pests and their principal modes of action. The success of some of these has promoted their commercial production. One of the most important examples is *Bacillus thuringiensis*, (commonly abbreviated to *Bt*), a bacterium that produces a toxin lethal to a range of insects. One strain, HD-1, has been successfully used for some time against lepidopteran pests. The protein-aceous toxin (δ-endotoxin) is produced at sporulation by the bacterium, in

Table 5.5 The range of microbial pathogens used against insects (from Payne, 1988)

Group	Example	Mode of action	Specificity	Advantages	Disadvantages
Bacteria	*Bacillus thuringiensis*	Produces toxic proteinaceous crystal (δ-endotoxin) which attacks the gut epithelium and cause muscular paralysis. Death between 30 min – 3 days	Generally order-specific	Easy to grow aerobically on cheap substrate. Spore-forming – easy to introduce	Non-infectious – requires reapplication. Bacteria killed by ultraviolet light. Needs to be applied when larvae are feeding actively
Viruses	*Baculoviruses*	Infection of gut – death in 3–4 days	Species-specific within the insects – some with a larger host range	No common biochemistry with plant or vertebrate viruses	Although infectious, normally require repeated applications – when larvae are feeding. Difficult to produce
Fungi	*Verticillium lecanii*	Insect dies as hyphal growth and toxin production spreads throughout haemocoel. Death after 4 days	Generally order-specific, others with larger range	Easy to produce through fermentation. Easy to distribute	Require high humidity for spore germination. Reapplication necessary
Protozoa	*Nosema locustae*	Parasitizes gut or fat bodies – slows development and lowers fecundity	Generally order-specific	Easy to apply as spores to food. Vertical transmission (parent to offspring)	Low lethality

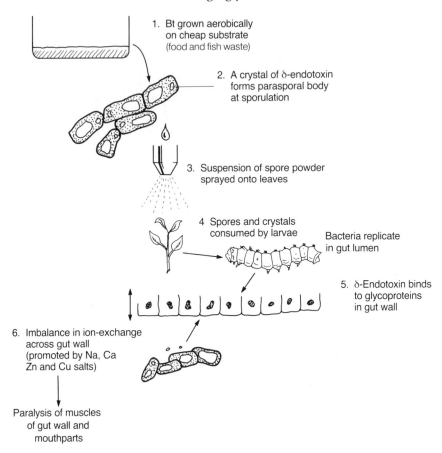

1. Bt grown aerobically
   on cheap substrate
   (food and fish waste)

2. A crystal of δ-endotoxin
   forms parasporal body
   at sporulation

3. Suspension of spore powder
   sprayed onto leaves

4 Spores and crystals
  consumed by larvae

Bacteria replicate
in gut lumen

5. δ-Endotoxin binds
   to glycoproteins
   in gut wall

6. Imbalance in ion-exchange
   across gut wall
   (promoted by Na, Ca
   Zn and Cu salts)

Paralysis of muscles
   of gut wall and
   mouthparts

Figure 5.10 The production of the insect pathogen *Bacillus thuringiensis*, its application and mode of action within the insect gut.

the form of a crystalline parasporal body (Figure 5.10). There are several varieties of this protein produced by around 20 serological types of the bacteria (Payne, 1988). Different varieties are effective against different insects – HD-1 is var. *kurstaki*; var. *israelensis* is effective against mosquitoes and blackfly larvae (Simulidae). Although they are easy to apply, these microbial insecticides are non-infective and need to be re-applied as with a chemical insecticide. Viral and fungal insecticides also suffer from low infectivity (Table 5.5). In contrast, *Bacillus popilliae* used against the Japanese beetle (*Popillia japonica*), is able to remain viable in the soil, probably because it is shielded from UV radiation.

Parasites can be important as vectors of infection in all organisms. *Heterorhabditis* is one of several species of nematode that spend one larval stage in the haemocoel of insects. The roundworm carries symbiotic bacteria (*Xenorhabdus*) which are released and proliferate in the insect tissues. The nematode feeds on the bacterial growth and the insect succumbs to septicaemia. Ten days after entering the host, large numbers of infective *Heterorhabditis* are released to the outside. These are available commercially against insect pests of the soil (Payne, 1988); *Romanomermis culicivorax* is sold as 'Skeeter Doom' in North America, used to control mosquito larvae in fresh water (Horn, 1988). Nematodes have been used as control agents in their own right – galls induced by roundworms in the weed *Acroptilon repens* are collected and crushed as a source of the parasite for control in the former Soviet Union (Harris, 1986).

The virulence of both the pathogens and nematodes mean they can be used as inundative control agents, although there is usually a delay between the application and the death of the pest (Figure 5.1). This can lead to less predictable results in the programme and is one reason for their slow uptake by agriculture (Payne 1988). They might also be used as a preventative measure as long as any chemical treatments do not reduce the pest population below a critical level for the pathogen to persist (Horn, 1988). The great value of these methods is their specificity and low toxicity to non-target organisms: the scope for improving the speed of action of these agents is large (Payne, 1988).

5.7.2 Sterilization techniques

Some very direct attacks on the reproductive potential of a pest are possible, lowering r for its population. The most impressive example is the release of artificially sterilized males to control the spread of the screwworm (*Cochliomyia hominivorax*) in central America, a fly which deposits eggs in the open wounds of mammals, including man. This represents a particular threat to cattle and, since each animal has a high value, the population level at which the screwworm passes its EIL is very low. The detection of a single screwworm usually initiates control measures.

This technique has been particularly successful with this pest because the female fly mates just once and has a low rate of dispersal. The flies are easily cultured in the laboratory and the smaller male pupae are readily separated from the females. These are then sterilized with a low level of radiation. This has no effect on their capacity to mate with wild females: the female does not distinguish irradiated males and there is therefore no selection against them. Large numbers of males are released to compete with the wild males for partners, resulting in a major reduction in the fecundity of the population. A total of 180 million sterile males has been released over an extended period, eliminating the screwworm in southern USA and Mexico by the mid-1970s (Horn, 1988).

Because of its potential impact on local economies, the appearance of small numbers of screwworms in north Africa in 1988 led to a release programme being initiated there. The fly was probably introduced in infected meat from South America and accounted for around 12 000 animal deaths in Libya. A release of over a billion sterile males, bred in a Mexican factory, lead to the demise of the screwworm in the area by 1991.

5.7.3 Semiochemicals

A range of chemicals is produced by some organisms to control the behaviour of other individuals or of other species. These are collectively known as **semiochemicals**: some are part of an organism's own defences against attack, others attempt to attract attention.

The secondary chemicals produced by plants to influence insect behaviour are the most obvious examples. The fragrant scent of a flower, and other, not-so-pleasant smells, are used to attract insects for pollination. But semiochemicals are also used to dissuade attention. Along with the physical defences against herbivory, many perennial plants protect their tissues with toxins and repellents. For slow-growing plants, it can be prudent to invest in a range of non-metabolic chemicals, preventing excessive grazing or browsing and extending the productive life of the leaves or other tissues. We often make use of their insecticidial or other properties – pyrethrum and nicotine are two examples. Such chemicals have a range of effects on insects – xanthotoxins in the Umbelliferae interfere with DNA synthesis, while bracken (*Pteridium aquilinum*) is one of several plants that produces insect moulting hormone. Nevertheless, some insects actually thrive on these secondary chemicals and can use them as nutrients, while others are concentrate the repellent in particular tissues as part of their own defence mechanism. The beetle *Chrysomela aenicollis* for example, produces a defensive secretion from the toxin, salicin, it derives from its food plants, the willows (Harborne, 1988).

In contrast, predators may use such compounds or their breakdown products to detect their prey. A chemical used by a predator or parasite to locate its prey is called a **kairomone**, and these have also been used in pest control. A breakdown product of tryptophan is produced by aphids in their honeydew. 'Artificial honeydew', (hydrolysed yeast and sucrose), is a synthetic kairomone used in California to attract natural enemies, including ladybirds, to fields of lucerne. These predators will remain in the fields even if no aphids are present (van Emden, 1988). Some natural enemies also have to detect a chemical characteristic of the host plant, a **synonome**, to be attracted to the site. *Diaretiella rapae*, a parasitoid of an aphid of Brussels sprouts, needs the stimulus of mustard oil from these plants to attack: in cultivars where levels of the synonome are low, biological control will fail (van Emden, 1988)

The **pheromones** produced by a female insect to attract mates can also be used to attract natural enemies to an infested area. More often they have been used to make population estimates of both pest and predator, and as an aid to judging the best time for insecticidal spraying. Occasionally pheromone traps dosed with pesticides are used – malathion traps are commonly used against the oriental fruit fly to protect the citrus orchards of California. In the few cases where an artificial pheromone is cheap enough to produce, it can be used to 'confuse' the male, saturating an area with the signal, and preventing the male finding the female. This is only likely to be effective at low population densities.

Insects' hormones, which are not strictly semiochemicals, have been used in control programmes to impair some aspect of the animal's development. Like pheromones, these have the advantage that they can be highly specific. Other insect hormones will disrupt moulting and some metamorphosis. A number of these are commercially available and have been used with some success against biting flies. Diflubenzuron is used to control the gypsy moth and cotton boll weevil in North America, by inhibiting synthesis of chitin, the main structural component of the insect exoskeleton (Horn, 1988).

Even with these treatments, there is evidence that insects have been able to develop resistance against excessive levels of hormones and pheromones. A tolerance to the signal from excess pheromones has developed in laboratory populations of cabbage looper moths and resistance to the effects of juvenile hormone is known in mosquitoes (Pimentel *et al.*, 1984).

5.7.4 Resistant plants

Pest control using resistant strains of plants has a long history. The grapevines of France were changed almost completely in the middle of the 19th century after many vineyards had been destroyed by the homopteran bug, *Phylloxera*. The various grape varieties are now grafted onto a resistant wild root stock imported from America.

Most cereal plants grown in the USA have some form of resistance bred into them, and there are cultivars that are resistant to over 25 different insect pests (Pimentel *et al.*, 1984). Improving plant resistance using conventional breeding techniques can now be supplemented by the newer techniques of genetic manipulation, providing a much larger genetic library from which to select characters. Novel defences against insect attack can be incorporated into the genotype of the plant (Table 5.6).

These new techniques can also improve the targeting of a pesticide. The δ-endotoxin gene has been incorporated into *Pseudomonas fluorescens*

Table 5.6 The range of defence mechanisms that have been incorporated into plants against insect attack

Plant	Mechanism	Author
Cucumber	Breeding the kairomone cucurbitacin out of the plant to prevent attack by cucumber beetle	Horn, 1988
Cucumber	Lack of hairs help parasite *Encarsia formosa* find whitefly larvae (pest)	Samways, 1981
Potato	An abundance of glandular hairs inhibits aphid attack without affecting natural enemy attack rate	van Emden, 1988
Tobacco	Incorporates trypsin inhibitor of the cowpea (*Vigna unguiculata*) to control herbivory by a wide range of pests – effective anti-metabolic agents	Hilder *et al.*, 1987
Tobacco	δ-Endotoxin gene incorporated into plant, effective in control of lepidopteran larvae *Manduca sexta*	Vaeck *et al.*, 1987
Rice, corn	Deficiencies in particular amino acids of particular varieties, or low levels of nitrogen discourage attack by several insects	Pimentel *et al.*, 1984

which can then be used as a soil dressing to protect plants from soil-inhabiting lepidoptera. *Pseudomonas* was chosen because it is non-pathogenic to humans and is relatively long lived. Inserting the gene into the bacterial chromosome ensured that there was less chance of it being transferred to other species (Payne, 1988). However, although the scope for further developments in this area is immense, so are the problems about regulation, release and the patenting rights of companies developing them (Meeusen and Warren, 1989).

Nor is resistance the complete answer. The monocultures of cereals and other crops which dominate much of western agriculture have little genetic variation. Under these circumstances, new races of pest can develop rapidly, overcoming any plant resistance. Varieties of wheat resistant to stem rust (*Puccinia graminis*) have perhaps five years of useful life before this resistance is overcome (Pimentel *et al.*, 1984). Similarly, microbial pesticides, applied continuously, can produce the same effect. *Plodia interpunctella*, a lepidopteran pest of stored grain, is known to develop resistance to δ-endotoxin within a few generations when exposed continuously (McGaughey, 1985).

Table 5.7 Control of pest population using different culture strategies

Technique	Mechanism
Tillage	Ploughing exposes soil pests to predation, desiccation and freezing kills weeds but can exacerbate soil erosion
Rotation	A two-year rotation of corn and soyabean prevents build-up of major plant-specific pests (corn root worm, wireworms, etc.)
Use of trap crops	Trap crops attract pests away from main crops which can allow for intensive insecticide spraying. The same effect can be achieved by planting an early crop, which is then sprayed
Weeding	Remove competitors for plant nutrients for crop by use of herbicides
Not weeding	Encourage natural enemies of pest species on other plants. Aphid numbers on cereals and brassicas may be reduced by leaving uncultivated strips at the margin of fields or leaving portions of it unsprayed. This can provide alternative prey for generalist predators to overwinter or survive periods of low pest numbers
Mixing plants	Planting two plants together can slow movement of pests between plants and lower overall pest population. In some cases, however, it may increase pest problems
Timing of harvest	Planting and harvesting can be timed to avoid the feeding stage of pests with some crops
Supplemental feeding	Spraying crop with nutrients (sucrose solutions) to sustain a natural enemy
Removal of rubbish	Removal of fallen fruit and other plant debris can reduce sites for pests
Balancing fertilizers with weeds	The level of fertilizer applied can determine the level of fungal attack in many cereal crops and needs to be matched to plant demand and the water content of the soil

5.7.5 Cultural control of pests

Most intensive agriculture attempts to maintain a monoculture during the growing season so that energy and nutrients applied to the crop nourish only these plants. The pest and the weed compete for these resources or feed directly on the crop, lowering yields. The farmer can modify his cultural methods to reduce the opportunities for the pest or increase those for its natural enemies (Table 5.7).

The methods used depend not only on the crop being grown, but also the scale on which it is cultivated. The need for mechanization might demand that fields be large and pest control itself then becomes mechanized. The retention of strips of uncultivated vegetation or parts of the crop which never receive pesticide treatment can preserve natural enemy populations. In the UK, uncultivated strips or grass banks have been evaluated as 'island' habitats to encourage generalist predators against aphids in arable fields. The density of predators was found to increase substantially over two years as the vegetation became established (Thomas *et al.*, 1991). On smaller plots, it is easier and more economic to adopt other approaches – weeding brussels sprout plots may increase the pest problem by removing refuges for the natural enemies (van Emden, 1988). One method of controlling the grape leafhopper *Erythroneura elegantula* in California is to encourage the native parasitoid *Anagrus epos* (Messenger, 1975). The parasitoid attacks the egg, but the pest overwinters as an adult, using an alternative leafhopper, which lives on blackberry, as prey. By allowing the blackberry to flourish in the hedgerows between vineyards, a large population of *Anagrus* is available at the start of each season and effective control is achieved.

An interesting example of how complex some of these relations can become is provided by Cook (1988). Fusarium root rot is a disease of winter wheat in the drier areas of the western USA, caused by *Fusarium culmorum*. Excessive additions of nitrogen, beyond levels that the wheat can use, encourage the pathogen. The first method of control is thus to limit fertilizer application strictly to that required. Additionally, leaving standing stubble after harvesting allowed colonization by other fungi which prevent *Fusarium* becoming established. In contrast, the control of *Pythium* root rot in the wheat of wetter areas is best achieved by burning or deep-ploughing the stubble (Cook, 1988).

5.8 CHEMICAL CONTROL AND INTEGRATED PEST MANAGEMENT

Biological control measures comprise a very small fraction of the total effort in pest control. Chemical techniques have clear advantages – rapid knock-down of the pest, cheap to produce and usually easy to apply. As we have learnt more of their mode of action and ecological properties, so we have produced increasingly selective pesticides, with a low persistence and low impact on other biota. A good example is the carbamate insecticide, pirimicarb. This is highly effective against aphids and some Diptera, but at normal levels of usage is non-toxic to several of their natural enemies including coccinellid beetles, lacewings (Neuroptera) and aphid parasities (van Emden, 1988). The specificity of established pesticides is also being improved. Lacewings, an important predator of aphids, are able to hydrolyse pyrethroids and a study of this process has improved the selectivity of the synthetic pyrethroids. Deltamethrin is over 1500 times more toxic to the aphid than the lacewing (Pickett, 1988) and the level of application thus determines the selectivity of the insecticide.

Set against the use of pesticides are less obvious costs, including their effect on the health of people exposed to excessive levels. There is a substantial number of fatalities each year from their misuse (Pimentel *et al.*, 1984). Beyond the expense of monitoring the residues from their production and use, there is frequently an environmental impact which is less easily quantified, especially on non-target species. Additionally, there are now over 400 insect species which are resistant to one or more insecticide (Horn, 1988), while to develop a new pesticide may cost up to £13 million (Meeusen and Warren, 1989). This figure takes no account of the hidden costs of its use, yet the pesticide may have a limited effective life. Alternative methods now compare favourably with these develop-ment costs – introducing a gene of δ-endotoxin into a plant variety would perhaps need £500 000 (Meeusen and Warren, 1989).

Unless a pest is rapidly eradicated or perpetually depressed by a natural enemy, there is unlikely to be a permanent resolution of many pest problems. A pest will use the genetic variation at its disposal as effectively as we apply these technologies to control them. Certainly many rapidly reproducing pests with *r*-selected life histories have developed resistance to a range of pesticides, and may well adapt to even the most complex of control strategies.

Nor is biological control always the answer. In contrast to the rapid control exerted by most pesticides, the time lag inherent in most biological methods make them unsuitable for emergency control. They are often perceived to be an ineffective measure for this reason (Jutsum, 1988). There is also the risk of introducing an additional pest which attacks non-target species, or a genetically engineered organism that has some unforseen impact on the ecosystem.

In reality, the choice is rarely between chemical and biological methods. Instead, an **integrated pest management** (IPM) programme is used which incorporates aspects of biological, chemical and cultural controls into one coherent strategy. Here the aim is to establish long-term control of the pest, based largely on natural enemy and cultural controls, but using pesticides when needed to aid regulation. IPM may require specific cultivation techniques to encourage the natural enemy and dissuade the pest – by avoiding extensive monocultures and introducing some heterogeneity into the habitat to provide patches and refuges. This can also extend to growing early season 'trap crops' to attract the pest away from a later main crop (Table 5.8).

The integration of the various methods is the key to the success of these

Table 5.8 Integrated pest management for the cotton boll weevil (*Anthonomus grandis*) (after Newsom, 1975)

1930s and 1940s
1. Use of early fruiting varieties to produce crop in shortest period
2. Use of early sown trap crops to catch overwintering weevils early in season, which were then destroyed with calcium arsenate dust
3. The insecticide was applied to spots to kill the highly localized patches in the main crop
4. Regular monitoring for rapid identification of action thresholds
5. Destroying crop residues to starve overwintering weevil population
6. Use of land clearances and controlled burning to remove hibernation habitats

1950s and 1960s
Use of chemical control by weekly spraying with DDT

1970s and 1980s
1. Lower overwintering population by
   (a) use of insecticide and defoliants at the end of the harvest
   (b) rapid harvesting of crop
   (c) trapping overwintering weevils using pheromone
2. Trap crops used early in season to catch overwintering weevils which are then destroyed using insecticides. This suppresses the population so only the third generation is likely to pass EIL
3. This lack of spraying during the growing season allows native pests of the cotton to be controlled by their natural enemies
4. Use of fast growing and resistant plant varieties
5. Use of insecticides in fields with high densities only (spot-spraying)
6. Chemically sterilizing weevils *in situ*

---

programmes. The experience of growers and past attempts at control become invaluable, such as those shown in Table 5.8.

The cotton boll weevil arrived in the USA in 1892 and by the early 1920s something approaching an integrated pest management scheme was in place. This exploited one weakness of the pest – its reliance on an overwintering population to re-infect the crop in the new season. A number of cultural and chemical controls were used to attack the overwintering population.

However, the use of calcium arsenate was not favoured by farmers, and the whole scheme was abandoned with the arrival of DDT in the late 1940s. This had the benefit of killing both the boll weevil and a number of other pests. Given its cheapness, DDT tended to be used to excess – many cotton growers were spraying their crops weekly in the 1950s. By the 1960s, widespread resistance to this and a number of other pesticides meant that control of the boll weevil was being lost.

The wheel turned full cycle: an integrated pest management pro-gramme was adopted from the 1970s, using chemical, biological and cultural controls to attack the overwintering population (Newsom, 1975).

Clearly the pesticides used should not interfere with the natural enemy – May and Hassall (1988) review models in which the natural enemy is affected by the pesticide used, and show how the timing of the spraying can determine whether pest numbers increase or decrease as a result. In some agricultural ecosystems spraying can be used to remove competi-tors or predators of the introduced natural enemy. For example, low doses of DDT were used to selectively kill coccinellid beetles feeding on the scale insect *Dactylopius* introduced to control prickly pear (*Opuntia megacantha*), increasing the effectiveness of the control agent (Samways, 1981).

An effective model allows us to define the time of year or the stage in its life cycle when a pest is most vulnerable to attack by a natural enemy. The model provides testable hypotheses which can aid our understanding and refine a control programme before it is initiated. A number of models are now used as predictive tools to aid decision-making: detailed computer simulations are used to control the pests of alfalfa, citrus, cotton, apples and soyabean (Horn, 1988).

Long-term pest management can only be achieved through extensive research into the life histories and population dynamics of the chief players. In controlling one pest, there is always the possibility of a second pest becoming more important. IPM requires treating the system as a whole to understand the implications of any action. Although reduction-ist methods will have been used to isolate the key factors and their interactions, a fairly complete description is needed for confident forecasting. A fully tested model can then allow the use of indicators to trigger control actions – for example, the appearance of webbing by the spider mite *Tetranychus urticae* in a glasshouse means that it has entered a dispersive phase. Then it is no longer under the control of the predatory mite *Phytoseiulus persimilis* and an acaricide is applied to the affected plants (Samways, 1981).

Summary

Pest and weeds are commonly alien species or sometimes native species released from the mechanisms that ordinarily keep their population growth in check. They achieve pest status when the costs of their control can be offset by the economic damage they incur. Many animal pests have a life history typical of *r*-selected species – small-bodied animals with short generation times and rapid reproduction (high *r*). In contrast, many weeds are a problem because of their competitive strategy. Insect pests typically show epidemic population growth when they escape the equilibrium density imposed by a natural enemy. The aim of biological control is to constrain a pest below its economic injury level, by depressing its population using a natural enemy. Long-term control requires that both the enemy and the pest should persist and the stability of their coexistence depends upon some pest individuals escaping predation. Simplifed models suggest that a specialist natural enemy (especially insect parasitoids) are likely to be most effective in relatively constant habitats, whereas more generalist predators will be more effective in ephemeral habitats which suffer rapid pest outbreaks. A range of other biological control methods have been used to control pests using various aspects of their ecology, including competitive exclusion, the use of resistant plants and a variety of genetic methods. These can all be used in conjunction with chemical methods in an integrated pest management programme.

Further reading

Horn, D.J. (1988) *Ecological Approach to Pest Management*, Guilford Press, New York.

Pimentel, D. (ed.) (1975) *Insects, Science and Society*, Academic Press, New York.

Wood, R.K.S. and Way, M.J. (eds) (1988) *Biological Control of Pests, Pathogens and Weeds: Development and Prospects*, Royal Society, London.

'Islands' of *Salicornia* in the mudflats of La Capellière nature reserve, part of the Camargue wetlands in the Rhône Delta in southern France.

6

Conserving communities

In the industrialized world, the perception of our role on the planet has changed dramatically since the end of the Second World War. In less than 50 years, man as pioneer and tamer of the wilderness has been replaced by an animal conscious of the limits to its environment. The rate of loss of natural habitats is the clearest sign of this species living beyond its means. Now we begin to understand that we have to be custodians of the planet's resources, and especially its biological diversity.

Extinction is part of the natural process of evolutionary change and some species would disappear from the globe with or without human help. But we have undoubtedly accelerated the process: counting only the large and spectacular, we are responsible for the loss of 63 species of mammal and 88 species of bird in the past 400 years (Diamond, 1986). Perhaps around 10% of the world's plant species are now threatened with extinction (Prance, 1991).

One of the central problems is deciding what should be conserved. People value particular features of the environment differently and each person will make different demands upon the available resources and habitats. The applied ecologist has to produce rational methods of deciding between competing claims. Very often our choices are limited as much by history and economics as ecology and geography. In Europe, for example, there is little scope or space for establishing extensive nature reserves, and instead we have to select and manage those areas that have escaped significant modification. Elsewhere, space may allow for a more inventive policy, where reserve size and location can be determined, at least in part, by ecological principles. Nevertheless, the ecologist is just one of many voices making claims on the wilderness, and the final decision is rarely based on ecological theory alone.

In an ideal world, the size and location of a reserve will be determined solely by the objectives of the conservation exercise. Setting up a reserve has two possible aims – to conserve an endangered species or to conserve a community. Usually, protecting a single species means that its habitat and community is also protected; indeed, decisions about conserving communities are often made according to the requirements of one or a small number of species, perhaps as indicators of the larger community. Although it is typically the large or attractive species that attract the public's attention, very often the community of which they are a part is of greater ecological significance. Ecologists continue to face conflicts of interest and difficult choices in conserving species and habitats and, in making these decisions, we need to be clear about the values we place on each.

The destruction of a habitat leads to natural communities becoming fragmented and confined to smaller and smaller areas. Every major terrestrial community has been affected in this way, with the distances between the isolated habitats increasing as the fragmentation proceeds. In some cases, setting aside a reserve may itself lead to further fragmentation if unprotected areas are exploited, creating a more hostile environment between reserves. These fragmented habitats have been compared extensively to the isolated communities on islands. Island biogeography theory has sparked much argument about the relationship between habitat area and the number of species it contains, and whether this can be used to determine the optimum size of reserves. The same theory has been used to predict how species number will decline as the area of a habitat shrinks. Certainly, a knowledge of the effects of insularization on populations and the species composition of a community is crucial for proper reserve management and selection.

This chapter reviews the theories that attempt to relate species number to the size of a habitat. We begin by briefly considering the designation of a rare species, and how fragmentation of its habitat may lead it to become endangered. Species–area relations can also be used to predict how species number might decline with a contraction of the habitat. The diversity of a community is also closely related to the scale and frequency of its disturbance, and we review the important role this can have in reserve management. Finally, several methods of selecting reserves are reviewed, concentrating on wetlands. These are species-rich habitats which naturally form isolated communities. As we shall see, a simple measure of habitat area or a count of species number is invariably inadequate. More elaborate methods have been developed, some of which attempt to quantify social and economic interests in conserving a habitat.

6.1 THE DISTRIBUTION OF RARE SPECIES

Prime factors limiting the abundance and distribution of a species are the availability of a suitable habitat, the species' capacity to move between habitats, and its rate of population growth. The term **'rarity'** implies a species with a small population, but it may also refer to one with a restricted distribution. Rarity is thus a composite term and it can take

Table 6.1 A classification of rarity based on the distribution and abundance of species (after Rabinowitz *et al.*, 1986)

		Geographical distribution			
		Wide		Narrow	
	Habitat specificity	Broad	Restricted	Broad	Restricted
Local population size	Somewhere large	58	71	6	14
	Everywhere small	2	6	0	3

different meanings for different species. A rare species can be abundant in just a single location, whereas some other species may be locally rare, although widespread.

In their survey of the British flora, Rabinowitz *et al.* (1986) used a three-way classification to partition plants into one of eight categories (Table 6.1). Of the 177 species, 160 could be divided according to their geographical distribution, their habitat specificity and their local population density. Common species are abundant over a large area, in various habitats. Rare species have some combination of a narrow geographical range, a restricted habitat or a locally small population, producing seven types of rarity. Most of the British flora are classified as rare according to this system, but this is not surprising – within most natural, undisturbed communities, most species are represented by just a few individuals, with only a small number of abundant species.

Some organisms have naturally low densities because of the area they need to supply their resources. This is most obvious with predators: the New Guinea harpy eagle (*Harpyopsis novaeguineae*) requires a large territory in which to feed (Diamond, 1986) and its population is ultimately limited by the size of its habitat. Some rare species are highly specialized, confined to very specific habitats, whereas others are locally rare, seldom occurring in some regions because they are close to the edge of their range. Both of these factors were true of the large blue butterfly (*Maculinea arion*) in England, where it recently became extinct. The butterfly requires a particular plant (*Thymus praecox*) and ant (*Myrmica sabuleti*), both of which are part of the diet of its larvae. Elsewhere, in the warmer parts of Europe, the ant is more abundant and so is the butterfly (Thomas, 1991). In other cases, the genetic history of a species may also reduce its capacity for population growth, irrespective of the size or nature of its habitat – as with the cheetah (Figure 4.21).

Some species are never abundant, even in the most established of communities. Tropical forests are one of the most diverse of terrestrial communities – occupying just 7% of the land area of the planet, they hold around 50% of its living species. Yet the population density for many species within an area of tropical forest is typically very low (Prance, 1991).

The abundance of different trees in these forests has been studied by Hubbell and Foster (1986) on Barro Colorado island, in central America. This was created by the digging of the Panama Canal in 1914 and has had over 70 years for the community to adjust to its isolation from the rest of the forest. They surveyed 238,000 trees (with a trunk diameter greater than 1 cm) on the island, and looked at the habitat requirements of both the rare and common species. Nearly 40 000 individuals belonged to just one species (*Hybanthus prunifolus*). The average density for most species was equivalent to 4711 individuals per hectare, whereas one-third of all species were termed rare, having less than 50 individuals in the 50 ha plot. Hubbell and Foster classified each species along two habitat gradients: firstly by their presence in five habitat categories, largely reflecting preferences in soil moisture, and secondly according to their 'regeneration niche' (section 7.1), scoring their occurrence in different gaps in the canopy – in recent gaps with full sun, or in older and smaller gaps with less sun (Figure 6.1). The rarer species usually had clear preferences, specialists that required specific conditions. Large gaps in the canopy are rare on Barro Colorado island and species favouring the open sun were relatively uncommon. In contrast, the most abundant trees were generalists, common in all conditions.

This survey represents a single snapshot of the forest, and of a community that is not in equilibrium (Hubbell and Foster, 1986): a number of species appeared to be rare generalists, perhaps because they were unable to regenerate under the prevailing conditions. There is evidence of a turnover in these species on the island, with rareness largely a result of low rates of colonization or of opportunities for establishment.

Along with the low population density of many species in tropical forests there is a large number of **endemic** species – native species found only in these locations. In the Pacific coastal forests of Ecuador, for example, between 40 and 60% of the plants are endemic, while the 0.8 km^2 of the Rio Palenque reserve has over 250 endemic plants (Prance, 1991). Islands also feature high levels of endemism; here isolated populations may evolve new species in the absence of significant migration and genetic exchange with other populations. Some islands, especially large tropical islands, represent the only significant habitat for many groups: Madagascar has levels of endemism of around 80% in both its plants and animals, although today just 10% of its natural vegetation remains (Prance, 1991).

A habitat need not be destroyed completely for species to disappear; some alteration of the interactions between species can have the same

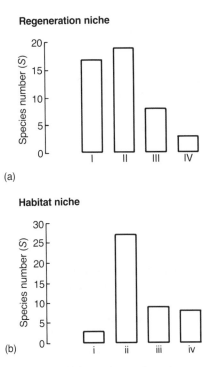

Figure 6.1 The incidence of rare tree species found in two niches on Barro Colorado island in the Panama Canal. Histogram (a) divides species according to their regeneration niche – the conditions necessary for their establishment: this is measured by their abundance in gaps of different sizes (I = full sun, II = partial sun, III = indifferent, IV = shade). In (b), species are scored according to their location on slopes (i = slope, ii = indifferent, iii = plateau, iv = swamp/ravine), probably reflecting soil moisture preferences in their habitat. For the most part, the rare species have specific preferences, especially in their regeneration niche. In contrast, common species are generalists which are predominantly scored in the indifferent categories. (After Hubbell and Foster, 1986.)

effect. The natural regeneration of the Brazil nut (*Bertholletia excelsa*) has been hindered by the decline of the one species of bee able to pollinate it, and of the agouti, a large rodent which buries the nuts in caches in the soil (Prance, 1991). Similarly, a distinct sequence of loss of Amazonian bird species directly follows the decline in their habitat quality (Terborgh and Winter, 1980). Particular guilds of birds are at greater risk than others: raptorial predators and large birds feeding on fruit and nectar disappear first as their habitat becomes fragmented. Notably, the more resistant bird species on both tropical and temperate islands are strong flyers, able to colonize islands after a local extinction. The same is found in the birds colonizing fragmented forest habitats.

Species at risk often appear to be specialists, K-selected species (section 5.1) with a small capacity for population growth and limited powers of dispersal. Although it may be found over a wide area, a species with a low population density throughout its range will suffer local extinctions. Then its capacity to move between patches and recolonize them becomes critical, especially if this habitat is fragmenting and contracting. Species with poor powers of dispersal are often amongst the first to be lost when their habitat becomes divided.

6.2 FRAGMENTATION AND INSULARIZATION

Fragmentation reduces the overall area of a habitat and increases the distance between the remaining patches. The smaller area lowers the carrying capacity of the habitat for resident populations (section 4.2), and the increased isolation limits the movement of individuals between the fragments. The net effect is a reduction in the amount of genetic variability within a population, itself leading to the problems of inbreeding depression. In this way, fragmentation can lead a susceptible species into an extinction vortex (section 4.7).

A minimum viable population is required for the long-term survival of a species and that implies a minimum area (Shaffer, 1987): in North America, 50 hectares are needed by a pair of red-cockaded woodpeckers to supply the food needed to raise a successful brood (Reed *et al.*, 1988). The minimum area also depends upon the quality of the habitat: a viable population of the large blue butterfly would be around 400 adults, with around 2500 usable *Myrmica sabuleti* nests; in an ideal habitat this would amount to about 1 ha (Thomas, 1991).

Habitat quality will deteriorate in smaller fragments where there is a large boundary relative to their area, reducing the proportion of undisturbed habitat. Such patches tend to be dominated by 'edge' communities, species that are atypical of the main habitat and which favour conditions at the junction of the two environments (Yahner, 1988). Very small fragments may support no substantial area of the original habitat at all, often because few of its abiotic characteristics, such as the soil or local climate, remain.

The quality of the habitat for many species will also depend upon the presence of other species, without whom the structure of the community may collapse. These **keystone species** play some crucial role within the community, for example, the pollinating insects. If these species are lost, much of the rest of a community will be unable to sustain itself. The breaking of other associations, including the loss of predators or mutualists, can also lead to changes in community structure, and a major loss of species.

The degree of isolation or **insularization** of the habitat is not simply represented by the distance between the patches. It is also depends upon the nature of the intervening environment and the species attempting to cross it. Unless other patches are very close, movement may be limited to the more mobile species, such as birds, insects or windborne seed. An oceanic island and a traffic island are each surrounded by inhospitable conditions, although different groups will cross these barriers with varying degrees of ease. In these examples, each fragment is well defined by its physical boundaries. Elsewhere habitat patches may be determined by less obvious discontinuities, such as a change in moisture, salinity or minimum temperature. Thus a pocket of soil in a rock crevice or a clump of *Salicornia* growing on a hot mudflat can also be represented as habitat islands.

Generally, larger oceanic islands tend to have more species, and a simple relationship between the area of an island and the number of species can be shown for various island types and for some taxa. There are two possible reasons for this: according to one theory, area itself governs species number because island size determines population size (in effect, the extinction rate) and the ease with which islands can be reached by new migrants (the colonization rate). A second theory argues that larger islands support more species because they have a greater variety of habitats. Of course both may apply. Area is indicative of the supply of a range of resources, including food or water, and it effectively summarizes the 'resource space' available for new colonizers or new individuals.

A population that grows slowly or requires a large area to support a viable population may find it difficult to maintain itself in a small habitat for any length of time. Many of our past predictions of the effects of fragmentation and insularization have drawn upon the theoretical studies of island communities. We briefly review these below.

6.3 SPECIES–AREA RELATIONSHIPS

We can demonstrate the effect of area on species number simply by sampling increasingly larger areas of a habitat and recording the number of new species found. Generally, in a uniform habitat, a plot of the total number of species (S) against the area sampled (A) produces a characteristic curve (Figure 6.2) – the increase in S is initially rapid as the most abundant species are recorded, but the rate of increase declines as the sample area gets larger and only the rareties are left to be found.

This curve is also found for particular groups of organisms, although not all. For some there is a straight-line relationship between the number of species and area, such as the passerine birds on the Cyclades Islands of

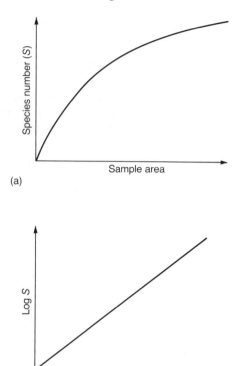

(a)

(b)

Figure 6.2 (a) The rise in the number of species recorded with increasing sample size. Eventually this tails off as fewer 'rareties' are left to be recorded – a large increase in sample size is needed to increase S. Plotted as logarithms (b), this produces a straight line.

the Aegean sea (Simberloff, 1986). In most cases, sampling over a larger range of areas will produce a curve and this tailing off suggests that there may be a maximum number of species for each group in an area. Alternative plots for different groups often include some type of logarithmic transformation – either by relating S to the logarithm of A, or more commonly by transforming both S and A to give the equation for a straight line:

$$\log S = \log c + z \log A \tag{6.1}$$

where S = the number of species at equilibrium; c = is the intercept, the number of species when $A = 1$; z = the slope of the line relating S to A.

The intercept represents the number of species found in the smallest sampling unit (such as single quadrat).

The slope is useful for comparing the rise in species number with area across different habitats. This is the most important term, predicting how many more species would be found by a given increase in area. The value z is specific to a particular group of organisms, say birds or flowering plants, although it commonly falls within a restricted range, between 0.2 and 0.35, for most groups on islands.

In theory, the degree of isolation of the island, the ease with it can be reached, determines z. Fragments of mainland habitats will be colonized relatively easily, so that most will quickly acquire species. The increase of S with area will thus be relatively slight and z small. In contrast, isolated oceanic islands will add to their S more slowly. They are also more likely to lose populations through local extinctions. Consequently we would expect oceanic islands to have fewer species than an equivalent area of mainland. As island size increases, these effects become less severe and large islands will be more able to maintain their species number. Thus S rises more sharply with A on oceanic islands and z is larger. Indeed, observed values for a range of organisms in fragmented mainland habitats are lower, around 0.13, compared with more isolated oceanic islands, with an average of 0.28 (Begon *et al.*, 1990).

In their theory of island biogeography, MacArthur and Wilson (1967) suggested that the relationship between S and A resulted from a dynamic equilibrium between rates of colonization and rates of extinction. This equilibrium is dynamic because there is a continual turnover of species, as established species become extinct and are replaced by new colonists. This leaves $\hat{S}$, the **equilibrium number of species**, roughly constant for an island of a particular size.

Larger islands will have higher rates of colonization, simply because they are bigger targets, more likely to accumulate species. The larger populations they support are less likely to become extinct. Thus the theory suggests that $\hat{S}$ will increase with island area.

Islands that are close to a source of immigrant species will be colonized at a faster rate. Being a short distance from the mainland and its pool of species, these islands stand a higher chance of acquiring sufficient individuals to establish a breeding population (Figure 6.3), even of those species with poor powers of dispersal. Rates of colonization will decline with distance from the mainland, and the theory therefore suggests that $\hat{S}$ will be lower on more distant islands.

The time that an island has had to acquire species will also be important. Early on, when few species have arrived, most new colonists will be the first representatives of their species on the island, so adding to S. Species number rises most rapidly at this stage. Later on, a greater proportion of the mainland species will already be resident on the island, so new arrivals add little to the total species count. The rate of colonization by new species thus declines with time. Similarly, a greater

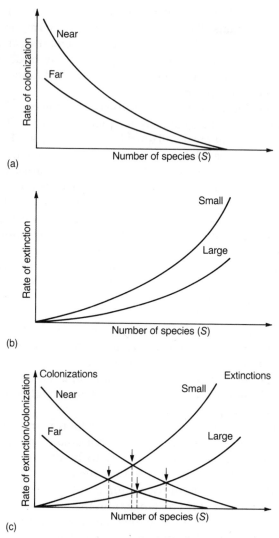

(a)

(b)

(c)

Figure 6.3 The principles of island biogeography theory. (a) Colonization declines as the number of species on an island increases. As species accumulate on the island, fewer mainland species remain unrepresented that are able to make the crossing. Islands near the mainland will tend to acquire more species including those with poor powers of dispersal, and at a faster rate. Larger islands offer a larger target, and so they also acquire species at a faster rate. (b) Extinctions increase with S, as there are more species on the island which could go extinct. Small islands will support smaller populations, and will thus tend to lose species more rapidly. (c) Where the rates of colonization and extinction cross, an equilibrium number of species will be established on the island (arrowed).

number of species on the island provides more opportunities for extinctions to occur. As time goes by, and S increases, the number of extinctions will rise (Figure 6.3).

Rates of colonization and extinction will also depend on the presence of other species. Those arriving later are less likely to find a vacant niche and will fail to establish themselves. Using models of interspecific competition, Case (1990) has shown that the chances of colonization decrease as both the number of competing species and the strength of competition rises. Equally, as species begin to compete for resources so the rate of extinction will increase. This competitive pressure will tend to be greater on smaller islands, with their limited supply of resources.

6.3.1 The species equilibrium

According to the MacArthur–Wilson theory, the equilibrium number of species on an island results from the effect of area and isolation on the rates of colonization and extinction. The intersection of the colonization and extinction curves gives $\hat{S}$ (Figure 6.3), maintained by the loss and gain of species. Such a turnover has been detected on both oceanic and habitat islands, even for highly mobile species – for example, Brown and Dinsmore (1988) demonstrated a dynamic equilibrium for birds nesting in wetlands in Iowa, and could show that $\hat{S}$ was a function of habitat area. There is also evidence that species colonizing an island reach an equilibrium within their niche class (Colinvaux, 1986) so, for example, an island would have a limited capacity for leaf-eating insects or large carnivores.

Given the rates at which species colonize and become extinct on an island, we can predict its $\hat{S}$. At equilibrium the colonization rate is balanced by the extinction rate:

$$C - E = 0$$

where C = the colonization rate per unit time and E = the extinction rate per unit time.

The colonization rate and the extinction rate can be derived quite simply. The number of new species colonizing the island per unit time is:

$$C = \lambda(P - S) \tag{6.2}$$

where P = the total number of species in the mainland pool; S = number of species already on the island (thus $P - S$ = those species yet to colonize); and λ = the average colonization rate per species.

The number of species becoming extinct per unit time is:

$$E = \mu S \tag{6.3}$$

where μ = average extinction rate per species on the island.

At equilibrium the number arriving is balanced by the number lost by extinction:

$$\lambda(P - \hat{S}) - \mu\hat{S} = 0$$

where $\hat{S}$ = number of species present at equilibrium.

This can be rearranged to:

$$\hat{S} = \frac{\lambda P}{\lambda + \mu}. \qquad (6.4)$$

The number of species at equilibrium can thus be calculated given the colonization and extinction rates and the size of the mainland species pool. Using instantaneous rates, the speed of approach to the equilibrium is:

$$\frac{dS}{dt} = \lambda(P - S) - \mu S.$$

This allows us to calculate the number of species present at any time since isolation:

$$S = \frac{\lambda P}{\lambda + \mu}(1 - e^{-(\lambda + \mu)t}) \qquad (6.5)$$

where e is the base of the natural logarithm.

As time (t) increases, so the second term $(1 - e^{-(\lambda + \mu)t})$ approaches unity and the acquisition of new species slows closer to $\hat{S}$. We can use the same method to derive rates of species turnover on an island.

The assumption that the average extinction and colonization rates are a fair representation for all colonists is highly unrealistic (Diamond and May, 1981). Also, this is a non-interactive model because it makes no allowance for the effect of established species on colonization and extinction rates. As such this gives simple straight-line relationships for C and E with S, whereas the concave plots of Figure 6.3 imply some interaction between the colonizing and resident species. Although this lack of interaction is unrealistic it does provide an adequate description of one or two real island faunas, where it has successfully predicted turnover rates (Colinvaux, 1986).

Only a limited number of species would thus be expected to persist on an island of a given size. If for some reason $\hat{S}$ is exceeded, the model predicts that the species number will have to fall back to its equilibrium number, a process termed 'relaxation'.

6.3.2 Relaxation

Islands that were once connected to the mainland may have a residual number of species which is larger than the area can sustain. Over a period

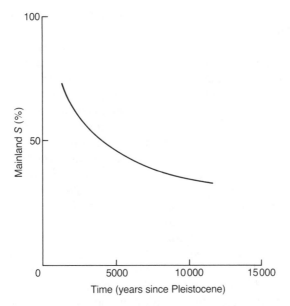

Figure 6.4 The loss of lizard species with time, as islands in Baja California have become isolated since the Pleistocene. The values are for islands with standardized size and latitude, so the rate of loss is purely an effect of the duration of isolation. (After Wilcox, 1978.)

of time this number will fall as the extinctions exceed colonizations, and the island relaxes to a new species equilibrium.

For example, Wilcox (1978) examined the important factors in the relaxation of the lizard fauna of the islands of Baja California. These were originally connected to the mainland of North America as a land bridge, but have become progressively isolated in the last 10 000 years. Not only was the size of the island important for the lizard fauna, so too was its latitude and the time elapsed since separation, the period of relaxation (Figure 6.4). In this case, distance from the mainland was unimportant, as colonization was thought to be insignificant compared with extinction.

Given that there are competitive interactions between the resident species, island biogeography theory predicts a rapid loss of species initially, decreasing exponentially as $\hat{S}$ is approached. By assuming no effective immigration, we can model this loss of species simply as an exponential function:

$$\text{rate of species loss} = kS^n \tag{6.6}$$

where k is called the relaxation parameter, the proportion of species on the island lost with time and n, the exponent, is an integer, usually from 1–4, used to scale the strength of interaction between species. A high level of competition would require a high value of n and thus imply a more rapid loss of species (Soulé *et al.*, 1979).

If we know how long an island has been isolated, the number of species originally present and the number remaining on the island, we can estimate the relaxation parameter (for a given value of n) for that island. With the data for a large number of islands, any relationship between the relaxation parameter and island size can be determined.

Terborgh (1974) measured relaxation for the birds of the West Indies, islands originally joined to the mainland of South America about 10 000 years ago. Terborgh set n at 2 to allow for interactions amongst the existing species, and then estimated k. He then went on to establish the relationship of k to island area. Using this estimate from the West Indies, he calculated the expected loss of bird species on Barro Colorado island in the Panama Canal. The predicted loss was of 16–17 species: the recorded loss due to relaxation was actually shown to be 15.

The same principle has been used to estimate relaxation for the number of large mammals on east African reserves (Soulé *et al.*, 1979). These rates are based on the extinction rates for large mammals on the islands of the Malay archipelago, since their creation by the rise in sea levels at the end of the last ice age. Obviously, applying these rates to the east African reserves has to be treated with caution, but these models predict a rapid relaxation for most of the reserves in east Africa. It is significant, however, that they predict similar values for $\hat{S}$ over the long term, irrespective of whether the interaction between species is severe or not. The average reserve of 4000 km^2 is estimated to lose between 5 and 6 of its 48 large mammal species in the next 50 years (Wilcox, 1980). The calculated values for two reserves are shown in Figure 6.5.

The danger with such estimates is the emphasis they place on habitat area, and there may be circumstances under which a reserve can retain a greater proportion of its species than theory predicts for its area. From his study of isolated sections of the Great Barrier Reef, Goeden (1979) demonstrated that the number of fish taxa associated with a 'patch-reef' was more closely correlated to its variety of habitats than its area. This is one of several examples where the diversity of habitats is found to be a more significant factor than area in determining $\hat{S}$. The evidence from different groups of animals is that area or habitat diversity may each be important in different groups (Connor and McCoy, 1979; Gorman, 1979). Part of the problem is distinguishing the effect of island area from habitat diversity as the two are so closely related. In experiments that have attempted to isolate each factor, the evidence suggests that area alone is still one determinant of the equilibrium number of species (Gorman, 1979). A range of other factors are known to determine $\hat{S}$ on islands, including topography and structural diversity of the island, as well as habitat diversity, all of which will increase with area.

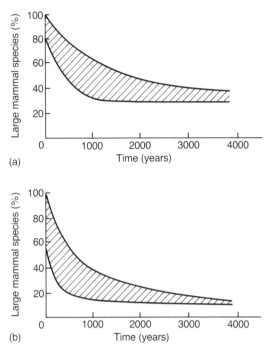

Figure 6.5 The predicted relaxation of large mammals in two reserves in east Africa. (a) Serengeti National Park, $14\,504\,km^2$; (b) Nairobi National Park, $2072\,km^2$. These estimates are based on the loss rates of large mammals with the fragmentation of the Malay archipelago at the end of the last ice age. The two lines show the limits associated with two extreme models, with different levels of species interactions. Over the long term, the predicted effect is the same in both cases: both reserves are too small to conserve many species. The smallest is predicted to lose four-fifths of its large mammal fauna. (From Soulé *et al.*, 1979, with permission from the publisher, *Biol. Conserv.*)

These complicating factors may explain why, in its simplest form, island biogeography theory is often a poor description of data from the real world (Spellerberg, 1991). Additionally, its emphasis on area and isolation limits the application of island biogeography theory to fragmented mainland habitats: the number of species in a habitat fragment will depend on a range of factors, including its history of disturbance, its species composition on insularization, and the conditions required by particular species (Spellerberg, 1991). Also, Spellerberg points out that few mainland reserves are isolated to the same degree as an oceanic island. As long as the surrounding habitat is not completely hostile to its species, many species will readily use the larger landscape, and will move relatively freely between fragments. Similarly, its community will be affected by populations in the surrounding areas. Taken altogether, this means that we should recognize that these theories should not be applied blindly to the ecology of reserves. Yet, in the past, these ideas have led to extended arguments about what constitutes the optimum reserve design.

6.4 INSULARIZATION AND RESERVE DESIGN

The relative ease with which small habitat fragments may lose species clearly suggests that large reserves are to be preferred to small reserves. However, a simple count of species may not always favour a single large reserve and, in some cases, it may be preferable to conserve a large number of smaller reserves. During the 1970s and 1980s a protracted discussion took place between proponents of each strategy. This became known as the SLOSS argument – whether it was preferable to conserve a single large reserve or several small ones.

Various arguments for large reserves were offered: theory suggested that large islands contain more species than an equivalent area of smaller islands because their size offered colonists a larger 'target', a wider range of habitats and lower rates of extinction. This would suggest that large reserves would be more able to sustain populations of species unable to disperse between fragments (Diamond, 1976) and, by maintaining their community structure, would relax to a larger $\hat{S}$ (Terborgh, 1976). On the other hand, some questioned whether smaller fragments do indeed have fewer species, and whether a number of reserves would increase the security of a species, should a disaster strike one of the reserves.

The effect of island size on the number of species was examined by Quinn and Harrison (1988) with a method similar to that used to derive Figure 6.2. Using data from previous studies of various groups on offshore islands, nearshore islands and terrestrial 'habitat islands', they plotted the cumulative number of species against cumulative area for each data set in two different ways. In the first method, they started with the largest island and then successively added smaller islands to get the total species count. This minimizes the fragmentation by scoring the largest islands first, with only new records from the smaller islands adding to the score. In the second, their count started with the smallest island and added successively larger islands (Figure 6.6). Now fragmentation is maximized. If fragmentation depresses species number then the total S should be achieved more rapidly with the count beginning with the large islands.

All islands scored for each study came from a single group to minimize changes in habitat type with latitude. 29 of the 30 studies they examined in this way suggest that more species are found on smaller islands of equivalent area to a single large one. Of the oceanic islands, 8 out of 10

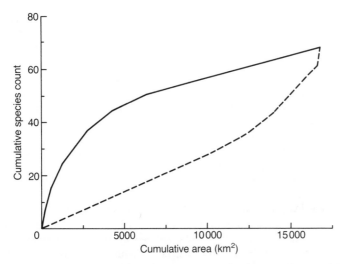

Figure 6.6 Two methods of plotting the number of bird species on the Hawaiian islands with area. The broken line plots the records for the largest islands first and so minimizes the effects of fragmentation; the solid line plots the smallest islands first, maximizing fragmentation. As the lines never cross, a number of small islands always have more species than an equivalent area in a few large islands. (With permission, from Quinn and Harrison, 1988.)

studies (including flowering plants, ferns, birds, bats and insects) show this pattern.

The essential feature of these groups is that few species are common to all the islands, so that some species are only found on one or two islands. Adding more fragments adds proportionately more species than an equivalent increase in area. Higgs and Usher (1980) also suggested that a reduced proportion of shared species for a given value of z could allow a collection of smaller islands to have a higher S. They were able to show this for higher plants in three habitat islands, in this case, disused quarries in Yorkshire. Similarly, in their survey of 22 heathland fragments in Dorset, Webb and Hopkins (1984) found that the number of beetle species actually declined as the size of the fragment increased. It also fell with increasing amounts of heathland within a 2 km radius of a sampling point. In effect, as the fragmentation of the heathland increased so did the number of species of beetles. This was probably because the smaller fragments were subject to more edge effects as non-resident species 'strayed' into the fragments. Quinn and Harrison provide a list of possible reasons why more species might be found amongst a number of small islands (Table 6.2).

Based on this evidence, a number of small reserves, in some circumstances, may be preferable to a single large reserve. In Quinn and

Table 6.2 Possible reasons why a collection of smaller islands may have a larger number of species than a single island with an equivalent area (after Quinn and Harrison, 1988)

1. When the sequence in which species arrive at an island determines the number of species (S), a number of islands gives the opportunity for more 'trials'
2. If a population can exist at various equilibrium densities, a number of islands offers a range of possible population sizes, which can in turn alter the species composition of a community
3. The large edge relative to the area of a small island may allow it to accomodate more species which prefer this habitat
4. Many ecosystems maintain their species diversity by undergoing periodic disturbance. Small islands may be more prone to disturbance
5. A large species pool will reduce the probability of any one species being the first to colonize. In consideration with points 1, 2 and 4, a larger number of islands provides a larger number of possible starting communities
6. Smaller islands may be smaller targets, but they are more spread out. The possibilities of landfall for a migrant may thus be greater than on a single large island
7. Perhaps the small populations that small islands support allow for more rapid differentiation of the population, although the evidence is rather in favour of higher endemism on larger islands
8. If different species go extinct on different islands with a variable species composition, smaller islands will between them hold a larger number of species. If the same species become extinct on an island of a given size then larger islands will have a higher species count

Harrison's study, however, the effect of fragment size was least marked in terrestrial habitats, probably due to a greater movement of individuals between patches, and this is likely to apply to mainland reserves. The essential consideration is the objective of the conservation exercise, whether the aim is to protect a single species or a community. A single species will require a minimum size of reserve to sustain a minimum viable population (section 4.6), or at least be able to move between a network of smaller patches. It is also important to protect a functioning ecosystem, large enough for a viable community to sustain itself. Larger reserves are less likely to suffer from 'edge' effects, where the boundary is dominated by uncharacteristic communities. Soulé and Simberloff (1986) suggest that watersheds make obvious units for conservation in terrestrial ecosystems, in which the inputs of nutrients or pollutants can be readily monitored (section 8.2), and sometimes controlled. What surrounds the reserve will also govern the exchange of species between islands or fragments: many reserves are not encompassed by wholly hostile environments and their isolation is only partial (Spellerberg, 1991).

For single species conservation, large reserves have several advantages.

1. They can confer the adequate long-term protection against chance extinctions which comes with large population sizes.
2. Large reserves are less likely to be destroyed by major environmental events like floods or fire.
3. Large reserves will support larger populations that will be less prone to a loss of genetic variability.

Where individuals can move freely between smaller reserves, some of these advantages may be less marked. Large reserves may also have several advantages where the aim is to protect a community.

1. They have a larger $\hat{S}$ than a single smaller reserve.
2. They have a slower rate of relaxation (Terborgh and Winter, 1980).
3. Some habitats can only exist on larger fragments.
4. Large reserves will have a small perimeter compared with its area. Species which rely on an intact community will be less likely to succumb to edge effects as one habitat gives way to another.
5. Large reserves are required to support larger predators which may be keystone species, necessary for the survival of the community, or species that rely on seasonal or patchy resources.
6. Large reserves will be better able to withstand variations in nutrient flow or pollution insult.
7. Large reserves will be less sensitive to disturbance, necessary in most communities to maintain high species richness.
8. Large reserves will ensure that a succession will lead to a range of habitats, rather than domination by a small number of species.
9. Large reserves are better able to support species which are poor dispersers, species that would be at risk amongst a series of small reserves.
10. The stability of the community in a series of small reserves may depend on the ability of predators and prey to move between unoccupied patches; unless a predator finds its herbivore prey, the latter might quickly decimate the vegetation of a small site (Lawton, 1987a).

A collection of small reserves does have some distinct advantages, especially for the conservation of single species. A number of reserves can protect an endangered species from the spread of a contagious disease or some other catastrophe. The decimation of a population in a single reserve will not then mean the end of the species. The problem of insularization and the movement of individuals between a number of reserves can be ameliorated if the environments separating small reserves are not too hostile to migrants. In an ideal world reserves might be organized with a particular geometry and configuration to maximize this movement, although the real world is usually less uniform than these ideas require (Westman, 1985).

The ecologist can also argue for a land use in the intervening areas which is sympathetic to the movement of particular species between fragments and which causes no further deterioration of the reserve itself. Keeping the reserves close together is most likely to encourage movement between these habitat fragments, as will corridors, such as hedgerows between woodlands (Spellerberg, 1991). However, some of these routes will be selective and will not allow free movement of all species (Soulé and Simberloff, 1986). Additional efforts to move individuals or their gametes between fragments may also be needed (Allendorf and Servheen, 1986). In practice, protecting a single species and a community amounts to the same thing: the integrity of the community has to be maintained if the endangered species is to survive. To some extent, any discussion about reserve design is becoming redundant – 'the historical phase of designing reserves is drawing to a close' (Soulé and Simberloff, 1986). Outside Amazonia there are few areas where the remaining habitats allow for a range of choices.

6.5 INSULARIZATION AND SPECIES COMPOSITION

Several species of bird in New Guinea move between two distinct habitats during their life cycle, breeding at high altitudes but spending their immature stages in the lowland forest (Diamond, 1976). They need habitats that span an altitude range of 16 000 feet and their conservation depends upon both habitats being preserved. Indeed, for many species reserve design has to allow for ease of movement between isolated patches. However, in setting aside reserves there is a danger of protecting only the interests of a few species, usually the most conspicuous members of the community. For others, designating the reserve may increase the risks of extinction, especially if this leads to further fragmentation, without due regard for their requirements or movements. We thus need to consider which species are most likely to be lost with increasing insularization.

A variety of plants and animals are able to travel long distances, but each will require a particular set of conditions if they are to establish themselves in a new habitat. Wilcox (1980) has made some general speculations about vertebrate colonization and extinction on islands. Endotherms have a high energy demand (section 2.3), requiring more food, and consequently have a lower population density in a given area. This is one reason why large mammals are the group most frequently endangered, irrespective of poaching or hunting pressure. In contrast, small ectotherms have a lower energy requirement, a higher carrying

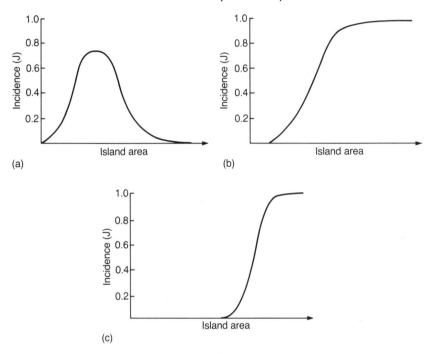

Figure 6.7 Plots of the incidence functions of species on islands, based on Diamond's work on the birds of southeast Asia. These represent the proportion of islands of a particular size with a breeding population. (a) Supertramp species are those able to colonize rapidly, but which will lose competitive battles on larger islands with more species. (b) Tramps colonize most islands and will be able to sustain themselves, perhaps by continual invasions. (c) In contrast, high *S* species are poor colonizers and need large islands and the presence of other species to survive. (After Diamond and May, 1981.)

capacity and, as a result, are less likely to become extinct. This would explain why islands at equilibrium characteristically have a vertebrate fauna dominated by birds and bats, and to a lesser extent by the reptiles and amphibians. Evidence from his study on the lizards in the Gulf of California also tends to confirm such effects (Wilcox, 1980).

Differences in the invasive power of species also emerge from within these groups. Diamond (Diamond and May, 1981) looked at the bird species in the New Guinea archipelago and noted that some species were absent on islands of a certain size. Three basic patterns were observed. Firstly there are those species only found on smaller islands which have a few species: these are termed **'supertramps'** – good colonizers, but with no competitive ability, and typically at the *r*-end of the *r-K* spectrum (section 5.1). As island area increases, $\hat{S}$ rises and the 'supertramp'

Conserving communities

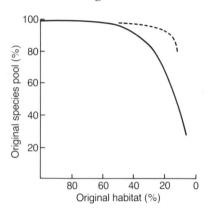

Figure 6.8 The proportion of species remaining in increasingly fragmented heathland habitats. Two extremes are plotted – one for a group of species that requires a large area and has poor powers of dispersal (solid line) and a second group that is more able to disperse, and has smaller area requirements (broken line). Although the latter group survives the initial insularization, species are predicted to be lost rapidly when fragmentation passes a threshold. (After Wilcove *et al.*, 1986.)

species are lost through competition. At the K-end, the competitive species are found on larger islands with a higher $\hat{S}$. In the middle are the **'tramps'** found on both species-rich islands and smaller islands.

These differences can be quantified as **incidence functions**. The incidence (*J*) of a species is the proportion of islands within a size-class that has a breeding population. A plot of these incidence functions against island area (Figure 6.7) indicates the probability of finding a particular species on an island of a particular size. An incidence function may thus give some indication of which groups are most likely to disappear first as habitats become fragmented.

This idea has been used to study the fragmentation of Dorset heathland, an area of semi-natural shrubland in southern England. Heathland developed on freely drained, sandy, acidic soils following forest clearance, a community dominated by shrubs adapted to dry conditions, especially heather (*Calluna vulgaris*) and heaths (*Erica* spp.). Through changes in land use, this once extensive habitat has been reduced to around 450 fragments, totalling just 5% of its original area. Wilcove *et al.* (1986) used a variation of the incidence function to model the loss of species as this fragmentation proceeds. The model allowed for re-colonization of unoccupied patches from both surrounding fragments and a mainland species pool. By using incidence functions, they specified colonization and extinction rates for different species. Two extremes are shown in Figure 6.8 – one group with good powers of dispersal and little chance of going extinct locally, and a second group of poor colonizers which suffered greater local extinctions. Not surprisingly, it is this second group that was the first to succumb to the effects of fragmentation, although beyond a certain point, the more resilient species were also lost rapidly. In the model, insularization was shown to increase extinctions over and above that due to a simple reduction in area, as the increased distances between patches slows colonization. Again, the least mobile species are the most likely to disappear with increased fragmentation. In addition, fragmentation breaks many of the interactions between members of the community which would lead to secondary extinctions.

Fragmentation may be less damaging in temperate communities such as the heathland, because many species exist at higher densities over wider areas than in tropical areas. Many of the species also have greater dispersal powers (Wilcove *et al.*, 1986).

6.5.1 The loss of species through introductions

Extinctions occur not only through the loss of habitat, but also because other species displace them. This has been a common phenomenon associated with the many intentional and accidental introductions made in different parts of the world.

Many intentional introductions have failed, sometimes for very obscure reasons (Simberloff, 1991). In other cases, a successful introduction has had little impact on the host community – overall perhaps just 10% of introductions have actually caused extinctions. Island communities however, tend to be more susceptible to the effects of an introduced species, especially of predators (Pimm, 1986) – rats, cats, mongooses, pigs, goats and snakes have all devastated vertebrate faunas on various islands. Around a half of all extinctions of island birds have been due to introduced mammals (Diamond, 1976), sometimes indirectly – the grazing by introduced goats has probably caused the loss of nectar-feeding birds on Hawaii (Pimm, 1986).

Many of these species have been introduced as part of a biological control programme, although even such planned introductions have a poor record of successful colonization. Of 679 animals introduced into Hawaii since 1890 (nearly all of which were insects), 436 failed to establish themselves (Simberloff, 1991). More than half of the bird introductions to lowland Hawaii since 1860 have failed (Pimm, 1986). This is a general trend found across many islands – successful introductions of birds declines as the number of existing bird species and other resident vertebrates on the island increases. This would suggest that competition is a major cause of failure. Similarly, introduced species which are morphologically similar to resident species are less likely to establish themselves, as are members of the same genus.

The most important determinants of success are the number of existing species and the strength of their interactions. It would seem that these community-level parameters govern establishment rather more than the ecology of the invading species (Case, 1990). Perhaps for these reasons the most devastating introductions are not of single species, but of communities which have evolved together over a period of time to produce a highly integrated community. This adaptive advantage may, for example, account for the spread of Eurasian temperate grassland around the globe, a community of grasses adapted to large-hoofed grazers (Simberloff, 1991).

6.5.2 Community stability and fragmentation

The loss of particular resident species can be of major significance: keystone species help to maintain the integrity of the community by their interactions with the rest of the biota, or by the habitats they provide for other organisms. For example, Terborgh (1988) suggests that composition of the plant community on Barro Colorado is an indirect result of the absence of top predators and their effect on the populations of seed-eating mammals such as the agouti, peccary and others.

The plant–herbivore associations are particularly important in regulating many communities. In an earlier study, Terborgh (1986) describes how the seasonal cycle of fruit production in tropical forests governs the populations of a number of animal species. When other fruit is scarce, vertebrate frugivores turn to figs and nuts, produced throughout the year, or occasionally consume insects. Palm nuts, figs and also nectar are key resources in the Peruvian forest, a reliable food source at times of shortage, even though their overall contribution to the energy budget of the forest is minimal. When other food is scarce, 60% of vertebrate frugivores rely on figs, and these plants probably fix the carrying capacity for the frugivores in this forest. The loss of this resource would cause the whole system to collapse.

The extent to which such interactions between species govern the stability of the community is still a matter of considerable debate. It was long believed that communities with the greatest species richness (S) had the greatest resistance to disturbance (McIntosh, 1985). Communities dominated by just one or two species were thought to be more likely to show instability, with large variations in the abundance of species and high rates of species loss. Models of theoretical communities indicate that this is too simplistic, and that both the extent and strength of the interactions between species are important in determining stability (section 10.2).

This becomes further complicated when we realize that a single species may have several functional roles within a community and their interactions with other species may also vary with time. One of the aims of conservation is often to increase community stability, and this is sometimes assumed to rise with S. Species richness has been commonly used to assess the conservation value of a site, even though theory suggests it would be a poor guide to the stability of that community.

6.5.3 Species richness and disturbance

Plotting the changes in species richness along an environmental gradient provides some indication of the factors that govern the diversity of a community. At the global scale the most obvious gradient is the increase in species richness found toward the equator. A number of theories have been suggested to explain this (section 10.4), including the idea that the tropics have more stable environmental conditions, or at least, conditions that are predictable for longer.

This is unlikely to be the whole story. For example, a smaller variety of polychaete worms is indeed found on the temperate (Mediterranean) coast of Israel, compared with its tropical (Red Sea) coast. Yet, despite their higher S, the tropical reefs are in a less predictable environment and are more likely to be disturbed by storms (Ben-Eliahu *et al.*, 1988). From such evidence it has become apparent that the degree of disturbance is a pre-eminent factor determining S in many ecosystems. Disturbance creates space for the invasion of new species and prevents the domination of the community by a small number of competitive species. What is important is the scale and frequency of these disruptions.

During the development of a community the competitive interactions between species lead to the replacement of opportunist species by a more persistent collection of competitive species (section 7.1). This provides fewer opportunities for new colonizing species and, as a result of competitive exclusions, S may be expected to decline with time (Figure 6.9). Maximum species diversity will thus be achieved when disturbance creates gaps for new species to colonize, without destroying the larger part of the community. This is called the **intermediate disturbance hypothesis** (Connell, 1978). The scale and frequency of disturbance has to allow fugitive or supertramp species to maintain a population, without damaging the integrity of the larger community.

In tropical forests, gaps of various sizes are needed to ensure the colonization by most tree species (Hubbell and Foster, 1986), but gaps which are too large limit recovery due to excessive loss of nutrients (Jordan, 1986). A disturbance should not be on a scale that reduces the productive capacity of the system over the long term, either through the loss of primary productivity or of plant nutrients. The agricultural practices of the Amazonian Indians, and natural disturbances such as tree falls, landslides and fires help to raise S. At the same time this increases

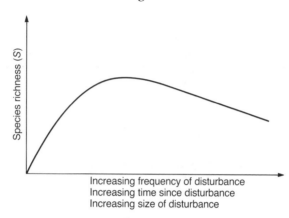

Figure 6.9 The effect of disturbance on the number of species in a habitat. Too much disturbance causes S to be depressed, but at intermediate levels gaps are created, which increase the rate of turnover of species and so raise S.

the adjustment stability of the forest by preventing some trees becoming locally dominant. Elsewhere, other species may promote a particular frequency of disturbance: some have suggested that the trees of the northern coniferous forests produce volatile oils to encourage fires, which will create gaps and regeneration niches for the germination of their seedlings (O'Neill *et al.*, 1986). This **induced disturbance** is a product of the community itself.

The frequency of disturbance, relative to the recovery time (elasticity) of the community, is crucial. Disturbances which are infrequent will not prevent competitive exclusions and S will be depressed as the competitive species dominate the community. Disturbances which are too frequent will displace species too quickly to be matched by colonization. The composition of the community may well reflect this frequency of disturbance. Also, the higher species diversity of regularly disturbed communities may provide a broader range of responses to subsequent disturbance, increasing their elasticity. A community with a mixture of pioneer and competitive species is then likely to show increased adjustment stability to future disruptions.

There is now considerable empirical evidence that disturbance is important in maintaining a high S in tropical forests and reefs (Connell, 1978; Ben-Eliahu *et al.*, 1988). Disturbance is probably important in maintaining species turnover in all habitats and so cannot account completely for the greater diversity of tropical forests. Other gradients in S can be found on regional and local scales, emphasizing the need to consider sampling area when attempting to explain these patterns, and the conservation value of a community. Indeed, we may wish to preserve some part of these gradients within one or more protected areas (Ratcliffe, 1986).

Including patches of disturbed and regenerating communities within a landscape may be crucial for the long-term prospects for the reserve (Bourgeron, 1988). As Ricklefs (1987) notes, the frequency of change in the environment means that the majority of natural communities are not in equilibrium, but instead following behind a range of continually changing abiotic parameters. We then need to consider stability at the larger, regional scales, rather than in one small fragment of a community in a small slice of ecological time.

6.5.4 Maximizing species richness within a reserve

A dominant species is able to secure a large proportion of the resources for its own growth. Amongst a community of sessile organisms, this is usually the species that can grow most rapidly with the resources available, and respond most rapidly to any increase in resources. Species-poor habitats are often nutrient-rich habitats, where a small number of species have become dominant. Management of reserves may attempt to reduce nutrient availability or increase the frequency of disturbance to prevent such dominance (Usher, 1989).

A nature reserve attempts to encapsulate in a small area features of the original habitat, before it had been fragmented. However, its capacity to withstand disturbance is much less than that of the larger area and our management has to be sympathetic to these differences. Managers often use a programme of smaller scale disturbance to mimic the larger scale processes – in this way fires and grazing are used to maintain prairies and old field communites in North America (section 7.4). Many of the habitats we attempt to conserve are semi-natural and would resort to a later successional stage without a particular frequency of disturbance. For example, a range of mowing and grazing procedures are used to maintain the floristic and insect diversity of grassland in England, which was, until recently, heavily grazed (Morris, 1991).

The effects of disturbance as a means of managing a habitat is readily reflected in the species richness of wetlands. Wood (1983b) contrasts two management strategies applied to adjacent reserves in the Camargue, the saline coastal marshes of the Rhône delta. Grazing by cattle and horses serves to enhance the number of bird species in the delta. A policy of no interference is adopted in the largest reserve, whereas in the smaller Tour du Valat reserve, a regular cycle of disturbance increases the predictability of the resources. In contrast with the *laissez-faire* approach, this programme of active management increases the variety of birds by reducing the population size of each species.

Grazing, burning and mowing are also part of some fenland management schemes in the UK (Wheeler, 1988). Opening up habitats by
creating edges will certainly introduce more species, but this poses
dangers to the flora and fauna which can only survive in the absence of
competitive invaders. The edge habitat can also be an effective barrier for
dispersal by some species (Yahner, 1988). This is one reason why reserve
management should include all fragments of a habitat, with regimes of
disturbance coordinated between a network of reserves (Morris, 1991). In
this way, a mosaic of patches can be created, each at different stages in a
cycle of disturbance and recovery.

6.6 SELECTING NATURE RESERVES

Conserving a community aims to preserve its character, its species
diversity and its capacity to accommodate disturbance. These are
attributes that are inherent in the structure of the community and its
relation to the larger ecosystem. In addition, there are value judgements
that we may hold about a site, such as its aesthetic merits or scientific
interest. We may also conserve a community because of some function
that it performs, perhaps in controlling our environment or handling our
effluent. Given a number of sites, and the prospect that only one or two
may be conserved, how should we choose between them? And what
weighting should we give the various attributes of each site?

The value that we put upon a site varies with the aim of the
conservation exercise; the problem is reconciling the different views or
objectives in conservation. Usher (1989) notes that 'nature conservation is
a blend of applied ecology and the social sciences' and in selecting a site,
we have to allow for the different expectations that people hold for it. The
major complication is that these criteria are scored in a variety of units,
with little or no common currency between them. Measures of the
economic worth of a piece of land and the number of species it contains
are not easily compared. It then becomes very difficult to reconcile the
scores that a site has for its different attributes.

6.6.1 Species composition and site selection

In selecting a site to conserve a single species, we are largely concerned
with a minimum viable population and the area needed to support it.
Where our prime aim is the conservation of the community itself, one
objective will be to maximize its species richness and diversity. This is
commonly rated highly as criterion for site selection (Usher, 1986a). We
might then use indices of diversity to select between sites, although their
measurement is not without problems (section 10.3). Not all groups can
be counted in these surveys and we may have to choose which organisms

are to be taken as indicative of the larger community. Also, the effect of area and of sample size has to be considered when comparing indices, along with the choice of the most appropriate index. Magurran (1988) describes the merits of various measures of diversity in detail.

At equilibrium, a community is assumed to have a full complement of species, so that each niche is filled (section 3.2). In habitats where this can be reasonably assumed, S may be used as a measure of habitat diversity, directly indicating the number of available niches. In choosing between potential sites for conservation, we would normally favour those with the greatest habitat diversity. Using S to measure this however, can be problematical – establishing that a community is at equilibrium can be very difficult (Connell and Sousa, 1983), and not all niches may be filled at any one time. Direct measurement of habitat diversity might be a preferable guide for choosing conservation sites (Goeden, 1979), although the ecologist is not always able to see the habitat from the plant's or animal's perspective.

Because of the relative ease of measuring the abundance and population growth of one or a small number of species, some ecologists prefer to use indicator species in site selection. Where they can be identified, keystone species would be the obvious choice, but very often, it is the most conspicuous and the most easily counted groups that are favoured. The stability of the community may depend on an entirely different group of organisms. Birds are easily identified and counted, but perhaps it is the soil microbial community that has greater significance for the long-term prospects of an ecosystem. Certainly, the birds may have greater popular appeal, and may be more advantageous in winning the conservation of a site. The ecologist, however, has to keep an eye on the ecological processes as well as the political processes in conservation.

6.6.2 Objectivity in site assessment

Unfortunately, we can rarely provide an inventory of all species or their significant interactions for a habitat in need of protection. Practicality dictates that the ecologist uses methods that can effectively identify the most valuable sites in a reasonable time. The intuitive sense of an ecologist with experience of a particular habitat will count for much in these circumstances. Ecological theory can only make the broadest speculations about how a site will change and develop, given the snapshots taken with a small number of surveys. In practice the ecologist may be obliged to use methods of rating sites that will not be strictly ecological. Then it is important to identify where the objective process stops, and subjective judgement takes over (Goldsmith, 1991b).

Equally, we have to be aware of the limits to our 'objective' scores of a site's worth. An assessment which is largely objective should produce the

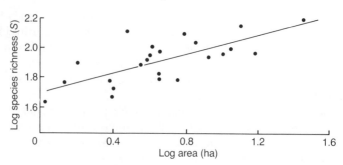

Figure 6.10 A plot of species richness against site area for the ground flora of fragments of woodland in Bedfordshire, England. The line of best fit (log S − 1.704 + 0.31 log A) predicts the number of species in a woodland of a particular size. A species count above that predicted may be used to select sites worthy of particular conservation. (With permission, from Dony and Denholm, 1985.)

same ranking from two independent assessors. This is most easily achieved with well-defined and easily quantified parameters that require no value judgement. As with indices of diversity, many of the scientific methods are often compromised by the practical difficulties of their measurement, and relatively few criteria will give a completely objective measure. Then qualitative criteria have to be scored against guidelines issued to the assessor.

The ten criteria used in the selection of sites of special scientific interest (SSSIs) in the UK (Table 6.3) were originally based on an ecological classification of communities (Ratcliffe, 1986). Some crucial parameters, such as area, are easily measured, and are absolute values. Others, such as 'typicalness' can only be scored relative to other sites taken to be the 'best example' of a habitat. The practical problems of assessing poorly defined criteria such as 'naturalness' usually require ranking, rather than some absolute scale of the degree of modification.

One wholly quantitative method was used by Dony and Denholm (1985) in their survey of vascular plants in the woodlands of Bedfordshire. They used the species–area relationship to isolate the effect of area on S. From this, they could identify the sites that had more than the predicted number of species for a given area (Figure 6.10). A higher species number is again found on some smaller sites, but as these authors point out, a simple count of S may not be sufficient to choose between sites: the quality of species list is also important, particularly whether the fragment has rare species for that area.

A large number of methods have been developed to rank sites according to several criteria, condensing these into a single score. This may not be the best way to choose between sites, and we should perhaps compare them by their profile of attributes (Goldsmith, 1991). This would

Table 6.3 Ten criteria used in the selection of Sites of Special Scientific Interest

Criteria	Score	Commentary
1. Diversity	S or some measure of community diversity	Practical difficulties in measuring S or applying an index of diversity
2. Area	Community area or minimum viable area for a protected species	Depends on objective of conserving site
3. Rarity	Abundance of rare or indicative species	Definitions of rarity – widespread and locally infrequent vs localized but locally abundant: depends on objective
4. Naturalness	Extent of human interference	Depends on objective – some sites require management to maintain conservation value
5. Representativeness (typicalness)	Usually based on plant associations representative of a biogeographical area	Rarity and typicalness may produce a conflict of objectives
6. Fragility	Sensitivity to environmental change and disturbance	Resilience may not be easy to measure in practice
7. Recorded history	Availability of long-term records	Of value for scientific research or educational purposes
8. Potential value	Scope for habitat creation	The prospects for restoration of a site or enhancement
9. Intrinsic appeal	Some measure of human appreciation of a site	Difficulties of measuring accurately, partially because it is an ill-defined criterion
10. Position in an ecological unit	Some measure of site ecology within a region	Significant gradients within a region may be reflected by appropriate site selection

help to highlight the particular assets of any one site, and make comparison between sites valid for each attribute. It is also a more sensible approach when a number of objectives of the conservation effort have been defined.

Many of these methods are specific to particular habitats. In the following section, we briefly survey some attempts to quantify comparisons of lowland wetlands. This is a broad habitat type including inland and coastal marshes, found throughout the world and invariably important for a range of wildlife, including plants and birds. It is a relatively unspoilt habitat largely because of the engineering difficulties in its development. Wetlands are naturally found as isolated habitats, often with well-defined boundaries.

We begin with simple ecological scores before turning to compound measures and the methods that incorporate economic valuation of habitats.

6.6.3 Evaluation of wetland sites

The rich-fen wetlands form a series of fragmented habitats throughout England and Wales often with a very diverse flora (Wheeler, 1988). These lowland marshes are surrounded by agricultural land and their floral diversity suffers at some sites due to nutrient enrichment from agricultural runoff, allowing some species to dominate. In a survey over 13 years, Wheeler recorded the herbaceous plants in over 2000 quadrats of 100 m^2, and demonstrated that this habitat was an important reservoir for about 160 species. Generally, the number of rare plants in a plot rose with species richness, so selecting a site according to S alone would be effective for protecting rareties in the fens.

More elaborate methods, requiring less taxonomic expertise, and a less exhaustive survey are also possible, weighted according to the rarity of the plants at each site (Table 6.4). These short-cuts are possible because of Wheeler's original extensive survey, defining rare and principal species for the fenland habitat in the UK. From this survey, four scores were derived, which considered particular species. The principal fen species score (PFSS) is a simple count of the presence of plants that are taken to be characteristic of the habitat. RWPFSS, the rarity-weighted principal fen species score is the PFSS score with species weighted according to their rarity in the fenland habitat. The richness indicator species (RIS) was originally used to predict S for a particular fen, using a small number of indicator species. Here it is used to compare sites. The final measure, the rare species score (RSS) sums weighted rankings of uncommon but characteristic fenland plants. When compared, no combination of the measures gave an identical ranking of 15 different sites, although

Table 6.4 Five alternative methods for evaluating rich-fen sites for their floral conservation value in lowland England and Wales (after Wheeler, 1988)

Score	Simple species score	(PFSS) principal species	(RWPFSS) Weighted principal species	(RIS) Richness-indicator species	(RSS) Rare species
Method	Count of plant S	Count of species mainly occurring in rich-fen habitats	Total derived from principal species weighted by their frequency in samples	Count of species shown to be indicative of species-rich sites	Total derived from rare species weighted according to their natural occurrence and frequency in samples

Wheeler suggests that the PFSS is perhaps the most indicative of their conservation value as this scores the main fenland plants.

Species richness and rarity have also been compared for wetlands in eastern Canada (Moore *et al.*, 1989). Not only does the number of plant species decline with the standing crop of the wetland, so does the representation of nationally rare plant species. In the most productive of sites, reeds and bulrushes (*Typha* spp.) dominate at the expense of most other plants. Indeed, the sites with the greatest conservation potential have slow-growing communities, with species of low competitive ability, classified as stress-tolerators (section 7.1). These plants are unable to compete with *Typha* as nutrients become more available and the infertile wetland undergoes eutrophication. Notably, the most diverse commmunity is here less resilient to disturbance than one composed of just a few species. As with many other natural communities, the productivity of a wetland is not a useful criterion for selecting conservation sites (Moore *et al.*, 1989). Rather we should favour the infertile sites, with their slow-growing plant communities and greater species diversity.

Using particular species as representatives of their guild is the basis of the **habitat evaluations procedure (HEP)** used by the US Fish and Wildlife Service. A **guild** consists of those species with more or less the same resource requirements, and a broadly equivalent role within the community: for example, a guild may include all the nectar-feeding insects. Usually, the species chosen are deemed to be the most sensitive members of that guild. Measurements of habitat area are used in conjunction with a **habitat suitability index (HSI)** to derive a rough estimate of the carrying capacity (K – section 4.2) for each species. Each critical variable in the species' habitat is scored on a linear scale between zero (completely unsuitable) to one (optimum); the index is the product of all these variables. The area, multiplied by the HSI, scores the habitat in terms of the carrying capacity for the indicator species; if all factors were optimum, the carrying capacity would be determined solely by the area, the resource space, of the habitat.

A version of this technique has been used by the US Army Corps of Engineers to evaluate various impacts on swamps, lakes and woodlands in the Mississippi region (Pearsall *et al.*, 1986). Not only can it be used in site selection, it can also help devise a management plan: the method measures the effect of future impacts by the reduction in the carrying capacity. A version of the procedure has been used to ascribe monetary values to any loss in habitat. However, the method is not without its critics. It may seriously oversimplify the interaction between factors and the assumption of a linear scale for the HSI contravenes much ecological theory (Westman, 1985). The assessment also requires an extensive initial survey to work out the habitat suitability for the indicator species. In his review of 15 HEPs, Williams (1988) criticized the lack of consistency in the choice of indicator species, or of attempts to validate models used to score habitat quality for each species. Overall, many of these studies used unjustified assumptions about how these habitats would change, and in doing so, lost objectivity. Williams concluded that this is not necessarily a fault of the procedure itself, rather its implementation.

Sometimes the simple measures are best. The government agency Environment Canada has taken the number of ducks to be an indication of wetland habitat quality on the prairies (Keddy, 1991). This claims that the duck population gives a sensitive indication of habitat quality relative to the area of the wetland, because this integrates a range of contributory factors such as pesticide usage and hunting. This rather pre-judges the objectives of the assessment, as these factors may only be important for the ducks.

The idea that the diversity of birds in wetlands are some general indicator of their habitat quality has been developed by Cable *et al.* (1989). Their **habitat assessment technique (HAT)**, uses the number of rare species as a measure of their conservation value and also includes a measure of its economic worth. In this procedure, the number of individuals of a wetland-dependent species is multiplied by a weighting based on their regional rarity. These are summed for all species to derive 'species points' for that site. An area factor is then calculated based on the species–area equilibrium for these species, in an obscure formula that ascribes the largest site to be uneconomic and the smallest site not to be ecologically viable. From this a composite faunal index is derived as the species points divided by the area factor. The authors considered their ranking of 11 Delaware wetlands to be succcessful: the species points accurately reflected the quality of the habitat of each site, at least as it is represented by the variety of the birds, while the composite faunal index helped to resolve the 'interplay of ecological and economic forces'. By using birds of course, the index benefits from an established database saving much time and effort.

A more elaborate procedure has been developed scoring wetland birds, but which could in fact score any environmental parameter in any habitat assessment. This is the **analytical hierarchy process (AHP)**, a much more general technique using matrix algebra. In this example (Figure 6.11), the method was used to assess communities of breeding birds in Swedish bogs (Anselin *et al.*, 1989). The method requires selecting some indicative measure (species of known national rarity) of a criterion (the guild to which the birds belong), which is taken to be important for the selection of a site. A number of criteria are combined into an index (the 'focus') as outlined in Figure 6.11.

Each stage of the process can have weighted elements. In this example rare species of wetland bird are scored more highly than the common species. At the next level, the weightings between bird guilds (criteria) are

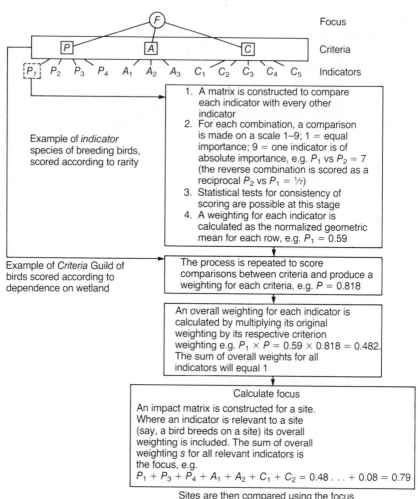

Figure 6.11 The analytical hierarchy process. This method was originally applied to wetland bird species but the principle can be used for any multifactorial analysis of a reserve. It is a composite measure, derived by scoring each indicator against other members of its group. A mean of these scores for each indicator provides a weighting for that indicator. Similarly the criterion to which it belongs is scored in the same way. An overall weighting for each indicator is derived by multiplying its original weighting by the weighting of its respective criterion. The sum of these modified weightings gives the overall score for all indicators at that site, the focus. In this way, different criteria can be compared and weighted according to the objectives of the site selection. One potential value of this method is the ability to include criteria scored in different units in the analysis. (After Anselin *et al.*, 1989.)

also scored and weighted in the same way. An overall score for each indicator is the product of its original weighting and the weighting of the criterion to which it belongs. The focus of a site is the sum of these scores for all indicators occurring at that site. Sites are then compared with these indices.

The AHP produced rankings of the Swedish wetlands which were consistent with previous measures based on multiple criteria. The method has several potential advantages:

1. Any criteria can be scored if appropriate indicators can be found of subjective attributes – for example typicalness, area, economic worth.
2. Different indicators can have different scoring systems.
3. All criteria do not have to be treated equally.
4. It formalizes the comparative process with some statistical check on consistency.

This method still retains much subjective judgement, but it does represent one possible way of scoring a site using multiple criteria which are measured in disparate units.

6.6.4 Monetary evaluation of sites

Even so, we should not be fooled into believing that a series of elaborate manipulations can convert a subjective assessment into an objective one (Ratcliffe, 1986). The real danger with these composite scores is being unable to identify their subjective elements. One alternative is to reduce the various estimates of value to an economic measure of the value people place upon them, by scoring them in monetary terms – how much we are prepared to pay for the utility a site provides. There are several difficulties here – people value the same attribute differently, a person's valuation may change with information and time, and the units of valuation themselves are subject to change. It also assumes that wholly ecological criteria can be accurately valued in monetary terms. Should we then simply rely upon the average or modal valuation that people place upon a site? And is our valuation (or knowledge) of a site today likely to be unchanged in 10 or 100 years time?

The capacity of an ecosystem to remove carbon dioxide from the atmosphere or to treat our effluent (section 9.4) is never traded and therefore never has a market price. We can, however, work out the costs of providing this service using existing technology, and thus value these services by using 'shadow prices' – the costs incurred if we needed to build equipment for the same purpose (Westman, 1985).

This is less easily achieved when measuring the aesthetic enjoyment that a site provides. The economic value of the pleasure derived from a conservation area can only be measured indirectly, perhaps by the amount people are prepared to pay as an entrance fee, or how far they are prepared to travel to enjoy it. One attempt to measure the valuation of a site was by Everett (1979) in his survey of the Dalby forest reserve in Yorkshire. Using the number of visitors to the site, their costs in reaching the site and the entrance fees collected, Everett scored the recreational worth of the forest. The economic value of the site was apportioned between its general recreational use, and that attributable to its wildlife according to visitors' reasons for visiting, and also their preferences within the park. The survey illustrated how very different people's valuation of the reserve were, and how such measures can only represent the average perception of its value at one particular time.

An additional problem with this approach is that it scores only the opinions of those visiting the site. A reserve may also be valued by those who are unable or choose not to visit it, even though for them it may serve some equally valuable function, such as the preservation of a single species. A site may even fail to be protected because it has too many visitors – Goldsmith (1983) describes the rejection of the Walthamstow marshes in London as an SSSI because of its degree of human disturbance. At other times, the decision rests with a very small number of people. The survival of an endangered species can sometimes be in the hands of a single individual (McDowell *et al.*, 1989). Then it is their knowledge of ecology and of the marketplace which is critical. Assuming perfect competition and perfect knowledge of the market place is a dangerous supposition when dealing with ecological resources.

An alternative approach has used 'supply-side' economics to value the input that a site may make to marketable goods or service. Ellis and Fisher (1987) considered the utility derived from a wetland as a means of treating effluent and supplying clean water, and also acting as a nursery for a fishery. In particular, they studied the wetlands on the Gulf coast of Florida, the sewage they treated and the blue crabs they produced for a commercial fishery. After piping the effluent to these sites, the quality of water drawn from such areas can determine the cost of any further treatment required, given the demand for the clean water. Similarly, if the price of shellfish falls with an increased supply then the cost of cleaning up or extending the wetland nursery to grow the food may not be merited.

There are many more methods of site evaluation which have not been considered here, and more general attempts have been made to produce a multifactorial index that can rate ecological worth with social and economic worth in an objective manner. Choosing between sites using these methods has to be highly qualified by the objectives of protecting a site, and used with a good pinch of experience.

Table 6.5 The main areas of management function in the conservation of nature reserves (after Pyle, 1980)

- The maintenance of successional stages
- The removal or mitigation of alien plants and animals
- The repair and prevention of vandalism, overuse, poaching and other human impacts
- Addressing the effects of climatic and hydrological impacts, such as flooding
- Addressing the effects of pollution
- Addressing other claims on the land, including rights of way, mineral or water extraction rights and so on
- Addressing disease and infestation in the reserve
- Mitigating the effects of past changes and past management of the reserve, including the absence of any key species. A reserve without a top predator may require culls of some species
- Applying a collection of methods to resist further extinctions

---

6.7 RESERVE MANAGEMENT

Not only is the selection of a site closely linked to the aim of the conservation effort, so is the subsequent management plan. As we saw earlier, one option is to create a network of fragments and to use a sequence of disturbance between the various sites. In a larger reserve this may not be necessary, though even here, some management of the disturbance will probably be required. Pyle (1980) has suggested that there are nine broad areas of management function (Table 6.5) common to the conservation of most reserves.

An initial survey is essential to site selection and for providing a baseline against which to judge its subsequent management. Compiling species lists and producing an accurate record of the status of all species is, for most large sites, impractical and the survey is has to be tailored according to the objective of the evaluation (Hellawell, 1991; Usher, 1991). Continued research and monitoring is used to judge the success of any management strategy, and to provide information to advance our understanding of its dynamics. This will also become a part of any impact assessment of a site – deciding what will be the effects of further insularization, development, or changes in management strategy. This is essential if the resources of the site are to be used to support it financially, either by encouraging tourism or perhaps by harvesting some population.

Small reserves are often used to protect fragile communities or rare species and need particular attention; larger reserves are commonly protected as typical examples of important habitats (Usher, 1986b).

Conserving a single species may require us to limit its population growth to that compatible with the size of the habitat. At one extreme, culling may be necessary to keep the population within the limits of the reserve, and to prevent the whole community from collapsing. This has proved necessary with several mammal species, such as elephant, on large African reserves. An alternative strategy is that of 'benign neglect' protecting the community, but doing little else (Soulé *et al.*, 1979). Even if there was no poaching, relaxation would argue against this approach, especially for large animals: with a low population growth, or where local extinctions are frequent, active re-introductions, from other reserves or from captive breeding, may be necessary. Such intervention is not without its own problems and there has been considerable argument between ecologists about using such methods to sustain endangered species over the long term.

In conserving semi-natural communities, management may amount to no more than maintaining the same frequency of disturbance, such as grazing or trampling pressure (Morris, 1991). The loss of the large blue butterfly from its special reserves in Britain followed because these sites were not heavily grazed, a regime which is essential for its food plant *Thymus* (Thomas, 1991). Some management practices may seem drastic (such as regular burning) or unnatural (such as mowing) but this prevents the community proceeding to a late successional stage, and the dominance of a small number of competitive species. Through this disturbance, reserve management seeks to maximize both plant and animal diversity.

The scale of development near a site also needs to be considered in its selection and management. A range of overlay techniques, now often based on satellite remote-sensing and computer databases, can be used to assess any such effects. These **geographical information systems (GIS)** may be run in combination with a defined model of a habitat to assess the potential impact of some change on the community. Satellites accurately map areas from space and provide spectral data for analysis at regular intervals. Image processing and other methods of data analysis extend this even further to measure rates of urban growth or pollution spread (Harris, 1987). Combined with conventional surveys of social and economic activity, we can then identify those areas most likely to be subject to disturbance, or detect particular conditions or particular times of the year when the community is most sensitive to disturbance. This information may be used to select reserves or can be included in any plans to restrict access. Equally it can used in an impact assessment of any future development, or in plans for remediation (section 10.5). This form of assessment is becoming increasingly important in large-scale industrial developments and urban planning and Westman (1985) provides a review of three schemes with a proven record.

Summary

The relationship between area and species number has coloured much of the debate about the effects of habitat fragmentation. A simple model of island biogeography, predicting a dynamic equilibrium number of species, has been derived from the interplay of extinction and colonization rates, based on island area and isolation. The empirical evidence does not always confirm this, and in some cases S may rather reflect the diversity of habitats associated with an increase in area. Nor is the model readily applicable to fragmented habitats on mainlands, although it has been used to predict the rate of loss of species in nature reserves caused by isolation. The insularization of any habitat decreases the average area of a fragment and increases the distance between patches. Islands may themselves be relatively fragile communities because of the small populations they support, and the high degree of endemism.

The rarity of a species is defined from a combination of population size, habitat specificity and geographical distribution. Species which have low rates of population growth and poor powers of dispersal are threatened by the fragmentation of their habitat. A number of other factors, including competition and physiology of the species, will determine which species are found in isolated habitats.

The conservation of a community aims to maximize the diversity of the reserve, and this may require an active management programme which includes a regime of disturbance. The species composition of many natural communities is perhaps maintained by a normal regime of disruption. Overall, habitats which are nutrient poor and with a natural cycle of disturbance have the highest diversity. A range of other factors needs to be considered in selecting a site, including the utility that people derive from it. A variety of methods have been used to try and condense these attributes into a single score, although these invariably include a large subjective element based on the qualities that a site is judged to have.

Further reading

Goldsmith, F.B. (ed.) (1991) *Monitoring for Conservation and Ecology*, Chapman & Hall, London.

Spellerberg, I.F., Goldsmith, F.B. and Morris, M.G. (eds) (1991) *The Scientific Management of Temperate Communities for Conservation*, Blackwell, Oxford.

Usher, M.B. (ed.) (1986) *Wildlife Conservation Evaluation*, Chapman & Hall, London.

Strip mining brown coal in southwestern Poland, producing complete devastation of the habitat. At other operations, where the topsoil has been properly stored and replaced, rapid restoration is possible.

7

Establishing ecosystems

Given time, most polluted or degraded habitats will be colonized by some sort of ecological community. The aim of reclamation is to accelerate this process, to speed up community development and to re-establish ecosystem function. To do this, the applied ecologist has to understand the relationship between the community of species and the processes of production, decomposition and nutrient transfer. Not only does colonization by plants and animals have to be encouraged, so also the interactions between members of the community that will drive these essential processes. Only then can a self-sustaining community be developed reasonably quickly.

Most patches of bare soil are rapidly colonized by vegetation, with a sequence of plants arriving as time passes. If the same succession repeated itself whenever the soil of a region was exposed, we would suspect there was some underlying mechanism governing the sequence – perhaps when the establishment of one species was dependent upon the presence or absence of another. This is termed an **autogenic succession**, because the sequence is determined by the interactions between the members of the community. Change derived from external pressures produces an **allogenic succession**; the gradual drying or flooding of a soil would, for example, encourage a succession of plants able to live at each particular moisture level. We should also distinguish between **primary successions** where there is no organic material for colonizing species to exploit, and **secondary successions** when some organic component remains from a previous occupation, before its disturbance.

The idea that communities follow a predictable sequence of development has been a dominant theme in theoretical ecology since the earliest days of the science, and disputed for almost as long. The debate has centred on the extent to which species within a community are interdependent upon each other. If a community was highly integrated, with the presence of each species dependent on the presence or absence of others, only certain combinations of species could form stable communities. By the same token, a large number of alternative compositions would indicate a low level of integration, and the collection of species would then depend, to some extent, on chance factors. Such a community may not develop a single, stable complement of species, but remain in transition, with a high rate of species turnover.

The suggestion that the composition of a community was more or less fixed was originally made by Clements (Colinvaux, 1986) and became known as the monoclimax theory: only one type of community would develop in a particular habitat and a particular climate. At the end of a succession this **climax** community would consist of a stable collection of species that underwent no further change. This deterministic view of communities has now been largely discounted as too simplistic. Today most ecologists recognize that any successional sequence is far from fixed, and chance events in the history of a habitat are of considerable importance (Usher, 1987).

Nevertheless, there may still be 'rules of assembly' that limit the possible combinations of species. Most attempts to describe such rules have concentrated on the mechanisms of interspecific competition (Connell and Slatyer, 1977). For example, Gilpin (1987b) describes how competition between various species of *Drosophila* within very simple laboratory communities only allow certain combinations of species to coexist: in this case, just three possible equilibrium states. The extent to which interspecific competition organizes natural and more complex communities is still a matter of debate (section 10.3). In addition, we can ask if these rules of assembly govern the sequence of a succession.

According to one definition, a community would be judged stable if, in the absence of disturbance, there is self-replacement of species with time. Then its composition remains more or less constant (Connell and Slatyer, 1977). Judging stability in this way would depend upon the scale over which the observations are made: a late successional community, dominated by trees may appear stable, simply because trees are long-lived and not likely to be replaced very quickly. If the measurements are not taken over an appropriate temporal and spatial scale, it would be impossible to demonstrate that a woodland shows inertial stability (section 1.4).

Late successional plants are well-described for most terrestrial eco-systems, but the species composition of a community can be a poor indicator of its stability. The variety of species changes as ecological processes change, and species themselves play some part in determining these functions. Our efforts at reclamation attempt to establish some degree of stability in these processes, to allow the community to develop and to regulate itself through its species interactions. In this way, the community can develop both inertial and adjustment stability, with the capacity to resist displacement and to recover from disturbance.

The practice of restoration often makes use of the existing adjustment stability of ecosystems, and their capacity for self-repair (Bradshaw,

1987a). In non-polluted freshwater habitats, for example, restoration may require only the provision of a suitable substratum. Colonization alone will then ensure the rapid establishment of an invertebrate community (Gore, 1985). Restoration attempts to return degraded habitats to their former state, often as a semi-natural habitat with some conservation or aesthetic value, or perhaps as useful agricultural land. Reclamation includes the range of techniques used to establish a functioning eco-system on a waste or spoil, where perhaps little was growing before. Derelict land is often a characteristic of the most economically depressed areas, and the establishment of a green space is itself a major improvement in the environment. In some cases, reclamation can also establish areas of conservation value, and occasionally sites of special scientific interest (section 6.6).

In this chapter we examine the current theories of succession and how these can be applied to the reclamation of degraded land. Restoration and reclamation techniques differ radically between habitats and we cannot hope to survey all the strategies used. In terrestrial sites, our effort is directed at developing a sustainable soil community, essential to establish vegetational cover above. Here we concentrate on the reclamation of mine spoils by accelerating soil-forming processes and plant community development. This is comparable to a primary succession and is later contrasted with the re-creation and restoration of semi-natural habitats by encouraging a secondary succession on disused agricultural land.

7.1 SUCCESSION AND DISTURBANCE

The capacity of plants or animals to colonize a site depends on their dispersive powers and their adaptation for life in that habitat. A lack of nutrients or severe physical and chemical conditions can make degraded soils very harsh environments – a spoil heap may have no physical barrier to colonization, but its conditions can isolate the site from all but the most tolerant of species. Succession on degraded soils will thus be drawn from a relatively small pool of potential colonizers.

Some definitions of succession only acknowledge change in the plant species (Miles, 1987), whereas others use the term to denote a chronological sequence of all organisms in an area (Lincoln *et al.*, 1982). Here, succession is taken to include the sequential changes in both species composition and ecological function, to distinguish it from a simple turnover of species. Three possible mechanisms could be responsible for determining both aspects of a succession:

1. change or disturbance in ecosystem processes;
2. the interactions between species;

3. the properties of the individual species, particularly key or dominant species.

For most communities, it will be some combination of these mechanisms that drives the changes during the development of the community. Even in an apparently stable community, these will continue to operate as its environment fluctuates and the prospects for individual species change.

Disturbance creates gaps and provides opportunities for new invaders (section 6.5). The intensity and scale of the disruption will determine whether regeneration begins with surviving species or with migrants from other areas: a decimated community will have to re-establish itself from colonists, initially those opportunist species with good powers of dispersal. A high frequency of disturbance prevents many species from becoming established and allows only a limited development of their trophic and competitive interactions. In contrast, a long-established community, subject to only infrequent and small-scale disruptions, will be largely dominated by competitive species (section 5.1). This offers few opportunities for colonizers, and species number declines as competition leads to a slow loss of species (Figure 7.1). Thus the greatest species richness is thought to occur in communities with intermediate levels of disturbance. With an established community, some form of disturbance, perhaps from an ant hill, mole hill or buffalo wallow, is important for opening gaps for pioneer species (Allen and Hoekstra, 1987).

A turnover of species is perhaps typical of most natural communities (section 6.3). In several habitats the dominance of one or two species may

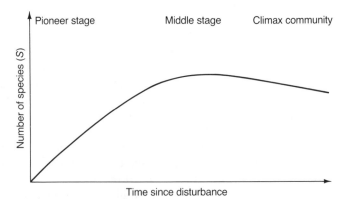

Figure 7.1 The rise in species richness during a succession and its decline toward the later successional stages. If the frequency of disturbance is shorter than its recovery time, a community may never develop past the pioneer stage. Theory suggests that a higher number of species are found at an intermediate frequency of disturbance as the competitive species of the later stages never become dominant in the community.

be relatively short-lived and a regular cycle of replacement follows, eventually returning to the original community. Several shrub and herbaceous communities are known to cycle through different stages as the dominant plants age and lose their vigour. This produces a composite habitat – a mosaic of patches, each representing a stage dominated by one or two particular species. Taken over this larger scale, the whole system may be regarded as a stable equilibrium, although different patches are at different stages in the cycle. These changes may be generated autogenically as, for example, in some heathlands, where the cycle is set by the loss of vigour in ageing heather (*Calluna* spp.), creating gaps which allow other species to dominate. Allogenic changes are also possible: on lowland heaths, *Calluna* only dominates where the soil has a relatively small capacity for absorbing phosphates. Where phosphates are less readily available, gorse (*Ulex europaeus*) or birch (*Betula* spp.) dominate (Chapman *et al.*, 1989).

A community subject to regular but non-catastrophic disturbance may show greater elasticity and a faster return time to an equilibrium composition. A mosaic of disturbance and change may allow a community greater adjustment stability, at least at the local scale. Some fugitive species rely on the creation of such gaps to sustain their population (section 1.4), but the whole community responds to the opportunities the gap provides – by growing over the gap, by exploiting the liberated nutrients in the soil or by setting seed beneath the new colonists. This is a form of secondary succession, dominated by the competition between plants for the resources available. In contrast, the scale of disruption and absence of nutrients in many mine spoils rarely allows for a rapid recovery, and only a primary succession is then possible.

7.1.1 Species interactions

An ecological community is a highly complex system and a large number of possible interactions, direct and indirect, will occur between its component species. The problem is to define the important interactions from all of these possibilities. Three basic types of interaction were identified by Connell and Slatyer (1977):

1. Facilitation: where the presence of a resident species makes possible or accelerates the establishment of the colonizing species by virtue of its own presence or its effect on the habitat.
2. Inhibition: the resident species prevents or slows down colonization by the new species.
3. Tolerance: late successional species are able to tolerate low levels of

resources (water, nutrients, light, etc.). Despite their slow growth they eventually dominate the community.

A succession is unlikely to result from just one of these mechanisms – as the community develops, facilitation may be replaced by inhibition, or some combination of all three processes may be operating simultaneously between different species (Table 7.1). Facilitation is more typical of primary successions where an inert mineral soil needs to acquire some biological activity before most higher plants will invade (Connell and Slatyer, 1977). As we see later on, the development of the microbial associations that fix the principal plant nutrients, nitrogen and phosphate, will facilitate the establishment of many higher plants in most terrestrial ecosystems.

In secondary successions, species interactions are dominated by competition for the existing resources, and inhibition and tolerance take on greater significance. However, in their review, Connell and Slatyer suggest there is little evidence for tolerance as a prime factor ordering the sequence of a plant succession.

A number of models of succession have been developed based on the chances of one species being replaced by another, or on the growth and population data for a limited number of species. These can often predict a community composition that compares well with existing climax communities (Colinvaux, 1986). Given the replacement probabilities between any two tree species, Horn (1981) predicted the variety of species in undisturbed temperate hardwood forests from earlier successional stages. Horn concluded that the general pattern of forest succession depends on the biological properties of the constituent trees, such as their shade tolerance. For example, a species of multilayered tree can grow rapidly in the open sun with its large photosynthetic area, but would fare worse in the shade where the respiratory costs of those multiple layers are a disadvantage. A single-layered tree will grow more slowly in the shade, but at a lower cost. It suffers less from crowding and will eventually oust multilayered competitors. For this reason mature forests are dominated by single-layered trees forming a canopy.

With no change in these replacement probabilities and no disturbance, these models predict that a forest community would eventually undergo no further change (Horn, 1981). Similarly, models of temperate woodland that include their abiotic factors predict a rapid successional change followed by a relatively consistent species composition for up to 2000 years, comparable to existing stands (Bormann and Likens, 1979). Specific models of the competitive interactions also suggest that communities become increasingly difficult to invade as the succession proceeds and the intensity of the competition increases (Lawton, 1987b; Case, 1990). Thus, with no major environmental change the stability of

Table 7.1 Three interactions between species governing a succession

	Mechanism	Example	Reference
Facilitation	The invasion by species B relies upon the presence or activity of species A	The mycorrhizal association of a fungus in the roots of some species of tree is essential to their colonization of reclaimed land	Miller, 1987
Inhibition	Species B is prevented from colonizing by the presence or activity of species A	Bracken prevents the establishment of heather on the Scottish Highlands by the density of its shade	Miles, 1987
Tolerance	Late successional species do not depend on predecessors – but may out-compete them by being more tolerant of overcrowding or a shortage of resources in the later stages of the succession	Single-layered trees can withstand shading and overcrowding and will eventually replace the more susceptible multilayered species	Horn, 1981

such commmunities must be maintained by inhibition. Much of the variation we observe between communities can be attributed to their initial species composition and the role of tolerance and inhibition as the succession proceeds. With some knowledge of the competitive inter-actions between species, we can thus make reasonable predictions about the successional sequence which follows.

7.1.2 Characteristics of species through a plant succession

An early successional community will typically have quick-growing, rapidly reproducing species, opportunists with highly dispersive seed (section 5.1). Late successional plants tend to be slower growing, devoting more biomass to non-reproductive structures, and flowering less readily. Competition between species only becomes significant in the later stages of a succession, when resources or space are in short supply (Usher, 1987). For example, during the secondary succession of old-field communities in Berkshire, competitive species are dominant as the woodland becomes established in the later stages (Brown and South-wood, 1987). This provides a more constant habitat for the insects and other animals living on the plants, facilitating the establishment of species with longer generation times (Figure 7.2). The succession thus marks the gradual replacement of short-lived opportunist species by longer-lived competitive species, reflecting the increase in the durational stability of the habitat (Table 5.1).

Pioneer plant species generally have small, wind-dispersed seeds, or are stoloniferous, invading a site using adventitious stems (Fenner, 1987). Seeds dispersed by animals arrive later when there are resources on the site to attract the animals. In the temperate zones these mid-successional species are frequently berry-bearing shrubs, grown from seeds that have passed through the gut of birds or other animals. Large-seeded plants are typically found later in the succession.

This sequence might differ in a secondary succession where there is a seed bank in the soil, ready to germinate after a disturbance. Gross (1987) has examined this experimentally by introducing seeds into the second-ary succession of old-field communities in Michigan. Early colonizing species such as *Verbascum thapsus* and *Oenothera biennis* produced small seeds, but were only able to colonize bare ground. Late arrivals, including *Daucus carota* and *Tragopon dubius*, had larger seeds and could establish themselves in vegetated soil. The difference in their time of arrival was simply their chances of getting there – the early colonizer seed is more readily dispersed and also viable for longer. Seed design, with its implications for seed dispersal, is thus one determinant of a plant's position in a successional sequence.

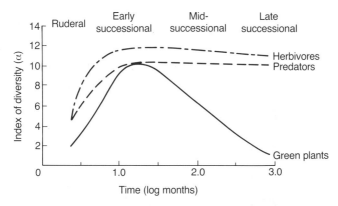

Figure 7.2 Changes in the diversity of green plants, herbivorous and predatory invertebrates (excluding Diptera) in the secondary succession of old fields at Silwood Park in southern England. Time is given as a logarithmic scale (months + 1) and the stage of the plant succession is given along the top. An index of diversity (α) summarizes species richness and relative abundance of the species. Plant and herbivore diversity both decline to some degree as the competitive plants dominate, but the same trend is not observed in the predatory invertebrates over the 10 years of this study. By the end of this period a woodland has developed. (Redrawn from Brown and Southwood, 1987.)

So too is the growth of the seedling following germination. The late arrivals studied by Gross produce seedlings with long stems and with a greater proportion of their biomass in their roots. This helps them to compete with other established plants. In contrast, early colonizers are typically rapidly growing plants that are intolerant of shade. As Grubb (1977) noted, the characteristics of a gap will determine which species are able to use it (Figure 6.1). The **regeneration niche** of a plant, that is, the conditions it requires to establish itself, may be the most important factor explaining its distribution in a habitat or in a succession. Grubb (1987) recognized two basic types of pioneer plants:

1. Long-lived pioneer plants that survive where resources are limited or the soil is unstable. As there are few competitors, rapid growth can be sacrificed to tolerate the harsh conditions.
2. Short-lived plants, rapidly exploiting abundant resources by quickly producing seed. These will eventually be replaced by competing species whose arrival may be slowed by their poor power of dispersal.

Where competitive associations between species go uninterrupted for long periods, we would expect niches to become highly differentiated, and the community to be dominated by specialists (section 3.2). In contrast, early colonizers should have a broad niche, able to accommodate a wide range of conditions. Thus, in general terms, a succession may

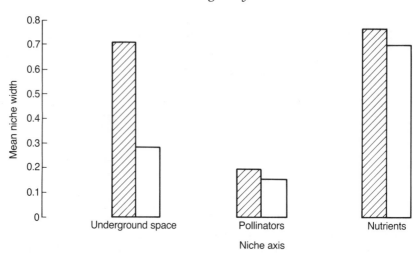

Figure 7.3 A comparison of niche width for three niche axes for early succession (shaded) and late successional (open) herbaceous plants. The early successional species tend to be found growing over a wider range of conditions, an indication of their greater adaptability. (After Bazzaz, 1987.)

be seen as a replacement of r-selected species by K-selected species. Evidence for this has been found by Bazzaz (1987) looking at the habitat requirements of plants grown up from the seed bank of early successional soils (Figure 7.3). In particular circumstances, potential colonists may, however, be limited by the nature of the soil. Some growth forms are suited to particular conditions – perennial grasses, for example, are invariably the only plants found on unstable sand dunes (Grubb, 1987).

The distinction between r- and K-selected species cannot completely account for the life-history strategies that we observe as a succession proceeds. Grime (1987) includes a third dimension that broadly measures adversity selection (section 5.1). His three-fold classification (Figure 7.4) is based on the plant's response to two factors, disturbance and stress, the latter measured as resource scarcity. The type of plant that dominates as a community develops depends upon the frequency and predictability of disturbances and resource availability (Table 7.2). Here dominance is measured by the contribution of a species to the overall biomass, as well as its interactions with its neighbours.

'**Ruderal dominants**' (R) are plants adapted to frequent and relatively predictable disturbances. Although they have little competitive ability, they can reproduce rapidly and have good powers of dispersal. '**Competitive dominants**' (C) have a high rate of resource capture enabling them to grow rapidly. These plants can modify their growth form to crowd out competing species, say by forming a tussock, but they are

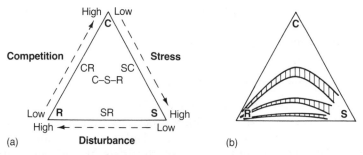

Figure 7.4 Grime's classification of three main plant strategies and their relation to a succession. The competition between species, the frequency and scale of disturbance and the stress resulting from nutrient shortage or the presence of toxins together determine which strategy is most likely to succeed. (a) The three main strategies are shown at each corner of the triangle, with intermediate forms between them. Stress tolerators (S) will survive in nutrient-poor habitats and are typically long lived and slow growing (e.g. sheep's fescue *Festuca ovina*). Ruderals (R) are rapid growing and short lived, quickly colonizing disturbed sites where there is no shortage of resources (e.g. chickweed *Stellaria media*). Competitors (C) are plants that crowd out other plants by their lateral growth and litter production, producing shade and difficult germination conditions (e.g. nettle *Urtica dioica*). (b) A succession is shown in three different habitats with low, intermediate and high levels of productivity. Each succession begins with the arrival of ruderals, followed by a stage dominated by competitors. These are replaced by stress-tolerant shrubs and trees able to survive in a nutrient-poor soil. In each case, the biomass tends to increase in each succession. (After Grime, 1974 and Hodgson, 1989.)

Table 7.2 Grime's classification of life history strategies in plants (after Grime, 1987)

		Intensity of stress	
		Low	*High*
Intensity of disturbance	Low	Competitors	Stress tolerators
	High	Ruderals	No viable strategy

unable to withstand disturbance. **'Stress-tolerant dominants'** (S) are able to survive in environments where resources are scarce by capturing and retaining the resources. These are typically long-lived plants, with well-developed defences against herbivory. Such species are found in unstable, unproductive habitats and therefore show high resistance to stress compared with the other two types. Although most plants will fall somewhere between the three types (Figure 7.4), in a succession there is a gradual shift from R through C to a final S-dominated community, when the stress of limited resources and the low frequency of disturbance favours the latter.

7.1.3 Development of ecosystem function

An ecosystem will only sustain itself by retaining or replacing the nutrients it receives. Decomposition processes are required to degrade dead tissues and make their nutrients available to the soil community and beyond. The storage and release of nutrients are crucial ecological functions that have to be developed in any soil before it can sustain a plant community.

The nature of a soil depends upon its parent material, position and climate, and also the plant community growing in it. As the constituent species change during a succession, so the chemical and physical properties of the soil develop. Most of the principal soil formations have taken several thousands of years to reach their present state, largely reflecting the dominant vegetation and prevailing climate.

Just as the plant community competes for incoming radiant energy, so the soil community competes for the chemical energy arriving in the organic debris falling on its surface. Decomposition processes are concentrated close to the top of a soil profile, releasing the energy that supports the whole soil commmunity – its trophic structure ultimately depends on the productivity of its **saprotrophs**, the bacteria, fungi and invertebrates that feed on dead and decaying organic matter. Aquatic sediments serve the same role, with an equally diverse community of decomposers wherever there is a regular supply of organic matter from the overlying water.

There is a decline in biological activity down the soil profile, moving away from the source of the organic matter and the aerobic conditions at the surface. Even in a soil with an established community, a process comparable to a succession occurs each time a pulse of nutrients arrives. The same processes of facilitation, inhibition and tolerance apply: for example, the colonization of wood by wood-boring beetles may be facilitated by the actions of fungi, while these same fungi produce antibiotics that inhibit bacterial invasions.

The sequence of arrivals is largely determined by the nature of the organic material and the ease with which it can be degraded. With plant material as the substrate, readily available sugars are exploited first by those saprophytic fungi with the fastest mycelial growth or whose spores can germinate quickest (usually phycomycetes and fungi imperfecti). Their metabolic waste products and the production of antibiotics will inhibit the arrival of other fungi and bacteria. As the simple sugars are used up, ascomycetes, fungi imperfecti and some basidomycetes begin to degrade the cellulose. Associated with these are the secondary sugar fungi, which use the sugars released from cellulose decomposition. Eventually only those fungi capable of attacking lignin dominate, mainly the basidiomycetes. Throughout the decomposition process, various proteins and amino sugars may become bound or chelated by organic acids, forming highly intractable compounds, humates, that are degraded very slowly. This has important consequences for soil quality – the soil texture, its capacity to hold water and bind nutrients, depends upon these and other undegraded organic matter.

Over a broad range of substrates, decomposition rarely follows such a simple linear sequence. Cellulose decomposers and many of the sugar fungi (such as yeasts) may be attached to a leaf or fruit before it arrives at the soil surface. Animal faeces also contain the germs of their own destruction. Decomposition is aided in all substrates by the action of invertebrates, comminuting the material and increasing the surface area for fungal and bacterial attack. Earthworms, mites, woodlice and millipedes all facilitate initial decomposition in this way and add bacteria to the substrate as it passes through their gut. The sequence is further complicated by coprophagous animals, such as woodlice, which consume their faeces, absorbing the nutrients released by microorganisms following a short period of decomposition in the outside world.

The bacterial sequence during decomposition is generally less predictable because of their highly localized distributions; unlike the fungi, they are unable to grow toward a substrate. There is also a general sequence of fungal colonization which follows nutrient depletion, particularly nitrogen, as the organic matter is degraded (Richards, 1987). The fungi are well adapted to low nitrogen substrates and are thus more important in the decomposition of plant remains low in nitrogen.

This succession in decomposition continually recurs in the soil. Large numbers of bacteria and fungi will be found where organic matter has recently been degraded, so the frequency of organic inputs governs the speed of decomposition. So does the distribution of organic matter, and the rate at which it is colonized: the soil is a highly heterogeneous habitat and bacterial populations can be highly localized. Thus, the rate at which a soil can degrade organic matter depends upon its recent history of exposure to that substrate.

Moisture, temperature and oxygen levels are the three principal abiotic factors governing the rate of decomposition. Generally, the cooler, wet soils of the temperate regions tend to accumulate more organic matter (Paul, 1989), whereas decomposition proceeds faster in the tropics with their higher average temperatures. Following clearance and the consequent release of nutrients from the soil, the increased productivity of a temperate forest can last several years. In contrast, this can be just one or two years in the tropics, due in large part to their higher rate of biological

Table 7.3 Functions of organic matter in the soil

Component	Function	Availability of carbon
Partially decomposed organic matter	Nutrient capital Improves water-holding capacity Improves soil structure and stability Energy source for consumers	High to medium
Humus	Nutrient capital Buffers soil pH Combines with clay to buffer soil pH and raise cation-exchange capacity Improves soil structure Energy source	Low
Microbial community	Nutrient cycling and capital Improves soil structure Energy source for consumers	High
Plant roots and non-microbial community	Nutrient cycling and capital Improves soil structure and mixing Energy source for consumers	High

activity (Jordan, 1986). The decline in plant productivity also follows a decline in plant-available phosphorus: this combines with aluminium and iron to become largely insoluble. The time taken for recovery is actually a problem of scale: in the tropics, the surrounding forest and its root system appear unable to maintain soil pH and fertility with clearances over $500\,m^2$, and the organic matter and its buffering capacity are severely reduced (Jordan, 1986).

In any soil, a sufficient amount of organic material is needed to maintain a decomposer community and to maintain the supply of nutrients to the higher plants. For this reason, establishing a high organic content to a spoil is one of the main aims of most reclamation efforts (Table 7.3).

7.1.4 Nutrient availability and succession

The speed of decomposition and level of biological activity are important determinants of soil quality. The cycling of two major nutrients, carbon and nitrogen, govern both processes. These are closely linked – carbon is important as the principal source of energy for the soil community and this energy drives the nitrogen cycle (Figure 7.5).

Together, carbon, phosphorus and nitrogen represent the most important plant nutrients and an established soil will have a significant

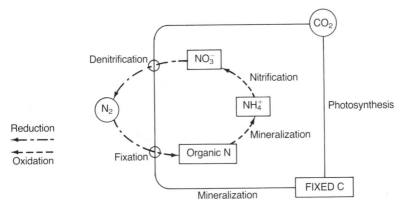

Figure 7.5 The interaction of the carbon and nitrogen cycles in the soil. The carbon cycle (shown by the solid line) provides the energy-rich compounds necessary to drive the nitrogen cycle through fixation and denitrification. (After Tate and Klein, 1985.)

capital of these minerals locked in its organic components. It will also have the microbial associations that make nitrogen and phosphorus available to the higher plants (section 7.3). All are readily lost from the soil community – carbon through respiration, phosphorus immobilized as insoluble organic compounds, and the mineralized nitrogen not absorbed by the biota being lost through leaching and volatilization (Table 7.4). The balance between phosphorus, nitrogen and the amount of organic matter in the soil is an important factor in the development of the plant community.

The degradability of organic matter depends upon the ratio of carbon to nitrogen (C:N) – microorganisms growing on this substrate need both elements to support their protein synthesis and population growth. For these decomposer bacteria, the C:N ratio in the organic matter needs to lie between 20:1 and 30:1, typically 25:1 (Richards, 1987). Fungi, on the other hand, are more efficient in their use of nitrates, and dominate soils with low levels of available nitrogen. The C:N ratio of the soil also governs the competition between higher plants and microorganisms for nitrogen: adding organic matter with a ratio above 30:1 to temperate soils will cause most nitrogen to be immobilized in the microbial community, leaving little available for plant uptake. At a lower ratio, below around 20:1, nitrogen begins to be in excess (carbon is now limiting microbial growth) and, if not taken up by plants, will be leached from the system.

In most soils, around 95–99% of nitrogen is held in the organic matter so that the size of their nitrogen pool is governed by its organic content. The rate at which nitrogen is released from this component depends on its rate of decomposition – high rates of turnover mean that little nitrogen accumulates as capital in the soil (Reiners, 1981). Its release also depends

Table 7.4 Sources of nitrogen available to plants

Process	Reaction	Organism	Controlling factors
Additions to available nitrogen			
Biological N fixation	Splitting of nitrogen molecule and its reduction to ammonium $N_2 + 6e^- + 8H^+ \rightarrow 2NH_4^+$	Symbiotic (*Rhizobium*) and free-living microorganisms, e.g. *Azotobacter*, blue-green algae	Inhibited by O_2, NH_4^+, NO_3^- Uptake by plants and microorganisms
Mineralization			
1. Ammonification	Breakdown of organic N to NH_4^+	Heterotrophic microorganisms	Binding by organic matter and clays Uptake by plants and microorganisms Loss by volatilization
2. Nitrification	(a) Oxidation of ammonium to nitrite $2NH_4^+ + 3O_2 \rightarrow 2NO_2^- + 4H^+ + 2H_2O$	Autotrophic bacteria *Nitrosomonas*	Available oxygen – water-logging inhibits oxidation
	(b) Oxidation of nitrite to nitrate $2NO_2^- + O_2 \rightarrow 2NO_3^-$	*Nitrobacter*	Leaching Uptake by plants and microorganisms Temperature pH (optimum 6.6–8.4)
Losses of N from soil			
Denitrification			
1. Dissimilatory reduction	Reduction of nitrate to gaseous forms of N: (a) $NO_3^- + H_2 \rightarrow NO_2^- + H_2O$ (b) $2NO_2^- + 3H_2 \rightarrow N_2O + 3H_2O$ (c) $N_2O + H_2 \rightarrow N_2 + H_2O$	Mostly heterotrophic bacteria using NO_3^- as a hydrogen acceptor, e.g. *Bacillus*, *Pseudomonas*	Anaerobic conditions – water-logging promotes denitrification. Nitrate concentrations pH (optimum 6.0–8.0) Presence of organic matter Temperature
2. Assimilatory reduction	Reduction of nitrate to ammonium in the synthesis of amino acids and proteins	Green plants, bacteria, blue-green algae, fungi	Inhibited by NH_4^+

upon the degradability of the substrate, and intractable substrates can immobilize nitrogen for long periods. Overall therefore, the quality (C:N ratio) and quantity of organic matter will be significant factors governing the supply of nitrogen to higher plants.

Phosphorus is also known to limit carbon and nitrogen assimilation by the biota – there is some evidence that phosphates are the prime factor controlling carbon and nitrogen immobilization, and it may govern the accumulation of organic matter in soils. Large amounts of available phosphorus are required by plants in symbiotic relations with nitrogen-fixing bacteria, such as legumes, so the addition of nitrates to the soil also depends upon available phosphates. Like nitrogen, phosphorus is assimilated by plants in a mineral form, although a large proportion is bound as unavailable organic forms. Together with its tendency to form insoluble inorganic compounds, this leads to a decline in its availability as a succession proceeds and is one reason why late successional communities tend to favour stress-tolerant plants. As we shall see shortly, several long-lived species solve this problem using a symbiotic association with a fungus.

In most soils, the increase in carbon and nitrogen and decline of phosphorus as the succession proceeds will eventually limit rates of decomposition, nutrient release and nitrogen fixation (Reiners, 1981). The lack of available phosphorus is thus likely to be the main factor limiting productivity in the later successional stages. Vitousek and Walker (1987) suggest that many terrestrial successions are typified by an increasing nitrogen capital and a reducing capital of available phosphorus with successional time (Figure 7.6). When nitrogen is deficient in the early

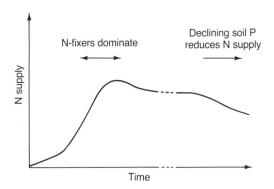

Figure 7.6 A hypothetical pattern for the supply of nitrogen as the soil develops during a primary succession. Nitrogen supply rises as fixation becomes established and while phosphorus is not limiting. Later the rate of fixation declines as phosphorus becomes biologically unavailable. (With permission from Vitousek and Walker, 1987.)

stages, the community will be dominated by legumes, able to flourish with the abundant phosphates. Their additions to the nitrogen capital of the soil are known to facilitate colonization by other plants, especially in reclaimed soils. In natural successions, woodlands become established when the nitrogen capital has reached around 400–1200 kg.ha^{-1}, and this probably represents a reasonable target for reclamation of a sustainable ecosystem on most wastes (Marrs and Bradshaw, 1982).

A succession may also be governed by the loss of nitrogen. **Nitrification** leads to the production of highly soluble nitrite and nitrate, which are readily leached from wet soils (Table 7.4). Several authors have suggested that nitrification is low or absent in soils beneath climax plant communities because of the nature of the leaf litter produced by the dominant plants. The tannins, phenols and other secondary compounds found in the leaves of the K-selected plants will strongly inhibit nitrifying bacteria. This has been offered as evidence that higher plants conserve nitrogen within the soil through the chemistry of their litter, but others suggest that nitrification may be equally limited by the shortage of phosphates or of ammonium (Richards, 1987). Either way, some nitrification can be shown to occur in late successional soils.

A greater loss of nitrogen results from disturbance of the soil. A range of disruptions are known to cause significant losses of nitrate, particularly where this increases the aeration of the soil (Table 7.4). A general model of the changes in nitrogen fluxes through primary and secondary successions has been proposed by Reiners (1981). As a succession proceeds, the biomass of the higher plants increases up to an equilibrium (Figure 7.7), after which additional nitrogen becomes fixed in the organic matter the soil. The mineralization of this detritus increases ammonia production, and a corresponding rise in nitrate production. These processes become balanced at climax. With a disturbance, such as a tree fall, a pulse of carbon and nitrogen arrives at the soil and both ammonium and nitrate production increase.

It is unrealistic to think that a succession can be reduced to a small number of causative factors. Interactions between species, such as facilitation and inhibition, may determine which plants colonize a site and in which order, but the availability of nutrients and space are themselves the product of other interactions between the soil, its community and the plant community. Chance and history will also have determined the sequential development of the community, and the initial floristic composition of a site will be crucial to its subsequent development.

A lack of predictability in a sequence of arrivals is no indication that there are no rules of assembly governing the formation of the community. But with so many interacting factors, and the additional elements of environmental variation and chance, we should not be surprised by the

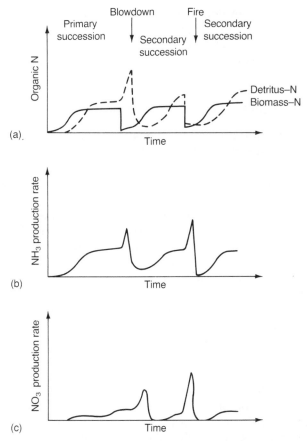

Figure 7.7 Reiners' model for the change in organic and mineral nitrogen during a primary succession, and following disturbances, in two secondary successions. In (a) the nitrogen in the detritus of the soil builds up after an equilibrium is reached in the standing vegetation (biomass). This itself reaches an equilibrium as input matches losses, and some fraction is mineralized to ammonia (b). Ammonia not absorbed by plants is oxidized to nitrate (c). Along with ammonia, any nitrate not taken up by plants will be lost from the system. A fall of trees or a fire adds nitrogen to the detritus, increasing ammonia and nitrate production, some of which will become fixed again in the succession that follows. The fire serves to mineralize the organic nitrogen more rapidly, hence the larger pulse of ammonia and nitrate. (With permission from Reiners, 1981.)

differences found between communities. Multivariate analysis, considering a number of related variables simultaneously, may be the only way in which these factors can be modelled in a sensible manner (Williamson, 1987). Although reclamation procedures largely concentrate on accelerating soil development, there is massive scope for understanding the organization of communities under experimental conditions (Harper, 1987). The variety of degraded soils, their location and area offer the prospect of experimenting with community-level phenomena on a scale very much larger than theoretical ecologists are used to.

7.2 THE PRINCIPLES OF RECLAMATION AND RESTORATION

The success of **restoration** and **reclamation** is judged by their ability to establish a self-sustaining plant community – in the first case, by attempting to re-create the original ecosystem, and in the second, by creating a community that meets the expected end-use of the site. Three ecological processes are needed for the community to maintain itself – primary production, decomposition and nutrient cycling (Visser, 1985).

In a secondary succession, some of these processes may already be in place. A soil with organic matter will have an established microbial and animal community, if somewhat depleted. Strip mining and similar operations can be arranged so that the displaced topsoil is replaced sequentially (Bradshaw and Chadwick, 1980) and a secondary succession, and restoration, will rapidly follow. Conversely, mining and quarrying wastes are effectively 'skeletal' soils, lacking both organic material or any significant life. Reclaiming this material will represent a primary succession, requiring that both nutrients and organic matter are accumulated to facilitate the arrival of higher plants.

Establishing a community needs to have clear objectives and a realistic appraisal of what is possible. These decisions may well have an economic or social dimension, particularly in urban areas, Amongst the factors that need to be considered, we need to include the following:

1. Our capacity to restore it: is there a realistic prospect of re-creating the original ecosystem, or are there physical or chemical constraints (for example, the presence of toxic materials) which prevent this? The site may be better used as a different habitat (old gravel pits are commonly turned into small lakes and ponds), or converted to productive agricultural use.
2. The financial constraints on the reclamation process: allied to this is the availability of other resources, including soils, water, litter or plants that may be used in the reclamation.
3. The local needs, ecological, social, or economical: we may choose not to reclaim a site because a valued plant or animal community has established itself. Alternatively, derelict land is often concentrated in areas of great social need where land for recreation or open spaces are preferred (Figure 7.8).
4. The pollution insult to surrounding areas: a site losing toxic runoff to local water courses, or material blown onto surrounding land may

Figure 7.8 A reclaimed car park in central London. This site, once formed of compacted rubble and a grossly polluted soil has been restored through landscaping and the use of imported topsoil to produce a much-needed green space and play area. The vegetation has flourished on the site for over five years.

require urgent treatment to reduce its impact. Unstable slopes or other hazards may similarly require attention. The impact of the reclamation process on local ecosystems and wildlife also needs to be considered.

Reclamation is an attempt to compress the time scale of a primary succession, rapidly establishing a nutrient base and soil structure in which a plant community can develop. In the first instance this requires a full appraisal of the physical and chemical characteristics of a waste, and the scope for ameliorating the unfavourable conditions of the site.

7.3 RECLAMATION

Primary successions in natural ecosystems can happen very rapidly. The skeletal soil of the moraines left by retreating glaciers will acquire both a

Table 7.5 Physical treatment of a skeletal soil

Feature	Problems	Treatment
Landscape	Stability of slopes	Rounding, flattening, compaction (prevention of underground combustion)
	Incongruent landform	'Blending' with natural landscape
	Hazards	Filling in shafts and pits Burying toxic wastes
	Drainage	Lagooning and treatment of runoff may be necessary Incorporating with natural drainage pattern
	Erosion	Landscaping of slopes and drainage, use of mulches and stabilizers to hold soil together
Soil	Consolidation and compaction	Ripping and scarifying of surface to break up surface crusts and open up soil. Vegetation and earthworms also help to improve porosity
	Particle size distribution	Incorporation of sludges or other wastes to improve range of particle sizes. May include modification of the soil chemistry in the same operation
	Surface drought	Incorporate organic matter to increase water holding capacity, especially mulches, sludges and manure
	Waterlogging	Improve drainage and rip surface
	High soil temperatures	Provide vegetation cover

soil and plant community in around 30–70 years and some wastes are colonized at roughly the same rate (Bradshaw, 1987b). However, a natural succession may be slowed not only by the prevailing harsh conditions but also by the rates of migration to the site. For example, plants tolerant of extreme soil pHs may need to travel long distances to colonize a waste (Bradshaw, 1987a).

The nature of the soil is the single most important factor in the reclamation of a site. Many spoils derived from mining and quarrying have poor physical and chemical characteristics which prevent rapid colonization by plants. This may be overcome by applying a topsoil that can immediately support a plant community, although this is usually an expensive solution. More often we initiate a succession on a skeletal soil. Typically, this has a poor range of particle sizes, as well little or no organic matter. This material includes a large variety of toxic wastes from the chemical industry, as well as the inert sands of a spent agricultural soil. Here we confine our attention largely to the problems of reclaiming colliery spoils, one of the commonest forms of wasteland in the industrialized world.

7.3.1 Physical problems of spoils

A soil has to provide anchorage, water and nutrients for a plant and all three are determined by the soil structure and texture. Following the landscaping and engineering of a site, and a period of settlement and weathering, work starts on improving the physical properties of the spoil (Table 7.5, Figure 7.9).

Any compaction of the spoil will require the surface crust to be broken open, by scarifying the surface or by ripping to a greater depth. Large rocks may need to be crushed or removed. The distribution of particle sizes is critical for the formation of a soil: spoils with predominantly large particles drain too freely, whereas very fine material compacts and waterlogs. The soil needs a particle size distribution that creates pore sizes which hold sufficient water for plant needs without becoming anaerobic. This can be partially achieved by mixing or crushing wastes. However, the water-holding capacity is most significantly increased by adding organic matter (Table 7.3). This improves the soil texture by causing fine particles to adhere to each other, and by encouraging biological activity. Organic matter aids the colonization of earthworms by buffering the pH and binding toxic metals. Earthworm activity increases the porosity and mixing of the soil, bringing substantial amounts of fine material to the surface each year (Bradshaw, 1991).

Sewage sludge, comminuted domestic refuse and farmyard manure have all been used to add organic matter to spoils. Peat and sewage sludge have the highest bacterial and fungal counts, and peat also contains large numbers of actinomycetes, an important part of the decomposer community (Visser, 1985). These different treatments add different amounts of nutrients, and they also differ in their ability to raise the water-holding capacity of the spoil (Figure 7.10). The same effect may also be produced using artificial soil stabilizers, gels that bind soil particles together. These have be used as a temporary measure to prevent the erosion of wastes with high levels of toxic metals, although their success over the long term is limited and under some conditions, they are known to limit germination (Bradshaw and Chadwick, 1980).

Figure 7.9 Reclaimed spoil heaps left over from limestone quarrying in the Peak District National Park, Derbyshire, England. Not only has the grassland community been restored, but the heaps have been shaped to follow the natural contours of this karst landscape.

On a smaller scale, however, stabilizers can aid seedling survival in the most unpromising of soils. Callaghan and his co-workers (1988) have examined the use of two synthetic polymers to aid seedling establishment in the arid soils of the Sudan (Figure 7.11). These are thin, sandy soils, with little organic matter to hold water during the period when the plant is establishing itself and producing roots. Not only do the polymers absorb water and buffer its supply to the plant, they can also bind plant nutrients. These chemicals are too expensive to use broadcast, but they can be used to improve the survival of single trees. Species of *Acacia* and *Eucalyptus* can be established with this technique, trees that will improve the local soil and provide essential fuel for the local people. These gels help to reduce the irrigation needs for re-establishing woodlands, and allow more marginal land to be used for this purpose (Callaghan *et al.*, 1989).

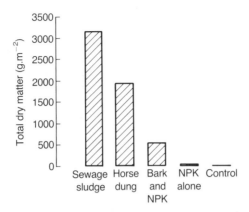

Figure 7.10 The minimum total yield of sown grasses with different amendments of organic material on iron mine spoil. The sewage sludge and horse dung were not supplemented with artificial fertilizer, yet have the highest productivity. The dry matter yield was correlated with the amount of phosphorus and nitrate in the soil. (After Borgegård and Rykin, 1989.)

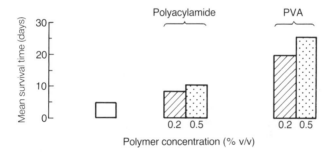

Figure 7.11 The mean period of survival (over 45 days) of *Acacia senegal* seedlings in a soil composed of Sudanese sand and Nile silt with and without the addition of a synthetic soil stabilizer. Each of two polymers were applied at 0.2 and 0.5% (v/v). The unfilled box shows the survival of the seedlings in the soil with no addition of a stabilizer. The polymers buffer the supply of water to the plants and, under a regime of the greatest water shortage, polyvinylalcohol (PVA) at 0.5% increased the survival of the seedlings five-fold. (After Callaghan *et al.*, 1988.)

7.3.2 Chemical characteristics of spoils

Chemical barriers to the colonization of a colliery spoil can take one of two forms – a deficiency in plant nutrients or an abundance of toxic materials (Table 7.6). Both can serve as an effective selection mechanism, allowing only tolerant species to colonize. Amelioration of these problems frequently goes hand in hand with the efforts at establishing a soil structure. Plant growth can be initiated in most spoils by the addition of

Table 7.6 Chemical and biological treatment of skeletal soils

Feature	Problems	Treatment
Toxicity	pH	*Too low* – ripping to increase leaching and liming in excess to mop up existing and latent acidity *Too high* – addition of organic matter Use of tolerant plants; leaching Mixing of acidic and basic wastes
	Salinity	Leaching during settlement period Use of tolerant plants
	Toxic metals	Addition of organic matter Inert barrier between waste and a topsoil Use of tolerant plants
Plant nutrition	Shortage of N, P	Organic and artificial fertilizers for initial treatment Use of leguminous plant for N-fixation, to create N capital in organic matter Keeping pH low prevents nitrification and loss of nitrate to leachates Keeping pH close to neutrality prevents fixation of P. Adding Ca and P or other wastes (basic slag) Encourage mycorrhizal associations

artificial or organic fertilizers, but sustaining this productivity without repeat applications requires that the organic fraction of the soil, and its nutrient capital, is built up.

Nitrogen

Plants take up nitrogen both as ammonium (NH_4^+) and as nitrate (NO_3^-) although the nitrate has to be reduced to ammonium in the plant to be incorporated into amino acids. The positive charge of ammonium causes it to be bound by clay and organic material in the soil which slows its release and its rate of oxidation to nitrate (Figure 7.5). Increasing the organic content of a waste will thus slow the loss of nitrogen from a spoil.

Inorganic fertilizers are often part of the initial treatment of a waste, when the aim is to quickly establish plant cover; but, in freely drained sites, nitrates are rapidly lost with the leachate from the soil. Highly alkaline wastes can also lead to rapid loss of ammonium through volatilization (Table 7.4). Farmyard manure, sewage sludge or other organic materials will slow this loss and improve the soil texture, aiding the formation of a decomposer community.

In natural soils, **legumes** represent the most significant source of additional nitrogen. These comprise a group of plants able to form a mutualistic association with the nitrogen-fixing bacterium *Rhizobium*. The bacteria infect the roots of the plant, which provides protection and a carbon source for its partner. Both the host and the *Rhizobium* will avoid the cost of nitrogen fixation if there is abundant available nitrogen in the soil, and so both ammonium and nitrate will inhibit fixation (Table 7.4). Otherwise, this association is the most significant source of nitrogen for the soil: most taken up by non-leguminous plants comes from the death of nodules formed when the bacteria infect the legume, although plants growing close to legumes benefit locally from fixation. On the thin sand left from china clay extraction, the most prolific of nitrogen fixers, the lupin (*Lupinus perennis*), can accumulate up to 295 kg.ha^{-1} of nitrogen per year within two years of being sown, about 76 kg of which becomes available to the companion grasses (Jefferies *et al.*, 1981). Over a range of other wastes, other legumes can fix around 100 kg.ha^{-1} per year (Bradshaw, 1983) and may thus remove the need for additional fertilizers.

The association of *Rhizobium* is essential if legumes are to be established on a waste, and inoculating a site with the bacteria will aid infection and nodulation. The species of *Rhizobium* is often highly specific to its host and the wrong species of the bacterium colonizing a plant will lead to ineffective nodulation, with no nitrogen fixation taking place (Day and Lisansky, 1987). This one of the major reasons for the failure of clover in mixed swards. There are, however, several commercial strains of *Rhizobium* which are able to form successful associations with a number of species of hosts. Inoculation usually applies *Rhizobium* in a fine peat, buffered to pH 6.8 with lime. Alternatively, it can be used applied as a seed dressing (Day and Lisansky 1987).

The variety of legumes provides a range of options for initial sowings, although the choice of plant will depend upon the availability of phosphates, soil pH, and whether or not the soil is freely drained (Jefferies *et al.*, 1981). A number of tree species, including *Alnus*, are able to fix nitrogen by virtue of their association with the actinomycete *Frankia*, which also forms nitrogen-fixing root nodules (Paul, 1989). This allows young trees to be planted in the early stages of a reclamation. In combination with grasses, legumes provide rapid cover, stabilizing the surface from erosion and initiating the development of a soil. Grasses can, however, inhibit the establishment of trees and legumes. For example, treating a metalliferous spoil with sewage sludge and burnt colliery spoil allowed grasses to thrive in the high-nitrate soil created, at the expense of the legumes *Trifolium* and *Lotus* (Johnson *et al.*, 1977).

It is important that the fixation of nitrogen should go hand in hand with the production of biomass, where the majority of the nitrogen is held. Given its rate of release from the organic component of a temperate soil,

Bradshaw (1983) shows how an annual requirement for 100 kg.ha^{-1} of nitrogen would need a soil capital of 1600 kg. At these levels, decomposition and nutrient release processes would ensure that the system would be self-sustaining. As a succession proceeds and organic matter accumulates, so this capital might rise (Figure 7.7).

The partitioning of nitrogen in the plant community appears to differ between reclaimed and natural soils, at least in the early stages of the succession. A larger proportion of the total nitrogen was found in the roots of plants growing on china clay wastes, rather than the soil, an indication perhaps that nutrient cycling was not yet mimicking natural soil processes (Marrs and Bradshaw, 1982). Further treatment would then be required to build up the nitrogen capital of the soil. Most sites require monitoring for some time after the initial plant community has become established, and may require subsequent applications of organic matter and artificial fertilizers to increase the nitrogen capital. Re-seeding of legumes might also be required.

Phosphates

The succession of a community from a skeletal soil broadly follows the accumulation of nitrogen, with late successional stages dominated by plants able to survive with low available phosphorus (Figure 7.6). Phosphates may be abundant in the early stages of a succession before

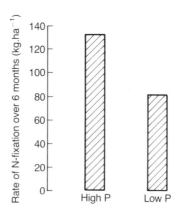

Figure 7.12 The effect of phosphates on nitrogen fixation by clover (*Trifolium repens*) on colliery spoil (Thorne, South Yorkshire) over the summer months. Plots were supplemented with phosphates at either 28.0 or 98.4 kg.ha^{-1}. Adding phosphates increased both nitrogen fixation and the growth of the legumes and the companion grasses. Other factors affecting the rate of fixation included spoil moisture and temperature, and the amount of light. (After Palmer and Iveson, 1983.)

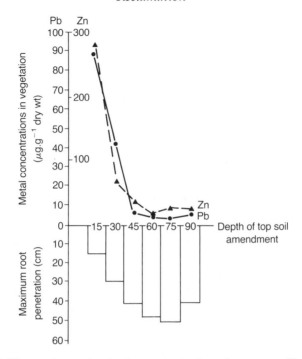

Figure 7.13 The maximum depth of root penetration of grasses and legumes after growing for six months on a lead/zinc spoil (Minera, North Wales) amended with topsoil to six depths. These are shown along the horizontal axis, from 15 to 90 cm. Root depth reflects the concentrations of toxic metals to which the plants were exposed, and the levels accumulated in the plants. Notice that larger soil depths are needed to provide an effective barrier to the metals in the spoil, and to prevent the topsoil itself becoming poisoned by the metals. (After Johnson *et al.*, 1977.)

they form insoluble complexes with organic components in the developing soil, whereas the soluble forms of nitrogen are readily lost without any organic matter to bind them. Phosphates are important for the early plant community: the rate of nitrogen fixation by legumes on colliery spoils is determined by the availability of phosphorus (Palmer and Iverson, 1983; Figure 7.12). Similarly, the nodulation of legumes was shown to be successful on amended metalliferous wastes, but the development of the clover was hindered by a lack of phosphorus (Johnson *et al.*, 1977).

Many wastes have abundant phosphate available for plant uptake, although phosphorus is largely unavailable in both acid and alkaline spoils. Such wastes require artificial fertilizers from the outset, or mixing with other wastes that can supply phosphorus. As the soil develops, plant litter adds phosphates to the soil, but the cycle of release and uptake causes a reduction in available phosphorus at each stage, and roots have to forage further in the soil to maintain their supply. This becomes especially important in sites rich in toxic metals, where non-tolerant plants may have limited root development and therefore are unable to meet their phosphorus requirement (Figure 7.13).

This demand for phosphorus has produced another example of a mutualist association in the roots of higher plants. In this case, a fungal association or **mycorrhiza** forms when the root cells of the plant are invaded by fungal hyphae. This extends the effective absorptive surface of the root and increases its rate of foraging for phosphorus. The fungus probably also aids ammonium, zinc and copper absorption by its host. Many woody plants will not establish themselves on mine wastes without this association: all the trees found to survive on colliery waste in Pennsylvania, including *Pinus*, *Betula*, *Populus* and *Salix* were shown to have mycorrhizae (Danielson, 1985). Mycorrhizal associations confer other benefits: they aid drought resistance in some plants and pine trees are known to be more tolerant of high soil temperatures when infected. They can also confer protection against plant pathogens: the capacity to produce antibiotics may be a common feature of those basiodiomycetes able to form mycorrhizae (Richards, 1987). Heathers and ericas are protected from the available toxic metals in acidic soils by the ability of their mycorrhizae to precipitate these metals on the fungal hyphae. In contrast, grasses are made more sensitive to copper and nickel due to the presence of mycorrhizae (Danielson, 1985).

The type of mycorrhizae depends on the host plant. Early ruderal plants may not require mycorrhizae, probably because they occupy open ground with abundant nutrients and little competition (Miller, 1987). In a secondary succession, the domination of ruderals can lead to a decline in the number and diversity of fungal propagules and this inhibits further development of the plant community. The community then has to rely on wind-dispersed seeds and propagules intercepted by the established shrubs to facilitate the development of mycorrhizae in a waste (Miller, 1987). As with *Rhizobium*, spoils may need to be infected with a range of fungal species to aid the formation of mycorrhizal associations in higher plants. Adding a topsoil can frequently supply both symbionts (Danielson, 1985). The growth of grass is improved with mycorrhizae, and legumes flourish when they have established both associations. The fungal component of the association significantly improves the soil texture, essential if the end-use is agricultural.

Other plant nutrients, including potassium, are required in much smaller amounts by most species, and demand can easily be met by an initial application of inorganic fertilizers. Potassium capital is established quickly as the organic content of the soil rises. Calcium is also needed by legumes such as clover (*Trifolium*) and a lack of this metal can hinder their establishment on an acidic waste.

Toxicity

Many collieries produce a spoil with considerable amounts of iron pyrite (FeS_2). This was formed by bacterial action in the anaerobic conditions under which the coal measures were laid down. On exposure to the atmosphere this is oxidized, again partially mediated by bacteria, to produce sulphuric acid (Gemmell, 1977).

Thiobacillus ferrooxidans oxidizes ferrous iron and sulphur leading to the production of sulphuric acid. This can cause indirect leaching of other metals: the production of the ferric ion serves as a powerful oxidizing agent which increases the solubility of various metals (Francis, 1985).

The extent to which the pyrite in a spoil has been oxidized depends upon the length of time it has been exposed to the atmosphere. In treating a waste, we have to buffer the acidity that has already developed, and the latent acidity – the acidity that will subsequently develop. A soil lacking any carbonate minerals will be unable to buffer this acidity and will suffer poor plant establishment, rapid loss of nutrients and a high availability of toxic metals.

Ripping a spoil increases its exposure, accelerating its weathering during the period of settlement and increasing the rate of leaching (Table 7.6). During this time, the drainage from the site may have to be collected and treated to prevent contamination of local water courses. Other wastes with high pH can be mixed with an acidic spoil although more usually, lime is applied (Figure 7.14). Up to 150 000 kg of $CaCO_3$ per hectare may be needed on highly pyritic spoils to mop up the latent acidity (Bradshaw, 1983). This has to be applied with the initial engineering of the site, as the surface is opened up, to incorporate the limestone below the surface.

In spoils that are excessively alkaline, the oxidation of sulphur compounds including gypsum (calcium sulphate) is encouraged to lower the pH. This is often the case in many arid regions: in the reclamation of alkaline soils spoilt by irrigation, sulphur-oxidizing bacteria, *Thiobacillus*, have been shown to significantly lower the pH when applied in combination with sulphur additions (Rupela and Tauro, 1973). A comparable technique has been used to mobilize and remove metals from digested sewage sludge (section 9.3).

Colliery spoils and semi-arid soils can also suffer a problem of salinity – with rapid evaporation, water predominantly moves up the soil profile and salts become concentrated at the surface. In temperate regions, this salinity can be leached away by watering, although without vegetation cover, all spoils can reach very high temperatures in the summer, promoting evaporation. Leaching may be possible on some semi-arid

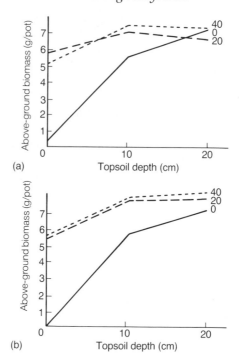

Figure 7.14 The effects of (a) fly ash and (b) lime amendments (g.kg^{-1}) on above-ground yields of barley (*Hordeum vulgare*) grown in pots of a pyritic colliery spoil (pH 2.7) from Glenrock, Wyoming, USA. Each amendment was applied with three levels of topsoil additions. All levels of of fly ash or lime amendments increased yields in the absence of topsoil, but these differences disappeared as topsoil depth increased. (After Taylor and Schuman, 1988.)

soils, although lowering the pH or displacing sodium by calcium using gypsum will also lower salinity.

Toxic metals are a feature of many wastes and will be more readily available to plants if the spoil is acidic. The organic matter of all soils is an important binding site for these metals, lowering their impact on the higher plants. Adding organic matter can thus reduce available toxic metals, although some treatments, particularly the addition of sewage sludge, may themselves be a significant source of toxic metals. Also some proportion of the metals will become available as the organic matter decomposes. Adding clay will provide a more permanent binding site for the metals. Alternatively, liming the waste reduces their availability by increasing the pH and encouraging the metals to form insoluble compounds.

Microorganisms also mediate the availability of metals, and a large microbial biomass can buffer the supply of metals to the plant community. Heterotrophic bacteria and fungi are able to bind metals to their cell walls (section 3.5), or increase their solubility by chelation with organic acids. These acids are released from the decay of the litter layer and will thus increase the mobility of certain metals. Such mobilization is important in the supply of major nutrients to both microorganisms and to higher plants, but can become a hazard if large concentrations of toxic metals are present.

One way to side-step some of these toxicity problems is to sow seed mixtures of metal-tolerant ecotypes (section 3.4). Although this limits the range of plant communities that can be created, at high metal concentrations it may be the only answer, short of applying a topsoil barrier. Even then, there is the danger of this expensive treatment becoming soured by the migration of metals up the soil profile, and into the rooting zone (Figure 7.13). Alternatively, a highly toxic waste may be isolated by covering it with a different, inert waste, which can then be reclaimed in the normal way (Bradshaw and Chadwick, 1980). Indeed, this is an appropriate treatment for a range of wastes, particularly from the smelting and chemical industries, where a variety of metals and poisons occur in the spoil.

7.3.3 Species interactions and ecological processes

Two species assemblages are crucial to the establishment of a functional community – those of the decomposer community and of the mutualist associations essential for the capture of nitrogen and phosphorus.

Without an efficient decomposer community, dead vegetation will form a mat on the spoil surface, which locks up nutrients and prevents the germination of seeds. The process of decomposition has a succession of its own, based on the nature of the carbon source: we have already seen how the release of nitrogen depends upon the proportion of carbon and its immobilization in the microbial biomass.

Early in the reclamation process, grasses and fast-growing herbaceous plants are useful for adding carbon and organic matter to the soil (Visser, 1985). Legumes such as clover enhance the nitrogen content and so aid the development of the microbial community, although these decompose more rapidly than grass and valuable nitrogen can then be lost through leaching. The rate of release from organic matter determines the amount of nitrogen capital that has to be established in a waste (Marrs and Bradshaw, 1982). In temperate soils, nitrogen is released at around 2% per annum from the organic matter: if the organic pool is small, this will not represent sufficient to match plant needs. If, however, nitrogen is released more rapidly, the soil can support a plant community with a smaller capital (Marrs and Bradshaw, 1982).

The rate of decomposition of the organic matter is also linked to the

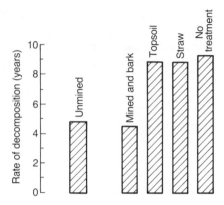

Figure 7.15 The estimated rate of decomposition of straw (measured as turnover time for complete degradation) buried in litter bags in strip-mined colliery spoils in the semi-arid conditions of New Mexico. One of three organic amendments was applied to the spoil, a control was given no addition and these were all compared with a natural soil. High numbers of soil microfauna were associated with the unmined soil and the spoil with the bark addition (especially oribatid mites, springtails and nematodes), and their grazing on the decomposer bacteria and fungi has enhanced the decomposition process. (After Elkins *et al.*, 1984.)

activity of the soil microfauna, especially those species that graze on decomposer bacteria and fungi. There is evidence that the fastest rates of decomposition are found in soils with abundant mites and other invertebrates that feed on the saprotrophs (Figure 7.15). These grazers can be effective regulators of decomposition on reclaimed strip-mine wastes (Elkins *et al.*, 1984). Such trophic interactions will thus control the availability of nitrogen – invertebrate consumption of bacteria and fungal hyphae increases the inorganic nitrogen in soil microcosm experiments (Ingham *et al.*, 1986a). On the other hand, feeding pressure by nematodes can significantly reduce the fungal colonizations in mycorrhizae (Ingham *et al.*, 1986b).

Grazing by both vertebrates and invertebrates will also affect these turnover rates. Large vertebrates accelerate decomposition by reducing the amount of standing dead vegetation, and by adding faeces and urea to the soil. The loss of nitrogen from the soil via this route may be relatively small – sheep only remove 5% (Marrs and Bradshaw, 1982) and grazing can be an effective method of narrowing the C:N ratio of the soil (Schafer *et al.*, 1980). Faeces add comminuted plant material that is already infected with microbial and fungal decomposers, and this fine-textured product is readily incorporated into the soil.

The invertebrates are particularly important for incorporating faeces and plant litter into a spoil. They can also be shown to undergo a succession of their own, following changes in the chemistry and structure of the spoil. Dunger (1989) describes a succession of springtails (*Collembola*) followed by earthworms as a plant community develops on reclaimed colliery tips in Germany. Springtails were also found to be important during the first three years of the reclamation of brown coal wastes when earthworms are absent (Abbott, 1989).

The species of earthworm varies with the quantity and nature of organic matter, as well as with a number of abiotic factors, especially soil pH. One feature of this succession is the switch from *r*-selective earthworms, such as the surface living *Dendrobaena octaedra* on the harsh soils of earlier colliery spoil heaps, to be replaced by deep-burrowing species such as the *K*-selective *Lumbricus terrestris* later on (Ma and Eijsackers, 1989). Earthworms thus extend the decomposition process gradually down the soil profile as the succession proceeds. They further affect the decomposition process by shifting decomposition from being fungal-based on most colliery tips to bacteria-based (Dunger, 1989), presumably as nitrogen becomes more available and acidity declines.

The arrival of herbivores with the development of the plant community can bring problems. A field of legumes represents a massive opportunity for population growth by aphids; the rate of nitrogen fixation by *Rhizobium* may be reduced by 86% by aphids attacking peas (*Pisum sativum*), reducing the carbohydrates that the plant can make available to its symbiont (Urbanek, 1989). The black locust tree (*Robinia pseudoacacia*) is commonly used in the USA to reclaim wastes because of its nitrogen fixing capacity, but significant tree mortality occurs with the arrival of a cerambycid beetle, *Megacyllene robiniae*, whose larvae bore into the tree and slow its establishment (Urbanek, 1989). The theft of seed from reclamation sites by ants has caused the failure of several reclamation programmes in Australia (Majer, 1989b). Other forms of invertebrate herbivory and grazing by vertebrates (especially rabbits) can seriously hinder the success of initial plantings. The weight of large animals can also compact a waste, hindering the germination of seed. Nevertheless, animals can play an important role in accelerating the decomposition process, in pollination, seed dispersal, and for maintaining an open community not dominated by a small number of species (Whelan, 1989).

The upper part of a soil profile may develop relatively quickly, taking just five years on a treated alkaline colliery spoil, although it requires much longer for activity to develop to the depth of a natural soil (Schafer *et al.*, 1980). The mutualist associations between fungi, bacteria, actinomycetes and higher plants are essential for the long-term integrity of the plant community, as are the interactions between decomposers and consumers in regulating the breakdown of organic matter and nutrient cycling. Although the emphasis in reclamation is largely on establishing these processes, the structure of the community that follows is often dictated by end use. In effect we have accelerated a primary succession in its earliest stages, when this process is most important, but there is much that the applied and theoretical ecologist can learn about how communities are constructed from the species interactions that follow these early stages.

7.4 RESTORATION AND SOIL EROSION

The agricultural landscape represents the largest area of modified habitat across the globe. This comprises a range of ecosystems, each reflecting the impact of agricultural practices on soil structure and processes. Under intensive grazing pressure the soil loses its structure, fertility and water-holding capacity. These compacted soils have no large soil animals and a shallow soil profile; being less porous they are more easily waterlogged. On the other hand, a regularly turned arable soil, without the protection of plant cover or regular additions of organic matter, can lose its upper layers to wind and water erosion.

Soil erosion is a global problem, not confined to arid or semi-arid areas, although these soils are more susceptible than others. The moist temperate zones have young soils high in calcium and organic matter, and with varieties of clay that raise their water-holding capacity. Such soils can sustain high productivity as long as they receive fertilizer regularly (Parker, 1989). In arid regions, soils are more fragile and more readily degraded to a non-productive condition. They are older, marked by a lack of calcium and high concentrations of sodium, and dominated by non-expanding clays (kaolinite). This, together with their low organic content, limits their water-holding capacity. These soils 'hard-set' under cultivation, forming a compacted sub-soil with a shallow profile.

Agricultural practices that are insensitive to the local ecology will quickly lead to a loss of productive potential – as in the dry wheat belt of Western Australia (Parker, 1989) and the Sahel of Africa (Le Houérou and Gillet, 1986). Tillage and disturbance have to be reduced to a minimum – without rapid stabilization, the thin, sandy, upper soils are readily eroded and it requires something close to a primary succession to restore them.

Similarly, the clearance of moist tropical forest can lead to the thin organic layer being washed away, exposing the subsoil to the sun. This is typically acidic, with low concentrations of silica and high concentrations of iron and aluminium oxides. When baked this will form laterite, a soil with the texture of brick, and with the same agricultural potential. This process of **laterization** is found throughout the humid tropics (Ramade, 1984).

Virgin forest binds and stabilizes the soil, and the biomass of the community also holds most of the nutrients of the ecosystem. Removing the trees removes these nutrients, leading to a rapid reduction in the organic content and biological activity of the soil. The loss of soluble phosphorus particularly limits the productive life of a cleared soil to just 2–3 years. In large gaps, the absence of the primary plant nutrients inhibits forest regeneration and allows grasses to invade, producing a savanna-type community, thereafter maintained by ungulate grazing (Jordan, 1986). In the drier areas re-establishing the trees is one way in which a living soil can be restored: Le Houérou and Gillet (1986) suggest that woodlots need to be established on the best agricultural land in each village. This will supply the people with fuel and timber and also begin the restoration of the soil. Within five years, the mature trees could provide browsing for the animals that have traditionally dominated African and Indian agriculture.

This '**desertization**' or reduction of soils to this skeletal form, is a product of human impact, rather than any long-term effects of climate change (Le Houréou and Gillet, 1986). It is a feature of the soils under intensive agriculture in northwestern Europe, North America and other areas of intense agriculture, as well as the arid and semi-arid regions. For example, the loss of the upper soil profile through wind erosion is a particular problem in flat regions such as East Anglia or the prairies, when the soil is left without cover for extended periods. Elsewhere, soils on exposed slopes are easily eroded by water, often with large masses of material moving with disasterous consequences. Rapid slippage has led to dramatic loss of life in Colombia, while long-term soil erosion following deforestation in the Himalayas depletes their agricultural potential and silts up hydroelectric dams.

Where political and economic conditions allow, endangered soils can be taken out of production – the United States Conservation Program aimed to remove 18 million hectares of cropland from production by 1990, in areas where soil erosion was above a specified rate. The over-capacity in some Western agriculture, allied to the costs of subsidising its production, has led a number of countries to adopt strategies to take land out of cultivation. The European Community has a policy to remove 20 million hectares from production and to improve its conservation value by the end of the century. This represents an oppportunity to restore habitats lost on an massive scale this century – 40% of broadleaved woodland, 110 000 miles of hedgerow and 25% of UK semi-natural habitats have disappeared under cultivation during this period (Green, 1989). Now the British Government has plans to create 12 000 ha of broadleaf woodland each year (Department of the Environment, 1990) and as much as 3–4 million hectares of marginal British farmland may be 'set-aside' by the year 2000 (Green and Burnham, 1989).

7.5 THE RESTORATION OF SEMI-NATURAL HABITATS

A principal aim of the set-aside scheme in the UK is to extend and enhance existing semi-natural wildlife areas (Newbold, 1989). Hedge-rows, broadleaf plantations and partially improved grassland are import-ant for the structural diversity of their habitat and the range of wildlife they can support. In the UK landowners can receive payment for taking at least 20% of their arable land out of production, so sites have to be chosen that make best use of the funds allocated. Even though not all sites will be protected, their value as wildlife habitats is enhanced by reducing their fragmentation (section 6.2). This can be accompanied by a management strategy that maximizes the diversity of species in a site by preventing the succession proceeding to a steady state dominated by competitive species. As Morris (1991) points out, a climax community is largely dominated by autogenic change, but a semi-natural habitat represents a **plagioclimax**, a stable community maintained by the activity of man. Consequently these communities will need continued management to maintain them.

In contrast to reclamation, restoration guidelines for semi-natural habitats in both Europe and the USA are aimed at lowering the fertility of soils: ecosystems that are floristically diverse generally have nutrient-poor soils. Intensive agriculture increases the concentration of nitrogen and phosphorus beyond natural levels, a process of enrichment termed **eutrophication** (section 8.4). Left to themselves, eutrophic soils will quickly sport a collection of weeds, ruderals (Figure 7.4) able to rapidly exploit the abundant resources. The weeds of arable fields are derived from wind-blown seed or from those persisting in the seed bank of the soil, able to rapidly colonize the soil in the absence of competitors (Table 7.2).

If cultivation ceases, the frequency of disturbance is reduced and competitive species begin to dominate the plant community. Most management strategies include some element of disturbance to prevent these communities proceeding to any form of climax, when the diversity of plants would decline (Figure 7.1). For example, a programme of burning, mowing and the selective use of herbicides has been used to suppress regeneration of woodlands in Connecticut, enhancing the wildlife value of communities that have developed on abandoned agricultural land (Niering, 1987). The same principle applies to coastal plant communities, such as salt marshes. Gray (1991) describes how grazing prevents the community being dominated by tall vigorous plants: wildfowl prevent typical salt marsh plants (*Salicornia*) being replaced by perennial grasses in the Netherlands. The effect of grazing will depend upon the successional stage, but cattle can reverse a succession in these habitats, giving a higher diversity of plant species. These interactions can be important at all levels within the plant community – the grass *Puccinella maritima* tends to flower later on grazed marshes, an ecotype (section 3.2) that maintains distinct genetic differences from populations on ungrazed marshes (Gray, 1991).

Disturbance creates patches in which both resources and space are available for new species to colonize. Only species typical of an early succession may be able to colonize bare ground (Gross, 1987), and the vigorous growth of weeds can limit the arrival of other species. Without a suitable regeneration niche, some species may never establish themselves, effectively slowing down the succession. Woodland plants are often poor seeders, typically long-lived, slow-growing perennials. They produce few, large seeds – common British woodland flowers, such as *Oxalis*, *Mercuralis* and *Hyacynthoides* also produce fewer seeds when grown in the shade, with a lower germination rate (Buckley and Knight, 1989). Their exclusion from the commercial seed mixtures used in woodland restoration is probably due to their low germination rate and the long time (up to 7 years) before some of these plants flower (Buckley and Knight, 1989).

7.5.1 Plant species enrichment

One technique of producing greater diversity has been borrowed from agriculture. Slot-seeding is used by the farmer to introduce clover into permanent pasture to increase its nitrogen-fixing capacity, and the same technique has been used to raise the floristic diversity of grasslands (Wells, 1983). Seed are drilled at the same time as a herbicide is applied, reducing competition from the established vegetation and creating a more favourable regeneration niche.

Herbicides have also been used in the restoration of prairies to control exotic weeds such as white clover, *Melilotus alba* (Kline and Howell, 1987). Prairies are characterized by an abundance of herbaceous plants, with less than one tree per acre (Kline and Howell, 1987). A natural succession in prairies can be encouraged by a particular schedule of burning, based on the life history of white clover, although burnings that are too frequent will allow trees and shrubs to invade. A prairie can be restored relatively quickly, within 3–5 years, although this depends upon plot size – a small plot will be unable to withstand frequent burnings. Other disturbances help to maintain the plant community – mound-building ants are needed to create patches to allow colonization and to improve soil structure.

In lowland Britain, one of the most diverse floral communities is the grassland associated with the shallow, poor soils of chalk downland. This is a community rich in herbaceous plants, a product of continued grazing over several thousand years. Changes in farming practice this century

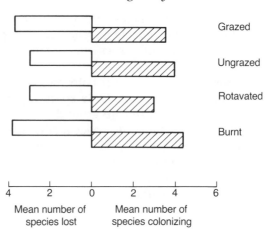

Figure 7.16 The rate of plant species colonization and loss from chalk grassland subject to four regimes of disturbance. These records are from the Aston Rowant nature reserve in the Chilterns, central England, between 1970 and 1982. Species turnover is higher in all the disturbed plots and greatest in the grazed plots. (After Ward and Jennings, 1990.)

have caused its decline and, as a result, chalk grassland is one target of the set-aside programme. It will develop spontaneously on abandoned arable land (Mitchley, 1988), but studies around Wytham Wood, near Oxford, have shown its dependence on seed migration from surrounding chalkland communities, rather than the seed bank in the soil. Two different strategies have been proposed to accelerate this secondary succession. Firstly, planting and the sowing of seed. This is expensive, but reliable and effective, even on nutrient-rich soils (Mitchley, 1988). Alternatively, a natural succession with active management can be used, including the removal of weeds. This is slower, less reliable, but cheaper. Either way, management in the form of grazing and disturbance is required to sustain the community.

The value of disturbance has been demonstrated for chalk grassland by Ward and Jennings (1990). Their study of several forms of disturbance show that sites grazed by vertebrates have the highest species turnover (Figure 7.16). A grassland that is allowed to form a continuous litter layer prevents colonization by other species, whereas grazing enhances colonization by trampling, creating gaps in the existing plant community. This has prevented an equilibrium plant community becoming established in the 14 years of this study (Ward and Jennings, 1990). Diversity is also increased by other disturbances (Figure 7.16). Where these habitats are fragmented, Morris (1991) suggests that management by grazing or some other form of disturbance should be rotated between a 'suite' of grassland sites. This is comparable, in some respects, to a mosaic of patches at different successional stages, with the whole community being at equilibrium (section 7.1).

Without management, the succession of plants on chalk grassland reflects the increased availability of nutrients facilitated by previous species, such as the leguminous shrub, gorse (*Ulex europaeus*) (Marrs and Gough, 1989). In a comparative study of semi-natural and agricultural soils, Marrs and Gough concluded that the major factor determining residual soil fertility was available phosphorus. This decreased as the succession proceeded to grassland, scrub and woodland. Although some sites showed a slight rise in the concentration of phosphorus with the development of the scrub community, the overall trend was down. Nitrogen was much more variable and any eutrophication was largely attributable to phosphorus.

Reducing the fertility of the soil can use two basic techniques – reducing the size of the nutrient capital in the soil or by binding nutrients in unavailable forms (Table 7.7). The first technique can be simply achieved by removing the topsoil (Berendse *et al.*, 1991), or by accelerating the rate of nutrient loss through leaching or by growing nutrient-demanding plants (Marrs and Gough, 1989). In one series of experiments, Berendse *et al.* (1991) compared the effect of hay removal, waterlogging and stripping of the topsoil as methods of lowering the nitrogen content of the soil and its impact on floral diversity. Removing the soil to a depth of about 5 cm was, not surprisingly, the most effective in lowering productivity, but it also increased plant diversity by creating gaps for new species.

Table 7.7 Accelerating natural processes of nutrient loss (after Marrs and Gough, 1987)

Encouraging ruderal growth to decrease available P
Encourage native species that can reduce soil fertility, e.g. *Bromus sterilis* on chalk gravel (the mechanism by which this happens is not understood)
Maintaining a fallow soil to promote leaching and little plant uptake

Removal by continuous cropping
Cropping with cereals has been used to reduce fertility of sandy arable soils to restore semi-natural heathlands, although this can take a long time
Grazing can also remove nutrients, although at a slower rate. Manure has to be collected: possibly a useful source of funds for nature reserves

Direct removal
Stripping of the topsoil can be effective and produce a saleable product
Deep ploughing sites can serve to mix topsoil with the less fertile subsoil and accelerate leaching

Soil transfer has also been used in restoration strategies. Turves may be transplanted for grassland communities (Wells, 1983) and whole soil profiles have been transferred to seed a succession in species-poor woodlands (Buckley and Knight, 1989). This is one method of introducing the seeds, mycorrhizae and some fraction of the other soil biota from a mature woodland into a newly established woodland community. At the other end of the scale, whole communities have been transplanted, including woodlands. Down and Morton (1989) describe the reconstruction of part of Darenth Wood in Kent, a site of special scientific interest (SSSI), to allow quarrying beneath its original location. In another example, grassland and marsh communities near Heathrow Airport, west of London, were transplanted, complete with their soil (Worthington and Helliwell, 1987). The profile of the soil was lifted in three layers, and replaced in the correct sequence. The donor site was needed to extract gravel to construct the M25 motorway, and the receptor site was a disused gravel works. The disturbance of the turves led to an initial increase in floristic diversity, although this was expected to revert to its original state if grazing was maintained at the same level (Worthington and Helliwell, 1987).

Hedgerows are transplanted in Germany, in a technique that places the hedge on a mound of soil which induces leaching of its nutrients. This helps to encourage the companion flora (Kaule and Krebs, 1989). Hedgerows and field margins are important as one type of corridor that connects fragments of woodland or other semi-natural habitats in an agricultural area, but they can also support sizeable resident populations of their own. Farmers in Britain are now encouraged to leave a 'headland' of unmown vegetation adjacent to the hedgerow, which is not sprayed with pesticides. This promotes a greater diversity of plants and the beneficial insects that can help to contain pest outbreaks in the crop (section 5.7). It also provides food for the birds and small mammals that inhabit the hedge. Hedgerows are commonly taken to be corridors for the dispersal of species between fragments of woodland, and Boatman *et al.* (1989) suggest that a headland of 6 m be left, producing a habitat around 12 m wide.

An understanding of the processes underlying succession, both of the interactions between species and of nutrient transfer in a community, underpin many of the techniques used in reclamation and restoration. Many ecologists have also recognized the potential role of constructing synthetic ecosystems to test theoretical ideas of succession. In established and older communities, it may be that the rules governing species assemblages become more important, and our management is geared to creating conditions that allow a greater variety of plants to colonize. In reclamation, it appears that ecological functions governing the availability of nutrients are of greater significance in the earlier stages of the succession.

Summary

A succession describes the progressive change in the species composition of a community and its ecological processes. Two types of succession are distinguished – a primary succession develops on a site that has been uninhabited and has no organic component; a secondary succession follows the destruction of a previous community. They differ in the availability of nutrients, and the latter will often favour ruderal plants able to colonize the site quickly. The colonization by a species may be determined by the existing occupants of the site – either facilitating or inhibiting their arrival. A number of species interactions are important in the development of the community, particularly the symbiotic association between higher plants, bacteria, fungi and actinomycetes. These aid the capture of nitrogen in the early phases of a succession and of phosphates in the later stages.

The succession of plants in a community appears to follow these changes in the availability of nitrogen and phosphorus. Reclamation aims to establish the processes of nitrogen fixation, especially through the use of legumes and mycorrhizae. Central to this is the production of organic matter which drives nutrient capture. This is crucial for the development of the soil community, as organic matter helps to regulate nutrient release. A range of plants and microorganisms may be used to overcome particular toxicity problems in a spoil, buffering its acidity and allowing for more rapid development of the soil profile.

Agricultural practices can accelerate the deterioration in soil fertility and structure. In tropical areas, loss of woodland cover leads to soil erosion and the formation of hard-set soils with little agricultural potential. In contrast, the restoration of temperate agricultural soils to a semi-natural habitat aims to reduce soil fertility, producing a species-rich community, more typical of the later stages of a succession. Grassland management requires a regime of disturbance that prevents the establishment of a climax community marked by a low diversity of plants.

Further reading

Bradshaw, A.D. and Chadwick, M.J. (1980) *The Restoration of Land*, Blackwells, Oxford.

Buckley, G.P. (1987) *Biological Habitat Reconstruction*, Belhaven Press, London.

Gray, A.J., Crawley, P.J. and Edwards, P.J. (1987) *Colonization, Succession and Stability*, Blackwells, Oxford.

Jordan, W.R., Gilpin, M.E. and Aber, J.D. (1987) *Restoration Ecology: a Synthetic Approach to Ecological Research*, Cambridge University Press, Cambridge.

Ravera, O. (ed.) (1991) *Terrestrial and Aquatic Ecosystems: Perturbations and Recovery*, Ellis Horwood, Chichester.

Sheep grazing in the shadow of the Hinkley Point nuclear power station in Somerset, southwest England. Modelling the contamination of food chains has established significant sources and routes of entry into the biota.

8

Modelling ecosystems

Ecological communities are highly complex. Not surprisingly, our models of the population dynamics of two- and three-species interactions have little capacity to predict the behaviour of a whole community. An ecosystem is even more complicated, when any model has to include a number of populations, their interactions with each other and their abiotic environment. Nevertheless, the model, as a formal description of the essential features of an ecosystem, is the most direct technique for understanding ecosystem processes.

In analysing ecosystems, we are principally concerned with the exchange of energy and materials between the community and its abiotic environment. We attempt to describe the processes that govern its species composition, productive capacity and stability. The simplest possible model of these processes is a budget, or mass balance, of inputs and outputs – such as the nitrogen gained and lost by an ecosystem. Here stability would be indicated by a lack of significant change in the balance over a number of years.

Ecosystems persist because their rates of production exceed their rates of decomposition, the processes of energy capture and release that drive nutrients through the system. Each function represents a collective measure of the processes of life within the system. The balance between the two provides an indication of how the system responds to stress and has been used for the early detection of pollution impact. Using this gross measure of a disturbance, we can then identify which part of the community has been affected by the stress.

Both processes are determined by a number of biotic and abiotic variables, and the complexity of natural ecosystems can make it difficult to define the role of any one factor precisely. This also limits our scope for experimentation – most ecosystems are too large and contain too many components for easy manipulation of these variables. Instead, the ecologist has to rely upon observational data, looking for repeatable patterns that can be correlated with changes in the abiotic environment. If, for example, the amount of rainfall correlates with the total loss of nitrate from a soil, we might suspect a causal link. In effect, we create a hypothesis, usually quantified, that we can test against further observations – perhaps predicting the amount of nitrate lost for a given amount of rainfall.

This is a simple model for one part of the system. More complex models will string several hypotheses together, attempting in each case to define the relationship between two or more variables. A model that repeatedly makes accurate predictions about the behaviour of the ecosystem will enjoy our increasing confidence, suggesting that the interactions between these components have been accurately described.

Viewing ecosystems in this way is termed **systems ecology**. It is a methodological approach rather than a body of theory about the ecosystem, but is very distinct from other ecological methodologies. It analyses ecosystems in a mechanistic way, treating the ecosystem rather like a machine processing materials and energy. In doing so, it attempts to isolate the main factors controlling and regulating ecosystem processes. Inevitably, this makes some presumptions about the integration of the ecosystem and, in its most extreme form, the method has been criticized because it tends to adopt a holistic view of the ecosystem (section 1.5). By looking at just one aspect of ecosystem function, this approach can also give a very partial view of the ecosystem processes. It collects species into functional groups, the components of the model, and this may distort their role, particularly when viewed from some other perspective.

Nevertheless, this approach offers a feasible method for testing hypotheses at the ecosystem level. A realistic model is an aid both to greater understanding and to decision-making for conservation and pollution management. We can also use models to compare very disparate ecosystems in some common process, such as the rates of nitrogen loss.

One alternative approach is to use a sample of an ecosystem to model processes in the larger system. **Microcosms** are meant to capture the essential features of both the abiotic environment and its community, but on a much smaller scale. These have the distinct advantages of allowing replication and direct experimentation. In a similar way, ecologists have partitioned parts of ecosystems in the field, especially freshwater habitats. Again, these **mesocosms** or enclosures allow replication, though fewer of the variables are controlled. We examine both methods later on.

This chapter provides a brief introduction to systems ecology and its use in assessing material transfer in ecosystems. We concentrate here on the movement of both nutrients and pollutants through temperate woodlands and freshwater habitats, and show how each, in excess, can affect ecosystem function and species composition. In particular, we look at the way such models have led to new strategies for restoring eutrophic freshwaters. From these models and the use of microcosms, a number of methods of ecosystem-level monitoring have been devised and these are reviewed briefly. We go on to review the use of models in the assessment of pollutant transfer and, more specifically, the question of bioaccumulation.

8.1 PRINCIPLES OF MODELLING

One measure of our understanding of the real world is the accuracy with which we can predict its future behaviour. This is most easily achieved where there are a small number of variables and relatively simple relationships between them. Physics enjoys this advantage, where some simple mechanistic relationship allows for highly predictable outcomes – such as the equations used to describe the motion of the planets.

All science involves some simplification, distinguishing spurious information (or 'noise') from that which is essential to the problem (the 'signal'). This is something we all do naturally, to hasten our understanding of a problem. Science has simply formalized the procedure: defining the problem, setting boundaries around the system and analysing the relationship between its components. We can then go on to quantify the relationships between these components.

In systems ecology, the components may be individual species, but more often they will be functional groups – such as 'decomposers' or 'leaf-eating insects'. The ecologist then treats each group like a 'black-box', the contents of which will probably remain undescribed or unquantified. All that is measured is its mass balance and its relationship to other functional groups. In this way, the role of each group in processing energy, material or information is described for the ecosystem, and a model of the whole system is constructed.

For modelling to begin at all, we need to state the objective of the study, to define what the model is meant to achieve. In this way we can decide which functional components and relationships are relevant, and are part of the signal. If the aim was to model the transfer of nitrogen through the system, two components might be the primary producers and the nitrogen-fixing bacteria, and their relationship defined by the rate of nitrogen uptake by the plants.

When explicitly stated and quantified, these relationships become the testable hypotheses of the model. Collecting several statements together in a logical sequence, as a formal representation of the system, produces the model. Much of what the scientific method has to say about the testing of hypotheses can be applied to the sequence of statements that comprise the model (section 1.2).

8.1.1 Types of models

Models fall into two basic categories. **Analytical models** aim to distil the principles that apply to a class of systems. In contrast, **simulation models**

Table 8.1 The main components of a mathematical model

The system

A system is defined by its boundary and consists of two or more interacting components. The boundary and components are themselves determined by the problem being studied. Most ecosystems are 'open' systems, where connections exist with the outside, across the boundaries

Sources or sinks for energy, materials or information are external to the system

Variables

1. External variables or forcing functions are variables outside the system, which influence its state, e.g. temperature, precipitation
2. State variables indicate the state of the system by their value or level. These comprise the basic components of the system and determine the essential features of the model, e.g. the level of phosphorus in the phytoplankton compartment

Processes

1. Rate processes are equations that quantify the flow of energy, materials or information passing between variables, e.g. rates of phosphorus uptake by the phytoplankton compartment

   Rates may vary with the size of the state variable; if there is negative feedback the system is regulated
2. Parameters are constants that describe fixed relationships between variables, e.g. maximum rate of population increase in phytoplankton

Both variables and processes change in space and in time

Conceptual diagram

These components are collected together as a flow diagram to illustrate explicit interactions that comprise the model

The diagram itself is usually a representation of a computer algorithm

---

attempt to predict the behaviour of a single, particular system. Analytical models favour economy, shedding detail to reduce the problem to a tractable size, encapsulating its essential features. These isolate the mechanisms of control, often without describing a quantitative relationship. Simulations require this detail to enhance the accuracy of their predictions – they sacrifice generality to give a detailed description of one system. This is especially useful in the management of an ecosystem and predicting its response to disturbance. The contrast between these two approaches was seen in models of the population dynamics of pests and their natural enemies (section 5.4).

We should also distinguish between **deterministic** and **stochastic models**. The former are analytical models that typically represent the relationship between components as a series of fixed equations. This allows only one possible output for a given input. The simple logistic population models (section 4.1) are of this type. Stochastic models, in contrast, admit the variability associated with most biological processes, and attempt to quantify this uncertainty. There is then a range of possible outcomes for a given input. Stochastic models may thus be more appropriate for simulating real ecosystems, although they can also be used to analyse the nature of the relationship between components.

Models differ according to the methods used to organize the components of the system, and the mathematics used to describe their relationships. A review of these applied to ecological modelling is provided by Jørgensen (1986). Here we concentrate on dynamic models that describe the relationships between well-defined functional compartments. There is, however, a number of basic terms common to most methods (Table 8.1) and also a diagrammatic notation associated with them (cf. Kitching, 1983). The basic process of model construction and refinement can be shown as a simple flow-diagram, itself a model or representation of the process (Figure 8.1).

8.1.2 Building models

Modelling begins by formulating the question to be answered. This will have arisen from a background of existing knowledge, experience, and perhaps information gained from pilot experiments. The question helps to set the boundary around the system, both in time and space, and to define the essential components. It may also suggest alternative modelling solutions which could be used if the initial choice fails to provide a solution.

Most models exist as a computer program and a series of equations summarizing these relationships. They start life as a system diagram, a flow-chart for the algorithm, which partitions the system and defines its functional groups (Figure 8.1).

The numerical terms for state variables or rate processes can be based on real data or reasonable guesses. A range of alternative equations may be tried for each relationship, perhaps using statistical techniques to choose an equation of 'best fit'. The model continues to be developed as further data are used to refine these terms, to identify other important components, and to evaluate the way in which the system has been partitioned.

Models may start with a minimum number of components – normally the approach adopted in constructing an analytical model – or in a more complete form, when state variables are removed if they are shown to be superfluous by later testing. If there are abundant data, the match between the observed and predicted results can often be improved by adding new state variables or subdividing the existing components further. However, any increases in complexity may make the model less

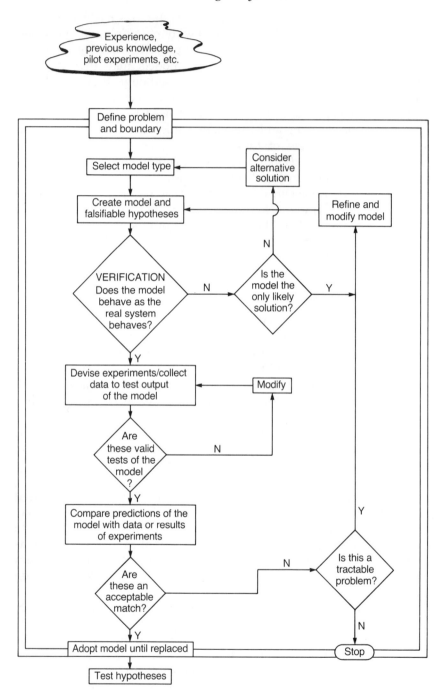

workable, and a balance has to be struck between further elaboration, and using the model for its intended task.

This balance can be judged through a process termed **sensitivity analysis**. Models based on a small number of state variables will be sensitive to inaccuracies in the measurement of any one component. This sensitivity is generally reduced in models that are more finely partitioned, but this requires measuring a larger number of variables and rate processes. Studying the operation of the model over a range of conditions, with different values for the external variables, serves to highlight the critical state variables and rate processes. A sensitivity analysis can then identify the components most significant to the performance of the model, and those which can be removed entirely.

As a basic principle of systems analysis, we would expect any regulation of ecosystem processes to be produced by some form of **negative feedback** – that is, a reduction in a rate process as the size of an associated state variable increases. This mechanism is found to regulate many processes in nature, such as body temperature or blood sugar levels. One example has already been described in population ecology (section 4.2): a population growing in a limited environment will achieve an equilibrium density, where the pressure for further growth is countered by the reduction in resources as numbers rise (Figure 8.2). In contrast, **positive feedback** tends to amplify the effect – the rate process increases as a state variable gets larger, driving the system faster and faster, pushing the system away from an equilibrium position. A population in an unlimited environment grows increasingly rapidly because the rate of population growth rises with the number of individuals.

If a model can be shown to follow the general behaviour of the system over a broad range of conditions, experimentation or the collection of further data to test the model begins. This process is termed **validation**, and compares the output from the model with that of the real system. This may include the response of an ecosystem to some form of stress or perturbation (section 1.7): experiments have the advantage of examining the system under controlled conditions. Otherwise, having measured the

---

Figure 8.1 A flow diagram showing the sequence of steps necessary to create and validate a model. A model usually seeks to answer a specific question or accurately mimic the behaviour of a complex system. It can be considered as a series of hypotheses or statements together in a logical sequence. Each must be falsifiable and testable, enabling the model to be refined or rejected. These are tested against its predictions about the behaviour of the system, using data from further observations or the results from specifically designed experiments. (Modified, after Kitching, 1983.)

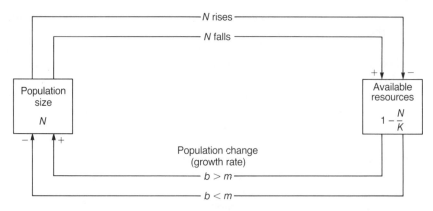

Figure 8.2 The principle of negative feedback, demonstrated by population growth in a limited environment. As the population grows, so the available resources diminish, reducing the birth rate (*b*) or increasing the mortality rate (*m*). A rise in the one state variable (*N*) reduces a second (the available resources) which in turn lowers the rate process (the growth rate). This negative feedback creates a stable, regulated system – a population below its carrying capacity will be allowed to expand up to it, but beyond this point, no further growth is possible. With positive feedback the opposite happens – an increase in a state variable leads to further increases in the rate process, and the system runs 'out of control'.

significant external variables, the predictions of the model are judged against further observations of the real system. If it proves to be 'robust' over a range of conditions, we accept the model and the hypotheses it embodies, that is, until it proves to be inadequate or false. We are now in a position to answer the original question directly, or use the model to provide a valid test of a hypothesis.

Simple dynamic models are used throughout this chapter. These have the longest history in ecosystems research: much of the work on mineral cycling in various ecosystems as part of the International Biological Programme used such modelling techniques (Waring, 1989). A range of other models, such as those based on matrix algebra, or multivariate techniques are not considered here: Jeffers (1978) offers a good introduction, with many applications.

Many established ecosystems show relatively consistent patterns of nutrient cycling over the long term, linked to a large storage capacity in the biota, and in the soil or sediments. A high rate of internal cycling is typically associated with a complex food web and a highly integrated community. One of the effects of simplifying this structure, that is losing a number of species, is to open up the system, allowing nutrients to leak away. The complexity of a community may also inhibit the return of the system to its original state after disturbance – the most rapid recovery is found in ecosystems where nutrients cycle easily, and where there is high efficiency in nutrient and energy capture – such as agricultural systems (Waring, 1989). The excessive loss of a nutrient may also indicate some disturbance to the system. These ideas have been tested in one of the most extensive efforts to model an whole ecosystem – Hubbard Brook.

8.2 HUBBARD BROOK

This classic study examined the movement of nutrients in an area of temperate hardwood forest in New Hampshire, USA, drained by the Hubbard Brook (Bormann *et al.*, 1974). One major advantage of this study was that it was replicated – six ecosystems were defined by the watershed associated with each tributary of the brook. Additionally, it was easy to quantify the losses from each catchment – the soil (a glacial till) overlies an impermeable bedrock so that besides volatilization, the only route of loss is via the stream draining each watershed. By measuring the levels of nutrients in these streams, and their volume of water, the total export could be calculated for each ecosystem. Precipitation gauges measured water and nutrient inputs from the atmosphere. Together with measurements of the nutrient capital stored in various compartments of the forest biota, a relatively complete budget could be calculated over several years (Figure 8.3).

Several important observations emerged from this study. The concentration of nutrients in the output was relatively constant, despite major variations in streamflow. This was due to the chemical buffering of the water, governing its capacity to carry nitrate anions. Consequently the mass of nitrogen lost depended mainly on the volume of water leaving the system. Additionally, losses through streamflow were greater than inputs from the atmosphere for nearly all nutrients; since rates of loss did not decline with time, the difference had to be made up by weathering of the bedrock. The exception was nitrogen, which was accumulating within the system (Figure 8.3): inputs from both precipitation and nitrogen fixation always exceeded losses. Most of this nitrogen was stored within the woody biomass, and harvesting trees significantly reduced the nitrogen capital of the ecosystem.

The response of this ecosystem to major disturbance was measured by clear-felling one whole watershed in 1965/6, and spraying it with a herbicide for the following three years. The vegetation was left where it fell. For the duration of this treatment there was thus little or no demand for nitrogen by primary producers. In a second watershed, strips of the forest were cut to study the changes associated with recolonization.

Within two years, the clear-cut had led to dramatic changes. The volume of water in this stream rose by over 30% compared with the undisturbed watersheds, and its particulate load increased nearly

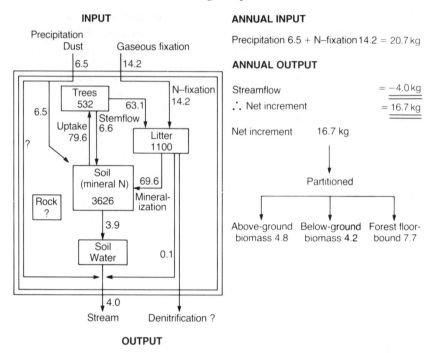

Figure 8.3 A model and budget for the flow of nitrogen through an undisturbed watershed – the Hubbard Brook experiment in the White Mountains of New Hampshire, USA. Note that denitrification was not measured here (it was not thought significant), so losses via this route are unknown. Nitrogen is accumulating in the watershed, largely in the biomass. Liberating this capital has drastic consequences for ecosystem stability. (After Bormann *et al.*, 1977.)

15-fold. In the absence of significant evapotranspiration, the rate of leaching of all nutrients was increased. Decomposition was accelerated within the de-vegetated watershed, although the ammonia and nitrates released by the nitrifying bacteria were no longer assimilated by a plant community. The result was a pronounced loss of nitrates – a 60-fold increase in the stream concentration within a year of felling. This was balanced in the stream chemistry by an increase in the loss of cations, most especially calcium. One consequence of these higher nutrient loads was an algal bloom in the enriched streams.

As the plant community re-established itself, so the scale of the nutrient loss declined, and some constancy returned. The models of the system describe how the movement of nutrients, soil and water are regulated by the biotic compartments of the ecosystem. Much of the nutrient capital of the catchment is held in these living components, and their activity and interactions limit the rate at which these are lost. The regulation of the system results not only from the competition amongst the biota for the available nutrients, but also the presence of plants and microorganisms, which together maintain the physical and chemical properties of the soil. In turn, the plant community effectively holds the soil and nutrients up the mountainside.

The soil profile develops from the biological activity at its interface with the atmosphere: the soil derives its structure and texture from the plant and animal community living in the first few centimetres. The porosity and crumb structure of the soil help to prevent it becoming waterlogged, or being washed away or drying out. The evapotranspiration of the higher plants pumps water out of the soil while their litter drives a decomposer cycle that increases its organic content, important for binding minerals and increasing its water-holding capacity (Table 7.3). The stability of the hydrological cycle within the catchment and the buffering of its water chemistry derive from a regulation imposed by the biotic components on the abiotic environment (Bormann *et al.*, 1974).

8.3 NUTRIENT CYCLING AND THE IMPACT OF POLLUTANTS

The Hubbard Brook experiment resulted in a simulation model of one ecosystem which is not readily applicable to other types of woodland growing under different climatic conditions (Blair and Crossley, 1988). Nevertheless it has led to several general ideas about nutrient cycling in terrestrial ecosystems. For example, the increased loss of nutrients following disturbance may be a common feature of a variety of eco-systems (Bormann *et al.*, 1974), and a number of studies have examined whether this might be used as an ecosystem-level measure of stress. The release of nitrogen from disturbed woodlands has been modelled by Reiners (1981) (Figure 7.7), and as with Hubbard Brook, this demon-strates how nitrate export is the end-point of a number of ecosystem processes. Nutrient loss under stress appears to be a feature of a range of terrestrial ecosystems from the tundra to hot, arid lands (Ausmus, 1984).

An increased export of essential nutrients has been found to follow a range of chemical stresses in different ecosystems, a response detected before any changes in population or community parameters. One common feature of stressed soils is an initial loss of nitrogen and other nutrients as the microbial community is disrupted. Early on, this may lead to an increase in primary productivity, as higher plants take up the available nutrients – a response typical of woodlands undergoing the first stages of acidification, when the trees show a short-lived burst of growth. At this time, nitrogen is being released more rapidly from the capital held in the organic or clay–humus component of the soil, and this can be readily detected in leachate collected from the site. If, in contrast, decomposition processes are inhibited by a pollutant, the amount of litter will increase in the soil (section 2.7), providing additional binding sites for both nutrients and pollutants. Then nitrogen loss would be a poor indicator of any displacement of the ecosystem. At persistently high levels of contamination, however, the disruption to the microbial community leads to increased nutrient loss, and the soil C:N ratio rises.

A dose–response relationship between nitrate loss and pollutant concentration has been derived from excised soil cores used as micro-cosms (Ausmus, 1984). Jackson *et al.* (1978) showed how the addition of toxic metals (lead, zinc, cadmium and copper), applied as a smelter dust and contaminated litter, increased nutrient loss in soil microcosms. The soil cores were collected from woodland around a lead smelter in Missouri, USA and received a total dose equivalent to the annual input at a distance 0.4 km downwind of the complex. Over the 20 months of the experiment, the losses in water leached through each microcosm were significantly higher in the treated cores compared with controls – for calcium (1.3 times), magnesium (1.3 times), potassium (1.5 times) and nitrate (6.5 times). Only the levels of phosphate in the leachate were unchanged by the addition of the pollutants. As Ausmus notes, this sensitivity offers a rapid method of measuring an ecosystem-level response – nutrient export rates increase with pollutant concentration until the nutrient pool is depleted or microbial processes in the soil are inhibited. Coupled with measures of the essential nutrients, such as C:N ratios, such responses also give an insight into the nature of the toxic impact.

Comparisons between the cycling of pollutants and essential elements have also been made for whole ecosystems. Van Hook *et al.* (1977) modelled the cycling of carbon, nitrogen, zinc, lead and cadmium within a watershed of mixed deciduous woodland (Walker Branch, Tennessee). The main inputs of pollutants were from the atmosphere, derived from three coal-fired power stations nearby. Outputs from the ecosystems were again measured as losses detected in the streamflow, although here there was the possibility of metals being lost with groundwater leaching out of the system. Given the measured loss rates in the streams, the **residence time** (or turnover time) of each element in the system was determined as:

Residence time (years) =

$$\frac{\text{the mass of an element in the ecosystem}}{\text{the rate of element loss from the ecosystem per year}}. \quad (8.1)$$

The same calculation can be performed for individual compartments within the system. In Table 8.2 these times are shown for both the whole ecosystem and for the vegetation component alone.

The watershed accumulated all of the elements – inputs were always greater than outputs. For both the metals and the nutrients, this

Table 8.2 Inputs and losses of nitrogen and three toxic metal pollutants in a mixed deciduous forest watershed, Walker Branch, Tennessee, USA

			Residence times (years)	
	Input	*Output*	*Whole system*	*Vegetation only*
N	13 000	3 000	2 145	4
Zn	538	140	2 546	3
Pb	286	6	18 932	3
Cd	21	7	108	4

The metals are largely derived from atmospheric inputs from three coal-fired power stations. For all elements the soil is shown to be the largest compartment and, in each case, the most significant process is the rate of decomposition of the litter. Recycling of nutrients and the pollutants between the soil, the litter and the above vegetation allows for an extended time in the whole system. This is considerably shorter in the vegetation alone (from Van Hook *et al.*, 1977).

accumulation depended upon the critical process of decomposition, the mineralization of organic matter. The single largest compartment for each element was the soil, and their retention here extended the residence time of each pollutant in the system. In contrast, the cycling through the vegetation was in all cases very rapid.

Soil was shown to perform a similar role in a woodland downwind of a metal smelter at Avonmouth in southwest England. Martin and Coughtrey (1987) modelled the soil as a series of compartments, represented by successive layers down the soil profile. By measuring metal concentrations in each layer over several years, they could derive the rate constants for the movement of each pollutant down the profile, from one compartment to the next. Predictions for 1979 corresponded well to the observed data. The residence times varied with the metal: in the first mineral layer of the soil, it took 0.65 years for a quanta of cadmium to move through the compartment, 1.8 years for zinc and 6.6 years for lead. These also changed down the profile: overall, cadmium and zinc were found to be relatively mobile, whereas lead was retained in the upper layers, confirming the findings from the Tennessee watershed (Table 8.2).

The same has also been found for lead in a polluted coniferous forest in Vermont, USA (Friedland and Johnson, 1985). This study showed that lead concentrations in the soil solution declined with depth, so that most of the lead is held in the upper part of the soil profile, further indication that the metal is being bound to the organic component. Despite this binding, and despite the accumulation of litter in heavily contaminated soils (section 7.3), the turnover time of the metal becomes shorter as contamination increases. The residence time for lead at the Vermont forest floor was 500 years, compared with 5000 years for Hubbard Brook, which receives only half the input of the Vermont woodland (Friedland and Johnson, 1985). Yet both sites have an equivalent output of the metal. It appears in both cases that the organic fraction of the soil and the decomposition processes are the most important factors regulating this rate of loss.

These studies also suggest that measuring the loss of any toxic metal from the soils would be a poor indicator of their impact on the system. We would need to consider some ecosystem-level parameter, such as nutrient export, or rates of litter decomposition (section 2.7) to determine if the system was stressed. Just as the death of individuals is used to measure the dose–response of a population, or the loss of species as the response at the community level, so nutrient loss has been used as an integrated measure of biological impact for the whole ecosystem.

The significance of such pollution effects has become apparent as a result of modelling of ecosystem processes. Decomposition has been found to be central to the regulation of many parameters in these temperate woodlands. The models have enabled us to determine normal rates of loss (and their variation) and the factors that regulate these functions. Nutrient export and decomposition rates also provide us with rapid techniques for measuring the response of the ecosystem to some stress.

8.4 EUTROPHICATION IN FRESHWATER ECOSYSTEMS

The Hubbard Brook experiment demonstrated the complex way in which components of an ecosystem interact, and how detailed models may be needed to describe its processes. Even in simple models, the outputs may not always conform to expectations. Briand and McCauley (1978) devised a model to examine the effects of nutrient cycling on the trophic structure of a freshwater system, based on just six compartments (Figure 8.4). When the totality of the interactions between compartments was considered, many of the responses were 'counter-intuitive'. For example, increasing the density of a predator left the size of the herbivore population unchanged.

This simple model was used to make several predictions about the impact of enrichment on freshwater ecosystems. It suggested that any addition of nutrients to the lake would largely go to the inedible component of the phytoplankton. This compartment mainly consists of the blue-green algae (Cyanophyta) which are avoided by planktivorous fish and the zooplankton because of their toxicity and taste. Where removing the source of the nutrients is not a feasible management option,

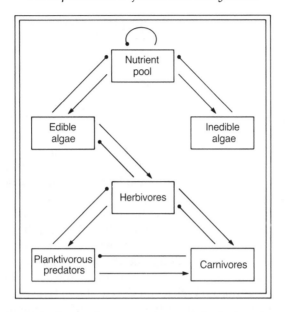

Figure 8.4 A model of a freshwater system in which the phytoplankton are divided into edible and inedible components. This model used loop analysis to examine the effects of perturbations. The interactions between components are represented as a loop allowing feedback, composed of both a positive feedback (shown by an arrow) and a negative feedback (a filled circle). For example, an increase in the edible algae allows herbivore numbers to rise, which in turn lowers the numbers of edible algae. These changes will have impacts on the rest of the system, sometimes with less obvious consequences – an increase in planktivorous predators has no effect on herbivore numbers but allows inedible algae to increase. (After Briand and McCauley, 1978.)

the model offers the alternative strategy of manipulating the balance between the herbivores and carnivores. Indeed, subsequent work suggests that this may be the only effective long-term method of controlling eutrophication.

Many natural water bodies are described as **oligotrophic**: clear-water ecosystems in which primary and secondary productivity is limited by a shortage of the major nutrients. The nutrient capital of the system typically increases during the life of a lake and so an oligotrophic lake may slowly become **eutrophic** with time. The process of enrichment, **eutrophication**, leads to turbid waters, mainly as a result of an increase in the abundance of phytoplankton. Very often these changes are marked by a shift from a clear-water community dominated by rooted higher plants or macrophytes to a turbid-water community dominated by a dense concentration of phytoplankton. With artificial eutrophication, following a large increase in nutrient supply, the process happens rapidly and the trophic structure of the lake becomes unbalanced. There follow reductions in dissolved oxygen concentrations and changes in the phytoplankton, higher plant and animal communites (Moss, 1988).

Temperate freshwater ecosystems are relatively stable communities, with highly predictable rates of community respiration (Uhlmann, 1991). They undergo a seasonal pulse of nutrients, and, if deep enough, become stratified into distinct layers during the summer. Then the colder, lower layers of the lake become distinct from the warmer, upper layers, and the deeper water may become depleted of oxygen. There is also little exchange of nutrients between the two layers, so that nutrients become limited in the upper layer, where most biological activity is concentrated.

A shortage of phosphorus limits the productivity of most freshwater systems, due both to its immobilization in the biota and the insolubility of its compounds. If phosphorus alone is increased, the cyanophytes, which are able to fix nitrogen, will dominate the phytoplankton community. Indeed we can predict their relative dominance from the balance between nitrogen and phosphorus (Welch and Cooke, 1987). The cyanophytes are more able to use low levels of carbon dioxide and become more buoyant when CO_2 levels are low and the water pH is high. This keeps them in the sunlit, upper layers of the water column. In addition, some species produce dense mats of vegetation that inhibit other phytoplankton and limit the swimming of the zooplankton. Together such factors mean that a slow-moving freshwater ecosystem can be rapidly dominated by blue-green algae, displacing not only other members of the phytoplankton, but some of the animal community as well. The reduction in the light reaching the lake floor also inhibits submerged and rooted macrophytes, and sediments become anoxic as large amounts of planktonic biomass are added to them.

A famous example of artificial eutrophication is Lake Washington, which became eutrophic as the city of Seattle expanded and discharged more of its effluent into the lake. Models accurately predicted the decline in the phosphorus concentration of the water when the effluent was diverted. These helped to demonstrate that phosphorus was the key nutrient in the eutrophication of the ecosystem (Welch and Cooke, 1987). Success here led many to believe that eutrophication could be solved by simply reducing nutrient loading. However, it quickly became apparent from other enriched freshwaters that the problem would not be cured so easily.

8.4.1 Modelling eutrophication and its management

Even models that have captured some of the complexity of the problem have failed to predict the behaviour of eutrophic lakes. Loucks (1985) and

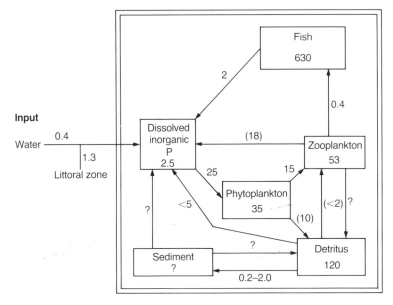

Figure 8.5 A simplified model of the phosphorus cycle in Lake Wingra, Wisconsin, USA, with the values found in late summer (mg.m^{-2}; rate processes are mg.m^{-2} per day). This is the basis of a computer simulation derived to manage a weed and a eutrophication problem. Note the small size of the pool of dissolved inorganic phosphorus in the water itself – this is the component with the most rapid cycling. The greatest source of this component is phosphorus in the zooplankton faeces. Estimated (indirect) values are shown in brackets. (Modified, after Loucks, 1985.)

his co-workers produced a model of the phosphorus cycle within Lake Wingra, a eutrophic lake in Wisconsin, USA, a summary of which is shown in Figure 8.5. This simulation was devised to aid the management of the lake, of which there were two principle aims – to remove excess phosphorus and to control an introduced weed, the macrophyte *Myriophyllum spicatum*. This dominated the shallow water of the lake and also represented a significant pool of phosphorus. One solution seemed obvious: harvesting *Myriophyllum* from the lake would not only remove the weed, it would also remove phosphorus from the system.

However, the interactions within the plant and animal community are highly complex. Macrophytes are important as a refuge for large zooplankton to escape predation and they also act as cover for the large predatory fish that feed on the plankton-feeding fish. Planktivorous fish select the larger zooplankton and their activity produces a zooplankton community dominated by smaller species. For these reasons, macrophyte cover is a crucial factor determining the trophic structure of a lake. One effect of removing macrophytes is to lower the average size of the zooplankton: without a refuge, the zooplankton face greater predation by the planktivorous fish, which themselves escape predation because there is no cover for their predators.

Through their grazing, zooplankton are one of the important regulators of phytoplankton abundance. By their excretion, they are also the most important source of mineral phosphorus for the phytoplankton (Figure 8.5). Significantly, larger individuals excrete phosphorus at a slower rate, so that a population dominated by small animals will have a higher rate of phosphorus turnover. The degree of zooplankton predation can thus regulate the available phosphorus in the lake.

The model allowed for each of these effects, and was used to consider a range of management strategies to control the eutrophication. One question was whether a single harvest of 50% of the *Myriophyllum* would significantly lower the phosphorus level in the water. The model predicted it would not. Instead, a better strategy appeared to be a repeated weed harvesting and measures to lower phosphorus inputs, reducing both the eutrophication of the lake and its weed problem (Loucks, 1985). Encouraging an increase in the average size of the zooplankton would cause an effective reduction in the rate of phosphorus cycling. However, this relied on the assumption that harvesting the *Myriophyllum* would lead to less macroinvertebrate predation of the zooplankton.

This strategy was never tested on the lake – the weed population declined dramatically, for unknown reasons, before the effects of harvesting could be measured. Obviously the model was an incomplete description of the real lake, and the shift to a new equilibrium was unpredicted. The model did suggest that the overall effect of an aggressive macrophyte, such as *Myriophyllum*, is to reduce the productivity of the phytoplankton relative to available phosphorus, and the whole ecosystem may switch between different stable states – either a community dominated by phytoplankton or by macrophytes.

Which state prevails may be determined by the consumers further along a food chain. Scheffer (1989) has modelled shallow freshwater lake ecosystems in the Netherlands, which were once oligotrophic. The switch to eutrophic conditions can be correlated with a change in the main fish population. Turbid lakes tend to be dominated by bream (*Abramis brama*), which has a higher growth rate in eutrophic waters. The bream feeds upon zooplankton and bottom-dwelling invertebrates. In doing so, they stir up the sediments, increasing both the turbidity of the water and the supply of nutrients to the phytoplankton. Macrophytes, on the other hand, find it difficult to photosynthesize in the cloudy conditions, and as they disappear so does the main predator of the bream, the pike (*Esox lucius*). Young pike require macrophytes to escape predation from adults.

A large bream population will thus inhibit recruitment to the pike population, indirectly lowering predation on itself. In the model, two stable states were found: one dominated by the pike – a clear-water, oligotrophic system, and one dominated by the bream, typically turbid and nutrient-rich. Sensitivity analysis of the model reveals a high degree of inertia for both states. For example, removing bream alone from the system will not allow it to switch to the pike-dominated state. Scheffer suggests that to induce oligotrophic conditions the food web and the nutrient levels in the water both need to be manipulated.

Changing the composition of the community to reduce eutrophication in this way has been termed **biomanipulation**. This strategy has been used to treat the eutrophic freshwaters of the Norfolk Broads in England, following a sequence of failed attempts to establish clear-water conditions using other methods (Moss, 1989). The broads were created by peat-cutting in the Middle Ages, but their water quality began to deteriorate with the expansion of the surrounding towns in the middle of the 19th century. Eutrophication appeared in different broads at different times, but was recognized as a regional problem in the late 1970s. Initially, the increased turbidity of the water was attributed to the many pleasure boats using the water, but eventually it was found to result largely from an increased input of phosphates.

The source of these nutrients included both domestic effluent and the runoff from agricultural land. Nutrient budgets were created to model the problem and these provided targets for reducing phosphorus inputs that were expected to improve water transparency. A number of experiments were performed, halting the input from rivers receiving sewage effluent, and also removing enriched sediments. The sediments represent the largest component of phosphorus in these ecosystems which, under anoxic conditions, will release the nutrient back into the water column.

These first attempts initially met with some success. Halting the phosphorus supply to one broad began to reverse its eutrophication, and in the first four years, phytoplankton biomass was reduced and water clarity returned. At this time, phosphorus release from the sediment slowed as a macrophyte, *Ceratophyllum demersum* began to dominate the broad. However, in the following three years, increasingly anoxic conditions in the sediments allowed the phosphorus release to resume, and a large phytoplankton crop followed, dominated by the cyanophyte *Anabaena planctonica* (Moss *et al.*, 1986). The macrophytes then became dominant again in 1987 (Moss, 1989). In a second broad, also isolated from a major source of nitrogen and phosphorus, but which had about 1 m of sediment removed from two-thirds of its area, the improvement was stable for a much longer period of time. However, since 1985 there has been some increase in the phytoplankton levels, probably due to the return of planktivorous fish, feeding on the larger zooplankton.

The average size of the zooplankton was again found to be critical on the broads: where the larger animals could escape predation – either by having a refuge from their predators or where there were fewer planktivorous fish, the zooplankton were able to control phytoplankton growth, restoring water clarity and allowing the macrophytes to survive.

The evidence from Lake Wingra is that a predominance of these larger species also slows the rate of phosphorus cycling. In turn, the growth of the macrophytes allows an increase in zooplankton size by providing refuge and cover for the top predators. In retrospect, the role of *Daphnia* as a large zooplankton may also have been central to the restoration of Lake Washington, resulting, at least in part,'from a reduced feeding pressure from an invertebrate predator (Uhlmann, 1991).

As Moss notes, the broads are more heavily enriched than those lakes where diversion of a nutrient input had been a successful strategy, as the broads have a larger capital of phosphorus in the sediments. One technique being tried in these cases is to lock the phosphorus in the sediments, by the addition of iron salts to form insoluble phosphate compounds. Many freshwater systems exist as either macrophyte- or phytoplankton-dominated communites (Figure 8.6) and there is some evidence that one community might suppress the other (Welch and Cooke, 1987). The key to the long-term restoration of eutrophic systems is thus to switch them from one stable state to another. Both the phytoplankton-dominated state and the macrophyte state could exist with a wide range of phosphorus inputs, so reducing phosphorus input alone is unlikely to be successful. In the broads, the switch to the algal-dominated state appears to have resulted from the use of organochlorine insecticides on the surrounding farmland in the 1960s, probably reducing the zooplankton community, and allowing the phytoplankton to flourish.

Where the phytoplankton community is dominated by cyanophytes, attempts at altering the trophic structure to induce a clear-water state may also fail. Some species of blue-green algae have particular strains that are toxic and rejected by zooplankton. Benndorf and Henning (1989) found that the seasonal increase in the toxicity of *Microcytis aeruginosa* in a reservoir was due to the selective consumption of non-toxic strains by *Daphnia*. The toxic varieties inhibit the invertebrate's filtering rate and so are avoided, leaving patches of the strain which can go on to dominate the ecosystem. Thus in a cyanophyte-dominated community, with eutrophic and calm waters, the grazing activity of the *Daphnia* may actually promote the growth of a toxic phytoplankton as the season progresses.

Filamentous forms of the cyanophytes will also inhibit grazing by the zooplankton by immobilizing them. Moss and his co-workers (1991) found that, in the absence of fish predation, larger *Daphnia* species could flourish only where cyanophytes were absent; in the presence of the filamentous *Oscillatoria*, only smaller species of zooplankton would be

Figure 8.6 Eutrophic freshwater systems can switch between alternative stable states according to their trophic structure. One state is a clear-water ecosystem, dominated by macrophytes, but this can switch to a turbid system, where phytoplankton, and especially floating mats of filamentous blue-green algae, dominate.

found. A range of methods has been used to control blue-green algae, particularly in reservoirs, where they taint the drinking water. A common technique is to artificially mix the water, preventing their domination of the upper layers, and the formation of filamentous mats.

Biomanipulation has been used with partial success to control blue-green algal blooms in some eutrophic lakes. Van Donk and Gulati (1991) altered the trophic structure of several enriched lakes in the Netherlands to encourage larger zooplankton, and to provide an effective predator of the Cyanophyta. They removed a large proportion of the planktivorous fish, increased the number of piscivorous fish and in some cases, added a larger species of *Daphnia*. Elsewhere they added refuges for the zoo-plankton (willow twigs), and also a freshwater mussel, *Dreissena*, as an additional consumer of the blue-green algae. Increasing the size of the zooplankton was again successful (Figure 8.7). Adding a large number of *Dreissena* (around $675 \, \text{m}^{-2}$) was also effective in lowering the cyanophyte populations. Interestingly, the main controlling factors in the system

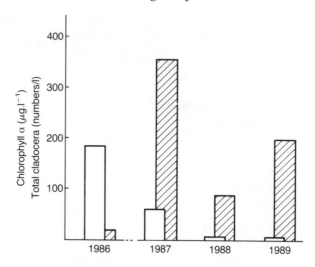

Figure 8.7 The amount of chlorophyll α (used as a measure of phytoplankton density) and the density of zooplankton (shaded boxes) in Lake Zwemlust, a small, shallow-water lake in the Netherlands. Values are shown for May for one year before, and three years after the onset of a management programme to control eutrophication. This included pumping out the water in 1987, removing the existing fish stock and replacing it with young pike (*Esox lucius*) and rudd (*Scardinius erythropthalmus*). Phosphorus levels in the lake did not change significantly after the treatment, but the transparency of the water improved by between 60 and 80%. The main controlling factors in the system after management were zooplankton grazing in the winter and nitrogen levels in the summer. (After Van Donk and Gulati, 1991.)

changed with its state. Before management, light was the dominant factor, but afterwards light and temperature were dominant only in the winter. In the early spring, zooplankton grazing was the main regulatory factor, whereas in summer it was the nitrogen content of the water.

These restoration methods were less successful upon the larger lakes in the same area. Part of the problem here is that the largest compartment of phosphorus in the water column is the fish (bream), representing around 50% of the total phosphorus in the system (Gulati *et al.*, 1991). Unless these are substantially reduced, their excretion and death serve to maintain phosphorus cycling within the biota.

8.4.2 Microcosm models of eutrophic systems

Several attempts have been made to isolate the essential features of the phosphorus cycle, and the role of the phytoplankton and the zooplankton in eutrophication. A mathematical model developed by

Borgman *et al.* (1988) was tested against data collected from laboratory microcosms containing *Daphnia magna*, to which different phosphorus loads were applied. The models were simple logistic models of predator–prey relationships, with additional equations modelling the sedimentation rates of phosphorus from the various components. Three hypotheses derived from the model were tested experimentally:

1. that the phytoplankton biomass was a function of the rate at which the zooplankton were removed (harvested), rather than the nutrient level in the water;
2. that the zooplankton biomass was a positive function of the phosphorus loading (and inversely related to the harvesting rate);
3. that the stability of the system depended upon the maximum phosphorus loading rate and the maximum zooplankton biomass.

From the experimental data collected from the microcosms, all three were shown to be correct. Phytoplankton abundance was more closely tied to the number of zooplankton, rather than phosphorus load: a 10-fold change in the phosphorus loading would only induce a two-fold change in the biomass of the algae. The presence of *Daphnia* lowers particulate phosphorus levels in the water by up to 50%, and they were found to be the main determinant of phosphorus sedimentation rates. Finally, if the addition of phosphorus became too rapid, the microcosms developed unstable behaviour.

Ordinarily, the populations of phytoplankton and *Daphnia* were shown to follow a series of damped oscillations in the microcosms, and this stability resulted from two controlling factors:

1. The rate of harvesting of the zooplankton was below their maximum rate of growth.
2. The feeding rate of the zooplankton was below the maximum growth rate of the phytoplankton.

Given stability in the plant and animal populations, *Daphnia* did not deplete its food, nor would itself disappear. Obvious points perhaps, but the tests on this model do suggest that the balance between the zooplankton and the phytoplankton could be critical both for the stability of the system, and the rate at which phosphorus cycles through it. In this case, the model's findings are amply supported by the empirical evidence from previous studies on eutrophication.

8.4.3 Other management techniques in eutrophication

Uhlmann (1991) has suggested additional treatments to immobilize or remove phosphorus from various compartments of the system and so reduce eutrophication. These include the addition of phosphate-precipitating chemicals to the water – iron chloride is added to the sediments of some broads to precipitate phosphorus. Elsewhere alum (aluminium sulphate) has been used to form an insoluble floc on the surface of the sediments, also effectively binding the nutrient. One alternative is to draw up deeper, calcium-rich sediments to bind the phosphates of the surface sediments. There is also the possibility of pumping out the nutrient-rich water and sediment to irrigate (and fertilize) farmland (Uhlmann, 1991).

Overall, the problem of eutrophication is a highly complex interaction between a limited number of players and one or two key abiotic factors. The degree of complexity is indicated by the features used to characterize the oligotrophic/eutrophic states – stable ecosystems denoted either by macrophytes/phytoplankton, by algae/cyanophyta or by bream/pike. These are not necessarily different sides of a three-sided coin, but can represent a variety of each state. In different systems, each condition is found to show a high degree of inertia, so that rarely will the ecosystem switch between alternate states by changing just one variable. Not only do we need to reduce pollutant loading, but perhaps also change the structure of the community. The complexity of these interactions are now beginning to be incorporated into our models of nutrient transfers through freshwater systems – a prime example of how a model is refined by comparing its predictions with the real world.

8.5 MICROCOSMS AS EXPERIMENTAL MODELS OF LARGER SYSTEMS

Rather than relying solely on some numerical abstraction of an eco-system, some ecologists have used a subset of the larger system to model its dynamics. A **microcosm** is a small community, isolated in some form of container, and taken to be representative of a larger ecosystem. This can only be an imperfect model of the original system, but the aim is to catch the principal interactions between its components and thereby mimic the behaviour of the whole system. In this way, it is hoped, the impact of a disturbance on an ecosystem can be assessed and replicated. An extension of this technique is the **mesocosm** or enclosure, where part of a natural ecosystem is partitioned off for experimentation but is contiguous with the original ecosystem.

The small scale of the microcosm limits the accuracy of its predictions about the real system: the laboratory aquarium, for example, cannot hope to capture the effect of wind turbulence stirring the sediments on a large, shallow lake. Nor do microcosms show the long-term constancy of natural ecosystems – the aquarium typically degenerates into some sort of green soup unless steps are taken to manage it. As Moriarty (1988) notes, microcosms commonly suffer from being too simplistic to accurately represent the real world, but too complex for easy interpretation.

Table 8.3 Advantages of microcosms in pollution assessment

1. Properly designed, they can reflect some part of the interactions of a real ecosystem
2. A microcosm represents a dynamic test measuring the response of a complex system over a time series compared to an undosed control system
3. Microcosms allow replication, giving some measure of the inherent variability in the system
4. Some variables can be controlled in a formal experimental design
5. Microcosms can be used to assess the ecological hazard associated with a pollutant, measuring both the duration and magnitude of the response of the system

Analytical difficulties of interpreting microcosm studies for pollution effects
1. Microcosms suffer the inherent biological variability of real ecosystems and this limits the level of response at which an effect can be detected
2. The response of a microcosm consists of two components:
   (a) the direct toxic effects of the pollutant;
   (b) the indirect ecological effects of the pollutant
   and these may not be readily distinguished either in the real system or a microcosm
3. The extent to which the microcosm reflects the complexity of a real ecosystem is rarely known
4. Standardized microcosms cannot accurately model real ecosystems and are therefore limited in their predictive powers
5. Response curves are often complex and are not easily compared between experiments or microcosms
6. Relating the measurement of an ecological hazard to the real system can be problematical – does the response of the smaller microcosm accurately reflect the real system?

Nevertheless, they have been used extensively to model ecosystems where simple population studies have little hope of producing a quantifiable model, especially in the microbial communities of soils or the planktonic communities of aquatic ecosystems. In addition, by comparison with control microcosms, they offer some scope for quantifying the response of an ecosystem to a disturbance, both in terms of its inertial and adjustment stability. If the objectives of the study are properly defined, and the limitations of the method are fully appreciated, these ecosystem-level responses can be important indicators of the factors governing the behaviour of the real system.

Microcosms also have the significant advantage of allowing replication (Table 8.3) and the testing of a perturbation against an undosed control. In this way they can represent a form of toxicity test, predicting the concentration of a pollutant that is most likely to have an effect, just as we might measure dose-responses at the population level. However, there are problems establishing their accuracy and reproducibility, and statistical comparisons between ecosystems are not straightforward (Conquest and Taub, 1989).

Some attempts have been made to measure the reproducibility of microcosm studies. Sheehan (1989) compared the responses of freshwater microcosms derived from different sources (including one lake sampled at different times of the year) to four different toxicants. The results from each microcosm ranked the pollutants in the same order of toxic impact, and to this extent they each identified the most significant pollutant. But more elaborate analysis was difficult: comparing the dynamics of the response of two microcosms is not simple, especially where these are not linear. Of course, the same is true of real ecosystems.

Microcosms do allow a comparison of the elasticity of ecosystems, the time taken to return to their original state, either between replicates, treatments or ecosystems. We can thus measure the variability in their response to different stresses and compare how quickly different ecosystems might recover from one disturbance. Experimentation also offers the prospect of isolating those factors that control these responses – essential information for managing stressed ecosystems.

8.5.1 Measuring microcosm responses

Three types of variables are commonly defined in microcosm studies:

1. abiotic characteristics of the system – for example, details of water or soil chemistry;
2. ecosystem level parameters – such as primary productivity or chlorophyll α level;
3. the population densities of organisms, their relative abundance, and their organization as a community.

Different measures provide different sensitivities to a perturbation, and examples from each of these categories have been used to monitor responses to a disturbance. Comparing replicate freshwater microcosms, Sheehan (1989) found that dissolved oxygen, a measure of community metabolism, was more than twice as sensitive as primary productivity to a range of toxic insults. Again, between standardized microcosms (constructed to a standard recipe of water, sediment and species), dissolved oxygen and pH had the least variability when perturbed with various biocides (Stay *et al.*, 1989). Other ecosystem processes, such as rates of primary production and respiration, suffer more variation between replicates and, the greater the perturbation, the more uncertain the response became, especially in the flux rates for materials. The most variable responses of all have been found in structural measures, including the density of the organisms. Overall the direct effects of various biocides on the standardized microcosms, such as a decline in zooplankton density, reflected responses observed at experimental field sites, at least over the short term (Stay *et al.*, 1989). Notably, the age of the microcosm had no significant effect on its response, at least in these standardized systems. A more extensive study is needed to assess their capacity to measure direct and indirect responses by the community over the longer term.

A number of variations on the principle of the microcosm have been developed to measure ecosystem-level effects. For example, artificial substrates introduced into rivers and streams are used to collect a community which, when recovered, becomes a replicate in an impact assessment. Trays of pebbles placed in the Adair run, a tributary of the New River in Virginia, USA, have been found to establish a nearly complete community of invertebrates within 14 days. Clements *et al.* (1988) collected colonized trays after 30 days, and then exposed them to three different levels of copper. While control trays in the laboratory continued to mirror trays left in the field, dosed microcosms showed reductions in both the total number of individuals and the total number of species. This was a dose-dependent response, being most marked in the summer. Once again, measures of any structural change in the community, such as indices of diversity (section 10.2), were found to be relatively insensitive to these changes.

In contrast, differences in the response of various groups of organisms were found to be the most sensitive measures of stress in an experiment comparing the use of microcosms and mesocosms. Pratt *et al.* (1988) measured the number of species of protozoa colonizing artificial substrates (polyurethane foam blocks) placed in laboratory microcosms and in enclosures established in Douglas Lake, Michigan. A number of other parameters were also measured, including algal and zooplankton density. Here the toxicant was various doses of domestic bleach. The most sensitive group in the mesocosm were the zooplankton (Figure 8.8), although the mesocosm community was less sensitive than a microcosm exposed to the same concentration. Some of these differences were due to the mesocosms being enclosed, reducing their rates of colonization from the main lake. The microcosms, on the other hand, were arranged in a flow-through system, allowing recolonization. Such differences will affect the nature of the impact of the toxicant and the response of the system, and serves to emphasize the need for the design of the microcosm to reflect the real system accurately. In this case, the colonization rate was always likely to differ between a closed and open system, and between ecosystems drawing from different species pools.

The nutrient status of a microcosm is also important. The response in static and flow-through aquatic microcosms to cadmium depended on their level of nitrogen and phosphorus; generally enriched microcosms

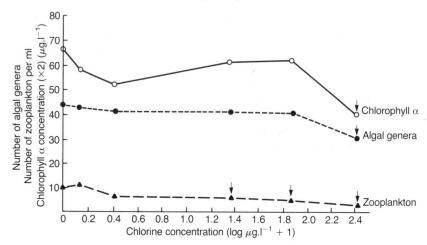

Figure 8.8 The effect of chlorine on chlorophyll α, algal genera and the number of zooplankton in mesocosms established in Douglas Lake, Michigan, USA, after 24 days. Each mesocosm received daily doses of sodium hypochlorate. Different measures showed significant responses at different doses (arrowed) – zooplankton numbers were the most sensitive to toxicant concentration. This differed however, from comparable microcosm tests. (After Pratt *et al.*, 1988.)

were less susceptible to the toxic metal (Hendrix *et al.*, 1982), although a significant proportion of the metal may have become immobilized by the abundant phosphates. Nutrient-poor microcosms tend to cycle their phosphorus more tightly than the enriched systems. Various parameters were measured in this study, but the most sensitive structural indicator of the cadmium level was the relative abundance of different taxonomic groups. The overall respiration rate and the rate of loss of nitrate were the most sensitive of ecosystem processes and their rise indicated some increase in decomposition as cadmium levels rose in the system.

Hendrix *et al.* (1982) suggest that microcosms could be used as part of a hierarchical testing procedure: having established that there are ecosystem-level responses, they suggest that microcosms would be useful in constructing models that describe how a pollutant moves through a system, its rate of accumulation and loss, and the possibility of comparing these processes between ecosystems. They also provide for replication and experimentation, to help to identify which components of the system need further study, the factors that regulate the response of the whole system.

8.6 BIOACCUMULATION

Modelling the movement of materials through an ecosystem allows us to derive a budget for each compartment. In this way we can measure the mobility of a pollutant within the food chain or its individual components – essential information for predicting the concentration in the diet of a consumer. From such information we can judge the extent to which a pollutant increases in concentration from one trophic level to another, a phenomenon called **bioaccumulation** or biomagnification.

The dynamic nature of these processes means that a pollutant's concentration in any compartment changes with time. In most cases, the rate at which a pollutant is lost depends upon its level within that compartment – the rate process (loss) rising as the state variable increases. With a constant rate of uptake, the level in the compartment will rise until the loss rate balances the uptake rate, and the compartment achieves a steady state. Then the amount of pollutant in the compartment is proportional to its level of exposure, the amount being taken up.

These compartments can be individuals, populations, trophic levels or some other functional group. Single and multiple compartment models were discussed for individual organisms in section 2.4, but these can be used to describe any component of an ecosystem, from a single tissue to the whole ecosystem. The loss rates in these models may be modified to include terms that allow for the loss of the pollutant through degradation or radioactive decay. Given these rates, we can derive its **residence time** – the average length of time a quantum of the pollutant remains in that component (equation 8.1). This measures the duration of exposure to the toxic hazard for that component.

Of the total amount of pollutant to which an organism is exposed, only some proportion will be assimilated into the tissues. This fraction is known as its **biological availability**, and will vary from one organism to another. Several factors determine the availability of different classes of pollutants, including their persistence (section 2.1) and a variety of ecological parameters. After its physical properties, such as size and means of dispersion, the chemistry of a pollutant is most likely to govern assimilation by the biota. Such factors also determine its degradability, either by its reactions in the abiotic environment or within the tissues of particular organisms. A pollutant is persistent if it has a long residence time in the ecosystem, and in the case of organic compounds, if it retains its original chemical form.

Within an organism, pollutants may be lost through excretion, through transformation by an organism's metabolism, or through radioactive decay. If the total rate of loss is low relative to the rate of uptake, the organism will achieve a level of pollutant in its tissues greater than in its diet. Similarly, the organism may concentrate a pollutant from the surrounding media, most especially aquatic species accumulating a pollutant directly from the water. It will show bioaccumulation; if this occurs over a number of compartments, organisms at the end of a food chain may be exposed to very high concentrations in their diet.

Bioaccumulation has been reported frequently for organochlorine insecticides, most notably DDT and its breakdown products, but the evidence is far from clear-cut (Moriarty, 1988). The principle is simple: a herbivore consuming a relatively large mass of plants with a low level of a pollutant assimilates some fraction of this intake into its tissues; the mass fixed in the tissues of the herbivore is subsequently available for a carnivore to assimilate. Because there are losses at each transfer, the overall mass of the pollutant may actually decline in each successive trophic level. However, each animal will consume many times its own weight of food, and thus will assimilate the pollutant from many individuals. A single consumer is thus exposed to a large mass of the pollutant and the concentration in its tissues rises. For bioaccumulation to occur, a consumer must retain a higher concentration of the pollutant in its tissues than its food, when in equilibrium with its dietary source. If this is repeated at each trophic level, the pollutant shows bioaccumulation along the food chain.

At steady state, the accumulation of a pollutant by an individual or a trophic level can be measured as its concentration factor (CF):

$$CF = \frac{\text{the concentration of pollutant in the consumer}}{\text{the concentration of pollutant in the diet}}.$$

This is an equation we met earlier (2.5). If the CF is greater than one at steady state, the pollutant shows bioaccumulation for that component or species. Notice that this includes only that assimilated from the diet, and not from the air or water. Non-dietary sources can be the major route of uptake for some pollutants, especially many organic compounds in aquatic ecosystems (section 2.4).

Bioaccumulation is most easily demonstrated for compounds that are not readily degraded or metabolized by any part of the ecosystem, especially the persistent organochlorine insecticides and elemental pollutants. Much of the evidence for the effect is, however, compromised by the details of the analysis (Moriarty, 1988). Dietary sources of the pollutant have to be distinguished from surface absorption or contamination, especially in aquatic habitats. Comparisons between compartments also need to be consistent – whole organism concentrations should not be compared with levels in a specific tissue or with a trophic level. Most importantly, comparisons between organisms can only be valid when they are both at steady state (Moriarty, 1988).

The close association of some organochlorines, such as DDT, with lipid, means that their concentration may vary between tissues according to the fat content of the latter. The proportion of fat also changes between species, between sexes and between individuals at different times of the year. Different organisms also have different capacities to metabolize insecticides, particularly where the insecticide has induced the development of resistant strains. The close relationship of metabolic rate to animal weight (equation 2.1) means that rates of uptake, loss and metabolic degradation of a pollutant can all change between animals of different sizes. Indeed, these may account for the different concentration factors found between trophic levels.

8.6.1 The effect of size on bioaccumulation

For a single individual, we can derive the steady-state concentration of a pollutant ($Q\infty$) as the ratio of pollutant uptake to pollutant loss:

$$Q\infty = \frac{fc}{k_{01}} \qquad (8.2)$$

where f = the weight of food consumed per day; c = the concentration of the pollutant in the food; and k_{01} = the rate constant for the loss of the pollutant by excretion or metabolism (Moriarty and Walker, 1987).

$Q\infty$ thus represents the rate of uptake divided by the rate of loss at steady state.

Metabolism is closely correlated with body weight and, as animals tend to get larger along a food chain, this alone may produce greater concentrations of a pollutant at the higher trophic levels. Moriarty and Walker (1987) have reviewed such effects for organochlorine insecticides. For example, food intake is commonly a function of body size – smaller species tend to consume more food per unit weight than larger animals. Consequently if all other things were equal – all species had the same rate of loss, were exposed to the same concentration in their diet and had the same rate of assimilation – the concentration of the pollutant would actually be greatest in the smaller organisms of the lower trophic levels. However, the rate of loss, k_{01}, is related to metabolic rate, and therefore to body weight. This rate differs between the major groups of animals, between different trophic positions, but again, tends to be faster in smaller animals. Larger animals with a slower metabolic rate will tend to lose the pollutant from their tissues more slowly. Animals at the end of a food chain could thus have higher concentrations simply because they are larger, rather than by virtue of their trophic position.

This possibility was shown with a mathematical model developed by Griesbach *et al.* (1982). They used a compartmental model of DDT in an ecosystem where no trophic structure was defined. Instead the 'animals' were grouped into five size classes, and each fed from a common 'food pool'. Since the compartments (the animals) were distinguished only by their size, any tendency for DDT to increase in the larger animals would suggest bioaccumulation was not due to trophic position alone. Rates of assimilation and loss of the pesticide were modelled as allometric

functions of body size (section 2.4). Following a pulse of DDT entering the
ecosystem, the model predicted that concentrations would rise most
rapidly in the smaller organisms. Over the longer term, however, the
insecticide would accumulate more in the larger animals. After a 'six-year'
run, the highest concentrations were predicted in the largest animals.

A sensitivity analysis of the model points to DDT being lost more
rapidly by smaller animals, effectively removing the pollutant from the
ecosystem. This loss contributed to the differences between the size
classes. Griesbach and her co-workers found evidence for this pattern in
the real world – higher concentrations of DDT are found in the larger
animals of aquatic ecosystems that have been exposed to the pesticide
over a long period. Based on the predictions from the model, and this
empirical evidence, the bioaccumulation of DDT may be a product of both
body size and trophic position.

Defining the trophic levels in a community, the compartments in the
model, is critical in predicting the trophic mobility of a pollutant.
Unfortunately, attributing an animal to just one compartment or one
trophic level can misrepresent the range over which it feeds and its
potential sources of the pollutant. An omnivore will feed at different
times on flesh or vegetation, causing its rate of pollutant uptake to vary.
Additionally, in a fluctuating environment, the steady-state condition
may be rarely achieved in most compartments, making it difficult to
confirm that bioaccumulation is occurring.

8.6.2 Modelling bioaccumulation in microcosms

Many early attempts at modelling the trophic mobility of pollutants relied
on simple microcosms. A pollutant might be introduced into an aquarium
for example, and the various components analysed after an interval of
time. Marine mesocosms have been used in the same way, but many of
these experiments have failed to mimic real ecosystems (Moriarty and
Walker, 1987).

Södergren (1973) used a freshwater microcosm where the trophic
compartments were physically separated, and where the budget for each
could be measured directly. Organochlorine pesticides were introduced
with the nutrient supply to the phytoplankton compartment (*Chlorella*),
which was then fed to various consumers. In an experiment examining
bioaccumulation of DDT, Södergren was able to show that the insecticide
was entirely bound by the algae, and in this microcosm the water was
only a minor source of the pollutant for the higher trophic levels.
However, these organisms had different capacities to degrade the DDT
(Figure 8.9). Unfortunately, the exposure time was relatively brief (eight
days) and no steady state is likely to have been reached at the higher
trophic levels. Even so, there was little evidence for bioaccumulation.

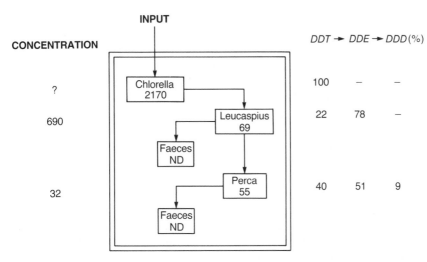

Figure 8.9 Compartmentalized microcosm model and budget of DDT assimilation and degradation in a simple freshwater community. Values refer to total DDT levels (ng) after the microcosm was exposed to the insecticide for eight days. This was introduced into the nutrient supply and was assimilated entirely by the phytoplankton. This meant that the herbivorous fish *Leucaspius* assimilated DDT from this dietary source alone. It was able to degrade this to DDE. The carnivorous perch (*Perca fluviatilis*) could further degrade this to DDD. There was little evidence of bioaccumulation of DDT between the herbivore and carnivore (concentrations in ng.g^{-1}): the CF for the perch was just 0.046. ND = no data. (After Södergren, 1973.)

A similar compartmentalized microcosm method was used by Ferard *et al.* (1983) to study cadmium bioaccumulation. Unlike DDT this cannot be degraded by the tissues. The microcosm had a food chain consisting of *Chlorella*, *Daphnia*, and the fish *Leucaspius*, which feeds on both the alga and the crustacean. The experimental design only allowed for dietary uptake by each consumer, and thus measured the concentration factors for each group. The algae were exposed for 10 days, reaching a steady state with the media. The metal was supplied in a highly soluble form and concentration factors up to 2200 were achieved, although assimilation rates were highest at the lower doses. For *Daphnia*, concentration factors also varied with levels of cadmium in the diet, but not in a consistent fashion, and only at one dose was it greater than one. *Leucaspius* showed very little accumulation of cadmium – over a four-day period, its total body concentration was just 1 µg.g^{-1} compared with the 570 µg.g^{-1} in *Chlorella* and 260 µg.g^{-1} in the *Daphnia* comprising its diet. There was thus no evidence of bioconcentration of cadmium along this food chain.

Cadmium will be readily accumulated by higher plants, such as *Elodea* but van Hattum *et al.* (1989) found that in *Asellus aquaticus*, a freshwater crustacean feeding on *Elodea*, direct uptake from the water was far more significant – the concentration factor was just 0.082 from the diet, compared with 17 500 from the water.

Studies on terrestrial systems suggest that cadmium may be one of the metals most likely to be concentrated along a food chain. Shrews, *Sorex araneus*, living near a copper refinery in Liverpool have been found to accumulate cadmium even though they regulate their concentration of copper (Hunter *et al.*, 1984). In a similar study, Andrews and Cooke (1984) described cadmium assimilation in small mammals from a reclaimed metalliferous waste site. As the soil is the principal source of cadmium for the overlying vegetation, this effectively eliminates respiratory uptake as a significant route into the animals. In this case, the insectivorous shrew had a concentration factor greater than one for cadmium, unlike *Microtus*, a vole feeding on grasses (Figure 8.10). In an established community such as this, organisms are more likely to be close to a steady state with their diet. Taken together, this all suggests that cadmium will indeed concentrate along certain food chains, although, with small mammals especially,

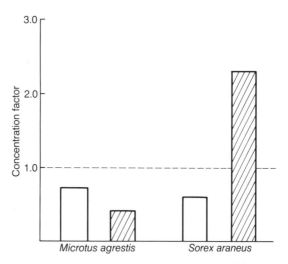

Figure 8.10 Evidence that cadmium may show bioaccumulation in one species of small mammal. Shrews (*Sorex*) and short-tailed voles (*Microtus*) were collected from established grassland on an uncontaminated soil (open boxes) and a metalliferous waste (shaded boxes). Inhalation was unlikely to be a significant source of the metal so these whole-body concentrations are probably derived largely from the diet. At the contaminated site, only the insectivorous shrew has a greater concentration factor than one, even though this species is perhaps half the size of the herbivorous vole. (After Andrews and Cooke, 1984.)

a series of samples through the year would be needed to show that bioaccumulation occurred in all seasons.

8.6.3 Modelling the bioaccumulation of radionuclides

Interpreting the evidence for bioaccumulation becomes more difficult when the pollutant can be degraded or undergoes radioactive decay. Models then have to include a term that allows for the reduction in tissue concentrations due to these transformations.

The decay of radioactive elements proceeds at a predictable rate and this term can be incorporated into a model to predict the concentration in any compartment at a particular time. Several models have been developed for both aquatic and terrestrial ecosystems, which attempt to describe the movement of radionuclides through each system, allowing for this decay. Brisbin *et al.* (1989) used such a model in their analysis of radioactive caesium in plants and animals near two nuclear reactors on the Savannah River in South Carolina, USA. An improvement in the effluent between 1971 and 1981 caused a larger fall in ^{137}Cs levels in the animals compared with the vegetation and, in both cases, this decline was greater than that attributable to isotope decay. There was no clear pattern of increase in the pollutant along the food chain, although here again it was impossible to ascribe some animals to a single trophic position. A number of correlations were found between ^{137}Cs concentrations in different groups even though they could have had no possible trophic connection. The correlations are probably a reflection of the overall level of ^{137}Cs in the ecosystem, rather than a feature of its trophic structure. Generally the results suggest that bioaccumulation alone could not account for the concentrations found in the higher trophic levels.

A compartment model for radionuclides in terrestrial ecosystems has been used by Jackson and Smith (1987) to predict the transfer of radionuclides between a simple soil–plant–animal system – a food chain involving sheep and cattle with man as the carnivore. The model thus assesses the potential exposure for anyone consuming contaminated sheep or cattle. Relatively fine compartments are defined in the model – for example, the plants are divided into roots, shoots, internal and external leaves and grains, with rate coefficients between each of these components. Empirical data has been used to refine these coefficients and to tailor the model to a particular site and soil. Using previous data for the movement of radioactive sulphur (^{35}S) into milk, a sensitivity analysis of the model suggests that critical determinants of the level of contamination are the rate at which ^{35}S is washed off the leaves and its transfer from root to shoot. Both are time-dependent and both serve to determine the amount of ^{35}S consumed by the cow.

These factors can differ between sites. Howard (1987) showed that surface contamination of the vegetation by soil or silt was the major source of ^{137}Cs in sheep, especially near the Sellafield nuclear processing plant in northwest England. Those grazing on salt marshes adjacent to the plant were more heavily contaminated than sheep on inland pastures – the main source of the ^{137}Cs was the outfall from the plant to the Irish Sea and some of this effluent returned in the silts deposited at high tide. Surface contamination of the vegetation was found to be a more significant source for the sheep, compared with that bound within the vegetation.

In contrast, surface contamination of the vegetation following the Chernobyl accident in 1986 was relatively unavailable to sheep and goats. In this case, it was the caesium incorporated into the plant which was the most important source of the pollutant (Howard and Beresford, 1989). The rate of transfer of ^{137}Cs into the milk of Norwegian goats actually increased in 1987–1989, despite the greater atmospheric deposition in the year of the accident. By these later years the radionuclide had found its way into the plant tissues, increasing its biological availability (Hansen and Hove, 1991). Levels of caesium in sheep on the uplands of Cumbria reflected concentrations in their diet but this relationship changed with time – most particularly with changes in the rates of consumption and the quality of the forage (Howard and Beresford, 1989).

The ecology of the areas where radioactive caesium was deposited after the Chernobyl accident also determined its availability to higher trophic levels. Livens *et al.* (1991) surveyed soils in central Norway, Scotland and northeast Italy, and found that plant uptake was generally greater if the soil had a high organic content and a low clay content. Such differences produced the greater contamination problems on upland fells and the increased uptake of ^{137}Cs by the Cumbrian sheep. The radionuclide is readily immobilized in the soil profile by clay minerals, but will continue to cycle through the upper, biologically active layers in soils with a large organic fraction. This was particularly true of the colder, more acidic moorland soils of Scotland. Even so there were differences between plants – mosses and lichens tend to accumulate more than higher plants, perennials more than annual plants. Concentration factors – comparing the levels in plant tissues with those of the soils in which they were growing – were larger than one in typical moorland species such as ericas and *Agrostis capillaris*, indicating bioaccumulation. However, values varied considerably between sites (Livens *et al.*, 1991).

Bioaccumulation certainly does occur under some circumstances with some pollutants. However, unequivocal evidence for it being a general property of many pollutants, applying in most ecosystems, is lacking. Other factors, such as body size, have to be distinguished from the effect of trophic position. For example, levels of ^{137}Cs in moose (*Alces alces*) in northern Sweden were shown to vary with age, sex and season immediately after Chernobyl (Danell *et al.*, 1989). Any attempt to measure bioaccumulation would need to eliminate or control each of these before comparing between species or trophic levels.

Even in controlled experiments, especially those based on models or microcosms, bioaccumulation is not often found, even after a steady state has been reached. For example, the woodlouse *Porcellio scaber* had concentrations of benzo(a)pyrene in its tissues 30–40 times lower than those in its diet, even after equilibrium had been achieved (van Brummelen *et al.*, 1991). Many more experiments are needed to establish whether bioaccumulation is actually the exception amongst the major pollutants.

Summary

Systems analysis offers one method of reducing the complexity of ecosystems to manageable proportions, a methodology that can extend the scientific process to highly complex ecosystems. With appropriate partitioning of the system, a model is used to generate predictions that can be tested against further observations, or experimental data. Systems ecology has been largely applied to studying the transfer of energy and materials in ecosystems, providing some insight into the factors that regulate ecosystem processes.

Disruption to nutrient cycling follows when ecosystems are enriched by pollution. Eutrophication of temperate freshwaters has been represented by one of several states, with the ecosystem switching between oligotrophic/eutrophic states – these alternative states have been characterized by the dominant members of the community – macrophyte/ phytoplankton, algae/cyanophytes or bream/pike. Models suggest that each state has a high degree of inertia. Based on these interactions, several efforts have attempted to manipulate the community structure, along with reducing the nutrient supply, to restore the ecosystem to a oligotrophic state.

Ecosystem-level parameters have been used to gauge the response of the system to stress, based on models of natural ecosystems and microcosms. Nutrient loss, particularly of nitrate, appears to be a sensitive indicator of disturbance in several ecoystems; others include rates of oxygen production, decomposition or respiration. Microcosms offer some potential for carrying out controlled and replicated experiments to measure ecosystem-level responses, although the details of their arrangement and of their analysis can be critical in the outcome.

Simple budgets for the levels of pollutants in various components of the ecosystem may suggest bioaccumulation, an increase in concentration with successive trophic levels. However, a range of factors needs to be considered to demonstrate this conclusively. The availability of many pollutants, including radionuclides, are determined by a number of biotic and abiotic factors, and elaborate models may be needed to predict concentrations in any component of an ecosystem.

Further reading

Harper, D. (1992) *Eutrophication of Freshwaters*, Chapman & Hall, London.
Kitching, R.L. (1983) *Systems Ecology*, University of Queensland Press, Brisbane.
Moriarty, F. (1988) *Ecotoxicology*, Academic Press, London.

Oxygenation ditch used to treat the sewage of a village of 450 people in the Languedoc, southern France. Using a timed cycle of agitation followed by resting, the tank goes through phases of aerobic digestion and sedimentation.

9

Exploiting ecosystems

Elaborate models of ecosystem processes are a recent innovation in ecology. Our exploitation of the productive or degradative capacity of ecosystems has a much longer history, relying only on the expert's eye and the model in the practioner's head. Experience has taught the manager to watch a small number of key indicators, such as water turbidity or the abundance of one group of organisms, rather than follow the whole system. These techniques have developed over many generations, sometimes being lost, and sometimes arising independently in different parts of the world.

In doing so, they have undergone a process of evolution of their own, surviving according to their utility. In sewage treatment, the activated sludge process is a well-documented example of how an idea can evolve, from its first appearance at the beginning of the 20th century (Berthouex and Rudd, 1977). With the selective pressures of formal scientific experiment and, more often, of trial and error, the method has been adapted and refined for its particular function – the rapid degradation of organic matter. Control is exerted through a number of variables, most especially by regulating the average age of the microorganisms within the ecosystem, to encourage a decomposer community which does not itself produce a large biomass.

Key variables are also used to indicate the state of the ecosystem, effectively summarizing the interactions within its community. These are most likely to be process variables, such as rates of nutrient transfer or respiration. Many of these processes are highly predictable at the community level and can be modelled easily using mass balances (section 8.1). One alternative is to follow the abundance of key species, those whose density or activity may control the state of the system, as in eutrophication (section 8.4). While this can be used for monitoring the state of microbial communities, it is often too slow for regulating an effluent treatment plant or a commercial fermentation. Most microoorganisms are known by their deeds – by the substrates they attack and the products they produce – and so we monitor the disappearance of a substrate, the rate of oxygen consumption or some other process variable which can be measured quickly. From experience, plant managers have learnt which factors are crucial to maintain high rates of decomposition or production.

In other cases, the appearance of particular species indicates the system shifting to an alternative state, often as the abiotic conditions change. Different effluents provide different substrates for the microbial community, causing major shifts in the dominant groups. By modifying the external variables we have been able to use the adaptive capacity of these communities to treat a large variety of wastes (section 3.6).

The central aim of managing these ecosystems is to sustain their productive or degradative capacity – maintaining them close to an optimum state where these processes are maximized. This can give some insight into the regulatory processes of the natural ecosystems which they mimic: a conventional sewage works consists of a number of communities, each accelerating some part of the decomposition process or the oxidation of ammonium as they occur in a freshwater system. These processes are central to most ecosystem functions, releasing energy and nutrients for the whole community.

In the previous chapter, systems analysis was introduced as a means of modelling ecosystems. Using this approach, it was shown that manipulating state variables, most especially the abundance of key species, could shift the ecosystem from one steady state to another. Here we consider an alternative approach to controlling ecosystems, the use of process control. This method, more normally applied to process engineering, manipulates external variables (Table 8.1) to control the processes within a system, based on simple mass balances. The performance of the system as a whole is judged by its output.

For the most part, we concentrate in this chapter on the ecology of a number of artificial ecosystems which have been created for specific purposes, mainly waste treatment. Here we look particularly at the methods used to optimize their function by controlling both the abiotic environment and the composition of the community. The criteria used to judge their performance include rates of decomposition, of gas production or, in some cases, biomass production. Although these systems are not as complicated as most natural ecosystems, many include highly complex communities which have yet to be fully described. Only now are we beginning to describe their ecology and how changes in some variables serve to control them. The same techniques can be used to exploit natural ecosystems for similar purposes, and we go on to consider how our understanding of ecosystem ecology is central to maintaining their capacity and ecology.

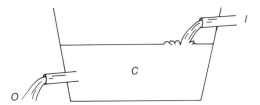

Figure 9.1 A representation of mass balance in a water tank.

9.1 SIMPLE PROCESS CONTROL

Many of the mechanistic approaches described in the previous chapter form the basis of our attempts to optimize artificial ecosystems. Process control uses a small number of variables, derived from some form of predictive model, to monitor and regulate these operations. In some processes the model may itself be part of a computer program used to control the system automatically, as with brewing and some other industrial fermentations. A number of key variables are monitored, simplifying the highly complex interactions within the ecosystem, but allowing us to consider its performance solely in terms of specific and quantifiable criteria. Using a mass balance (section 8.6), we can judge its performance against some key output, such as a nutrient loss or a desired product. Again, the method can be applied to the whole system, or to a single compartment:

total mass entering the compartment	−	total mass leaving the compartment	=	total mass accumulated in the compartment
(I)		(O)		(C).

We can envisage this balance as the level of water in a tank (Figure 9.1).

At steady state there is no net change in C – the level remains constant because the input is balanced by the output:

$$I = O.$$

The rate at which water is lost from the tank might depend upon the amount of water in it. A greater depth of water will exert a higher pressure, driving the water through the outlet faster:

$$O = k_1 C.$$

O increases in direct proportion (k_1) to the size of C. For our purposes we assume the amount lost is a constant proportion of the volume of water in the tank.

At steady state the input must balance the output:

$$I = k_1 C$$

and so

$$C = \frac{I}{k_1}. \tag{9.1}$$

Thus at steady state, the volume of water in the tank is set by the ratio of the input to the rate of loss.

Quite simply, with a constant rate of loss, C will rise as the input increases (until the tank overflows). Conversely, an increase in the rate of loss will lower the volume in the tank as long as there is no change in the rate of input. This is equivalent to equation 8.2 set out in the last chapter when we discussed the accumulation of pollutants in organisms.

Achieving steady states in continuous cultures of microorganisms can be used to measure their growth rate in a reaction vessel (or **bioreactor**). Again the volume in the culture is kept constant, so that loss of the media through the outflow is matched by new media being added. If the population of microorganisms is also to remain constant, cells lost at the outflow must be balanced by new cell growth. The input supplies a limiting nutrient at a rate low enough for it to be fully used in bacterial growth. At steady state, the rate of growth is thus determined by the rate of input, with the population under density-dependent regulation (section 4.2). A continuous culture is capable of regulating itself, achieving a steady state over a wide range of substrate inputs, except where the growth rate is close to its maximum. Altering the rate of input can thus be used to control the bioreactor, or more specifically, the growth rate of the culture. Because of the control we have over the quality and quantity of the substrate entering the vessel, this system (commonly called a chemostat) has been used extensively for studying microbial biochemistry and population dynamics (Figure 9.2).

A simple mass balance can also be applied to substances that undergo chemical transformation, as long as we have fully quantified the stoichiometry of the reaction. A budget can then be constructed for each element – as in the aerobic respiration of glucose:

$$C_6H_{12}O_6 + 6O_2 \rightarrow 6CO_2 + 6H_2O$$

or the fermentation of glucose to ethanol:

$$C_6H_{12}O_6 \rightarrow 2CO_2 + 2CH_3CH_2OH.$$

Both chemical oxidations yield energy, and the proportion of each element is maintained on either side of the equation. This allows us to construct a mass balance for carbon, hydrogen and oxygen as they pass through different compartments and organisms within the ecosystem. In the same way we can balance the energetics of the whole ecosystem.

The capacity of different bacteria to attack very specific substrates is used to degrade a variety of wastes and produce useful products – either

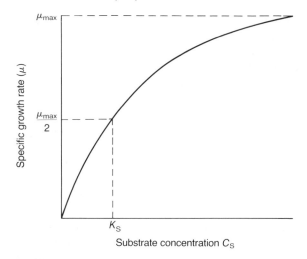

Figure 9.2 The monod relationship describes the response of a bacterial population to an increase in the limiting substrate. The population will grow until it reaches its maximum growth rate (μ_{max}), beyond which an increase in the available substrate has no effect. Below this level, the constant input of new substrate will displace a constant number of bacterial cells. At steady state, the growth of new cells supported by the new substrate is matched by losses at the outflow. A culture can be maintained indefinitely in this condition. The rate of population growth differs between species, producing different yields for the same rate of input. The growth rate of a species (μ) is a simple function of the substrate concentration at half the maximum growth rate:

$$\mu = \mu_{max}\frac{C_S}{K_S + C_S}$$

where μ and μ_{max} are the actual and maximum possible growth rates; K_s is a constant relating μ to C_s, the concentration of the limiting substrate. K_s is termed the monod constant and is defined as the concentration of the substrate at which the growth rate is exactly half the maximum rate.

by simultaneous reactions, where the activity of one organism facilitates the presence of another, or by step-on-step reactions where one organism provides a substrate for a second, again through facilitation. The plant manager aims to maximize a process without endangering the stability of these interactions, even though the community itself has rarely been completely described. In much of effluent treatment, the ecosystem is treated as if it were a black box and its performance is regulated using an external variable, such as the input rate. Little attempt is made to monitor the species interactions and the stability and persistence of the ecosystem are judged simply by its output.

Ideally, the bioreactor will retain a high degree of inertial stability, accomodating a range of variations in most external factors without becoming soured or switching to some alternative state. A community will respond to a combination of external variables – rates of decomposition, for example, are dependent upon pH, moisture and temperature, as well as the nature of the substrate. At low temperatures or low pH, or with certain substrates, a particular collection of decomposer micro-organisms will be present, though in most cases these go undescribed and uncounted.

The balance between the inputs and outputs from a system is the principle criterion used to judge its efficiency. Using mass balances in this way is not confined to artificial ecosystems: the method has been used to assess pollution and restoration problems in large natural ecosystems – such as the Great Lakes of North America, the North Sea and the Mediterranean (Uhlmann, 1991). This is the only realistic option when systems of this size are faced with a major pollution threat, and when there is little immediate prospect of modelling the details of their biotic interactions. In some cases, altering an external variable may not have the desired effect, as with the attempts to reduce nutrient flow into some eutrophic lakes (section 8.4). Here biomanipulation, altering the structure of the community, may be needed to produce a more permanent change in the ecosystem. On a different scale, biomanipulation forms part of the strategy in regulating several of the artificial communities reviewed in this chapter, usually by introducing a key species for some particular function.

9.2 SEWAGE TREATMENT

Decomposition is a central process in all natural ecosystems. The degradation of the complex compounds in organic matter liberates both its chemical energy and its mineral nutrients, making these available to the rest of the community. The size and diversity of its decomposer community determines the capacity of an ecosystem to degrade different organic fractions. This, in turn, is set by the regularity of the supply of organic matter – most natural ecosystems receive largely predictable types and quantities of these substrates. A large input of some organic effluent into the system may substantially perturb the balance between decomposers and the rest of the biota. Freshwater ecosystems in particular can become severely de-oxygenated by the activity of decomposers receiving a sizeable pulse of untreated sewage.

A sewage treatment plant concentrates the decomposition process in a series of controlled habitats, each of which optimizes conditions to accelerate some phase of the process. The aim is to convert as much waste as possible to carbon dioxide and water. In this way, the water leaving the plant has a low organic content, and will support little further bacterial growth. Its nutrient content, particularly nitrogen, will also have been lowered. To minimize disposal costs, the solids produced in treating the waste have to be reduced to the smallest bulk possible. Within the plant, this organic matter is derived from two sources – the bacterial populations and waste that arrives at the inlet pipe and the growth in the biomass that occurs within its ecosystems.

The efficiency of a treatment works can be readily measured by the difference in the organic matter in its inflow and its outflow. This is determined as the **biochemical oxygen demand** (BOD): the amount of oxygen required to oxidize the organic matter in a sample by biochemical processes (Table 9.1) – measured as the reduction in the oxygen content of the water over a standard incubation period. Most of this demand arises from the presence of carbonaceous compounds, although some are attributable to nitrogen-rich compounds – proteins, ammonia and so on – which have a BOD of their own. As part of the organic matter will be converted to new cells, a BOD determination does not represent a complete oxdiation of the effluent. This is determined chemically, as the **chemical oxidation demand** (COD), the amount of oxygen needed to completely oxidize the substrate to carbon dioxide and water. Consequently the COD is always somewhat larger than the BOD. In addition most sewage works attempt to oxidize ammonium to nitrate, using nitrification. This not only lowers the COD of the effluent, it also reduces

Table 9.1 Measuring the strength of domestic sewage – the five-day biochemical oxygen demand. A five-day BOD measures the amount of oxygen used by microorganisms under aerobic conditions when oxidizing organic pollutants at 20° C

An accurate test needs to ensure that:
- there is a diverse bacterial community present in the solution
- no essential nutrients are in short supply (N or P)
- extreme conditions of pH, metal concentrations or the presence of disinfectants are avoided (or removed)
- the method of determination is specified, including any dilution factors used, inoculations or nutrient additions

Additional considerations
- dilution may more closely reflect the impact of an effluent on a receiving water course – when low pollutant levels are accompanied by low microbial densities
- a BOD is not a complete measure of the oxidizable pollutants in an effluent and many need to be used in conjunction with a COD determination
- nitrogenous compounds will also exert an oxygen demand if nitrifying organisms are present

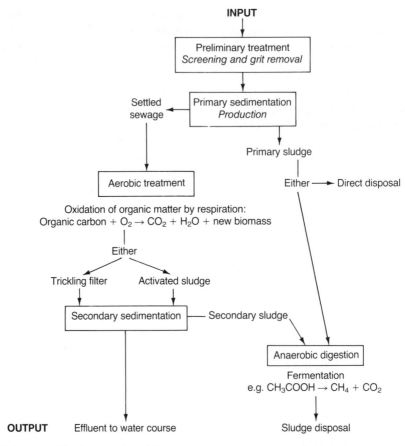

Figure 9.3 The layout of a conventional sewage treatment works. This process can be divided into a number of stages, not all of which may be present in each plant. The early stages are largely physical treatments, removing non-degradable components, and sedimentation to remove excess water. These are followed by biological treaments, usually commencing with an aerobic, oxygenated phase, and then perhaps an anaerobic digestion phase.

the risks of eutrophication, and the toxicity problems associated with ammonium for the freshwater community. In some works, denitrification may be used to remove high levels of nitrates from the discharge, to protect groundwater and drinking (potable) water supplies.

Sewage treatment also has to reduce the threat of disease from the pathogens and parasites carried in the waste, and these are also monitored in discharging waters. A count of coliform bacteria is commonly used as a measure of faecal pollution in freshwaters. These bacteria are normal inhabitants of the digestive tracts of mammals and are used as indicators of the likely presence of other pathogens.

These health risks have led most industrialized nations to make a considerable investment in sewage systems over the last 100 years. The conventional system consists of a pipe network that uses water to transport the waste to the treatment plant. Here the contaminated water has to be separated from the gross solids, and treated through a series of sedimentation and degradative steps (Figure 9.3). Nowadays, with the demand for water continually rising, this strategy is being questioned and alternatives are being considered. For many developing nations, the large investment required in the pipework and the treatment plant, as well as a shortage of water, make this an inapproriate technology. In drier parts of the world, and in the poorer nations, it may be better to treat the waste near to the source – perhaps with each home having a dry-treatment facility. No water is used in transport, and no water needs to be removed again during treatment. In effect these methods operate by increasing the C:N ratio of the waste *in situ*, so that a composting process begins, with a humus-rich soil conditioner being the end-product.

9.2.1 Conventional sewage treatment

The sewage arriving at the treatment plant will be derived from a variety of sources, and its quality will reflect the proportion of domestic and industrial sources in its catchment area. Effluents from a food processing factory or from an electroplating works have very different pollutant burdens; similarly the quality of the sewage varies with the time of day and season of the year.

Here we concentrate on domestic effluent, consisting largely of faecal and nitrogenous waste, with a preponderance of carbohydrates, fats and proteins, and which arrives as a very dilute solution. Raw sewage invariably contains a variety of other materials, each with different degrees of degradability. Rags, cans and other debris are easily separated out at the inflow using screens. Similarly, grit can be collected in troughs after screeening. Some components of the sewage, such as urea, will rapidly oxidize in contact with the air, and their degradation may be complete before they reach the plant. However, the greater part of the decomposition process is carried out by microorganisms within the works.

After screening, the velocity of the inflow is reduced in the primary settlement tanks to allow the suspended solids to settle out. The settled sewage or 'grey water' is drained off for aerobic treatment, and the sludge scraped and collected either for direct disposal (at sea or in landfill sites) or for further treatment by anaerobic digestion.

The settled sewage provides a range of substrates that the microbial

Exploiting ecosystems

community can use, including mineral compounds which drive the respiration of **autotrophic** microorganisms (those using inorganic compounds as their energy source) as well as the organic matter used by **heterotrophs**. Each step in the breakdown of organic matter requires a specific enzyme, and if this degradation is to run to completion, a full sequence of enzymes needs to be present, or at least inducible. The extent to which any organic compound is degradable depends upon the range of microorganisms within the community and the enzymes they synthesize; those with the greatest diversity of genera tend to have the greatest degradative capacity (Mudrack and Kunst, 1986).

Both aerobic and anaerobic metabolisms are exploited in the sewage works, and most treatment plants have at least two ecosystems – one using aerobic respiration to reduce the BOD of the settled sewage, and another digesting the sludge by anaerobic digestion or **fermentation**. Whereas respiration uses an inorganic compound as the oxidant, fermentation relies on an organic compound. Both processes entail the breakdown of the high molecular weight organics in the effluent, producing smaller compounds that have a reduced capacity to combine with oxygen.

Following primary sedimentation, the settled sewage undergoes aerobic treatment by one of two methods – biological filtration or activated sludge. The first method uses a fixed-bed bioreactor, the trickling filter, in which a microbial and invertebrate community is established on an inert support medium. In the second, a microbial community is used as a continuous culture in suspension. Both methods aim to maximize the supply of oxygen to the decomposers to accelerate the reduction in the high BOD of the solution.

The microbial populations growing on the settled sewage generate a biomass which itself forms a sludge, and a substrate for other microorganisms. This biomass would, in turn, demand large amounts of oxygen were it released into the environment. Consequently this has to be separated from the outflow or digested before leaving the plant. A community of anaerobic bacteria is commonly used to ferment this secondary sludge along with the primary sludge collected from the first sedimentation (Figure 9.3).

9.2.2 Biological filtration

The bed of a trickling filter is responsible for little physical filtration of the settled sewage (Figure 9.4). Instead, the clinker, slag or gravel within the bed is used to support a complex community of bacteria, fungi, algae and a range of invertebrates, which collectively reduce the BOD of the sewage. The open structure of the support allows a free circulation of air for the aerobic degradation of the organic compounds. It also provides a

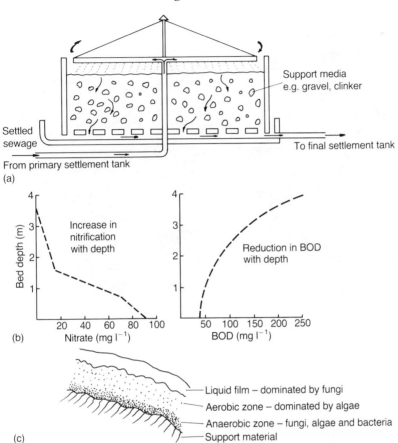

Figure 9.4 Biological filtration. (a) The trickling filter or percolating filter is a fixed-bed bioreactor over which the settled sewage is slowly trickled. (b) In its passage through the bed, the BOD of the sewage is gradually reduced by the microbial activity of the bioreactor. (c) The ecosystem consists simply of a support medium upon which a decomposer community develops – bacteria, algae, fungi and invertebrates form as a biological film on the support. In addition, there is a community of grazing invertebrates living on the production of the lower trophic levels. The interactions between these species help to regulate the development of the film and, in doing so, maintain the efficiency of the bioreactor. The film is up to 2 mm thick and is stratified into various layers, with oxygen becoming depleted inwardly with increasing film thickness. Similarly the whole bed is stratified towards the base, where a shortage of organic matter encourages the growth of autotrophic microorganisms, especially the nitrifiers responsible for the oxidation of reduced nitrogen compounds.

Figure 9.5 Biological filtration beds using clinker as a support material, on which a microbial community develops. This represents a fixed-bed reactor. The rotating arms trickle the settled sewage over the bed, and updraft through the clinker ensures rapid aerobic degradation of its organic content.

large surface area, concentrating the decomposer community into a small volume. In some works a special light, plastic support is used to enable very deep beds to be constructed, further reducing the land area necessary for proper treatment. Clinker beds are often circular or rectangular and up to 1.5 m in depth, above which moves a pipework trickling the settled sewage over the bed (Figure 9.5).

The microbial community forms as a biological film on the support, and is stratified down through the film (Figure 9.4). Several trophic levels form the complete community within the bed, including invertebrates that are important in regulating microbial density. The dominant bacterial genera include *Pseudomonas*, *Zooglea*, *Flavobacterium* and *Alcaligenes*. In the surface layers of the bed, where light is abundant, a number of algal genera are also found, amongst them *Stigeoclonium*, *Ulothrix*, *Amphithrix*, *Anacystis*, *Euglena* and *Chlorella*, and these are important in absorbing nutrients, especially nitrogen, which become available as the organic material is mineralized. Towards to the outer layer of the film are yeasts and moulds, including *Aureobasidium*, *Subbaromyces* and *Fusarium* as well as some fungi imperfecti. Throughout the bed, the fungi form the largest part of the biomass.

The nature of the community also changes with depth through the filter bed: as organic matter becomes depleted towards its base, autotrophic bacteria predominate – especially the nitrifying bacteria that use the nitrogenous compounds in the sewage as a source of energy. Within the bed, where light is limited, some algal genera may become heterotrophic (Dart and Stretton, 1980).

A small proportion of the oxygen needed by the organisms within the biological film is provided by the algae during the day, while at night algal respiration lowers oxygen levels. The main supply is provided by a draught of air through the bed generated by the sewage percolating downwards, and by a temperature differential established with the outside air. It is critical that the film does not grow too large to hinder this circulation, inducing anaerobic conditions. In particular, a dominant fungal population can produce a large biomass, increasing both the chances of the bed becoming blocked and the amount of sludge generated by the reactor.

Excess growth in the filter bed is lost with the flushing action of the sewage, and also through the grazing activity of the invertebrates, most especially worms and fly larvae. A variety of adult arthropods, especially springtails (*Collembola*), feed on the fungi and algae, and the efficiency of the ecosystem can be correlated with the activity of these consumers. The development of this grazing community is part of the maturation process of the bed, essential to prevent excessive growth of the film. For this reason, beds in northern temperate latitudes are best started in the spring and summer, allowing them three to six months to acquire a range of nematodes, annelids and arthropods. Without these species, an excessive growth in the film will lead to ponding of the bed in the wetter winter months (Dart and Stretton, 1980).

Most biological filters have a degree of self-regulation, simply due to the physical breakdown of the film – as it increases in thickness, so it becomes more likely to detach from the support material. The film grows most rapidly where there is a high nutrient load, toward the top of the bed (Figure 9.4). Because it is flushed away more rapidly in this region, the top of the bed tends to favour microorganisms with a short generation time. At the base, where the film is shed less frequently, longer-lived nitrifiers can become established (Mudrack and Kunst, 1986).

Compared with other methods of biological treatment, biological filtration is highly tolerant of shock loads of sewage or toxins. The presence of *Pseudomonas* and *Vibrio* species means that they have some capacity to degrade phenolic compounds (section 3.6), and thiosulphates and thiocyanates can be oxidized by *Thiobacillus* spp. At least in these artificial ecosystems, the stability of their function in the face of these stresses can be directly attributed to the diversity of their microbial community (section 10.3).

Table 9.2 The reduction in the pollutant burden of sewage in its passage through the various stages in a conventional treatment plant

Stage	Sewage input (% sludge)	Removal (%)		
		Suspended solids	*BOD*	*Coliforms*
Primary sedimentation	0.1–0.5	40–95	30–35	40–75
Biological filtration	0.1–0.5	20–90	60–95	85–95
Activated sludge	1–3	70–97	70–96	95–99+

It is also perhaps the most effective method of reducing nitrogen within the sewage, as oxygen is freely available throughout the filter bed, allowing the oxidation of ammonia. This is a repetition of the nitrification process that occurs naturally in many soils (Table 7.4). However, it is difficult to determine the capacity of a bed to process nitrogen because the biomass of its nitrifiers is not easily measured, and its efficiency can only be judged using a mass balance of inputs and outputs (Mudrack and Kunst, 1986).

An efficient trickling filter can remove up to 95% of the BOD and suspended solids of the incoming sewage (Table 9.2). Similar levels of reduction occur in the numbers of bacteria and pathogens. The sludge recovered from the sewage draining off the bed is collected in a secondary sedimentation tank (called a humus tank), usually to be fed to an anaerobic digester.

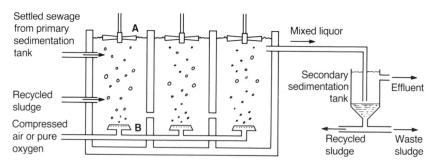

Figure 9.6 The activated sludge process requires part of the sludge which has already passed through the aeration tank to be returned, to enrich the microbial culture. Most aeration tanks have either mechanical aeration (A) or are aerated and mixed by bubbling air through them (B), rarely both together as shown above. If pure oxygen is bubbled through the tanks, this may be progressively decreased away from the raw sewage input, so that the added oxygen is matched to the reducing BOD along the tank.

9.2.3 Activated sludge

This alternative method of aerobic treatment uses a type of continuous culture in which an inflow of sewage (the substrate) into one side of a tank displaces treated effluent out of the other. Unlike the chemostat, however, some part of the outflow is returned to the bioreactor – the decomposition process is enhanced because the inflow contains sludge that has already passed through the tank, and is 'activated', or enriched with decomposers. This inoculum helps to maintain an active microbial biomass in the bioreactor (Figure 9.6), and imparts a stability to the community over a range of nutrient concentrations. The culture is continually stirred, or has air or oxygen pumped through it, to maximize the rate of oxidation of the organic matter (Figure 9.7).

Figure 9.7 Activated sludge tanks, under aeration from pumice blocks at the bottom of each tank.

The process relies upon the formation of **flocs** – clumps of micro-organisms and abiotic matter, that aid the clarification of the sewage and the reduction in its BOD. They also accelerate the settlement of the microbial biomass. A floc is effectively a suspended ecosystem of organic and mineral matter, with a community of bacterial and fungal decomposers and protozoan consumers. The bacterium *Zooglea ramigera* secretes extracellular polymers that form a gel, and is probably the main agent responsible for floc formation. A range of physical and chemical factors contribute to successful flocculation, including the availability of carbon and nitrogen, and of metal salts, such as calcium and magnesium, as well as the the presence of other species.

Because sludge is returned to the culture, the process cannot be controlled simply by the rate of inflow – bacteria are added to the culture with the returning sludge, so their mean residence time is longer than the period they spend in the aeration tank. Instead, the process is controlled by the sludge wastage rate – the growth rate of the bacteria being equal to the rate of sludge loss.

The retention time for sewage in the aeration tank is set by the dilution rate:

$$\text{dilution rate } (D) = \frac{\text{rate of inflow of sewage}}{\text{tank volume}}. \tag{9.2}$$

The mean retention time in the tank is simply the reciprocal of the dilution rate ($1/D$): if one-fifth of the tank volume is changed every hour ($0.20\,h^{-1}$), then the mean retention time of the sewage is five hours (equation 8.1). Although fluids will pass through the system relatively rapidly, perhaps within a few hours, part of the sludge is recycled, and the average retention time for the microbial biomass in the whole system is much longer (between 4 and 10 days). The steady-state conditions mean that the overall growth rate in the bacterial community is controlled through the rate of loss of sludge – the suspended microbial biomass lost with the effluent from the system. As with the dilution rate above, this is simply the reciprocal of the **mean cell retention time** (or **sludge age** – $1/\theta_c$). The process will remain self-regulating as long as a dynamic equilibrium between bacterial growth and substrate supply is possible and the maximum growth rate is never achieved (Figure 9.2).

With the formation of flocs, clarification of the effluent can occur very rapidly. However, the oxidation of the organic matter, and subsequently of the ammonia, leads to the BOD of the sewage declining more slowly. When these nutrients are used up, the cells begin a final phase called autodigestion, where the dead bacterial cells become a substrate themselves. The rate of BOD removal and quality of the effluent depends upon the phase of the bacterial community: a rapidly growing population will quickly lower the BOD, but will also produce large amounts of biomass.

An older community lowers the BOD more slowly, but produces less sludge. Most conventional activated sludge plants operate with the inoculant introducing flocs that are in this later, senescent stage. This is set by the rate of inflow giving a longer mean residence time for the sludge in the aeration tank, and an older average sludge age. By encouraging long generation times, autotrophic nitrifying bacteria also become established. Their oxidation of reduced nitrogen compounds is essential for the minimum possible BOD in the outflow.

Compared with biological filtration, activated sludge has a much lower diversity of species, probably reflecting the low diversity of habitats in this continuously stirred and relatively uniform ecosystem. However, there may be a large variety of heterotrophic bacteria using the sugars and fatty acids: most bacteria can only use one or two carbon sources and the variety of substrates within the effluent provides a range of niches. For the same reason, the dominant groups will vary according to the nature of the waste – effluent rich in antibiotics, for example, tends to have high numbers of *Pseudomonas*, almost certainly because of their plasmid-linked resistance (section 3.5).

There is also a large variety of protozoa in an aeration tank – over 200 different species have been identified, with the most common forms being ciliates. The composition of the protozoan community can be a useful guide to the state of the system, and indicator species are used as a measure of the long-term BOD loading on the bioreactors (Table 9.3). Most protozoa feed on the free-swimming bacteria outside the flocs,

Table 9.3 Indicator species of protozoa in activated sludge (after Mudrack and Kunst, 1986)

Overloaded system *(Poor oxygen supply)*	*Normal system* *(Good oxygen supply)*
Rhizopoda	
Naked amoebae	Testate amoebae
Ciliates	
Colpidium campylum	*Paramecium caudatum*
	Asidisca costata
	Euplotes affinis
Vorticella microstoma	*V. convallaria*
	V. campanula
Opercularia coartata	
Flagellates	
(typical of heavily loaded system at start-up) e.g. *Trigonomas*	

including some, such as *Vorticella*, which remain attached to the flocs. These consumers are important in maintaining a balance in the microbial culture and, again, if they are reduced in the culture, the quality of the outflow from the aeration tank declines. In the absence of protozoa, the turbidity of the effluent increases with the abundance of free-living bacteria, with corresponding increases in BOD and suspended solids. The protozoa, especially species such as *Balantiophorus minutus* and *Paramecium caudatum*, may also be important in enabling the formation of flocs, by secreting mucus.

A number of invertebrates are also found in activated sludge, especially rotifers and nematodes. Algae are mostly absent. The fungal community is usually small, although if their biomass increases, the sewage may 'bulk' – that is, increase its suspended biomass, keeping the sewage turbid and preventing settling. This is often found in systems where the level of nitrogen is low: fungi such as *Geotrichum* will thrive if both nitrogen and phosphorus levels are low, as do filamentous blue-green algae. A number of bacterial species can also develop filamentous forms under these conditions, and the communities then become dominated by a different collection of bacteria, including *Sphaerotilus, Microthrix* and others. This is one of several shifts that can be observed in an activated sludge community: others are associated with changes in nutrient and oxygen levels, and with temperature effects (Mudrack and Kunst, 1986).

In an efficient activated sludge plant, the total bacterial numbers, BOD and suspended solids of an effluent can all be reduced by up to 98%, and bacterial pathogens are reduced by a similar amount (Table 9.2). With a particularly intractable pollutant in the system, such as phenolic compounds, the retention time may need to be increased to allow complete decomposition. In some cases this might also require the development of a community with the degradative capacity and the use of an additional stage or treatment tank.

Relatively little work has been done on optimizing the aerobic treatment of wastes – typically this is achieved by altering the balance between the major nutrients, nitrogen and phosphorus, as well as the oxygen demand of the culture (Hamer, 1990). With the prospects for strain improvement by genetic engineering, a much greater variety of wastes could be treated aerobically. As Hamer notes, these techniques stand to be improved by our understanding of species interactions and community dynamics. This could allow us to introduce non-indigenous strains into a bioreactor, without the risk of them being lost through competitive interactions. Such methods could improve the prospects for greater use of **cometabolism** – the degradation of a compound by an organism which is not using it directly as an energy source. The initial dehalogenation of a chlorinated hydrocarbon (section 3.5) is an example.

Cometabolism offers the possibility of degrading many organic pollutants without any increase in the microbial biomass.

9.2.4 Nitrogen removal from sewage

Besides the carbon compounds in the sewage, the other significant source of a high BOD is the nitrogen-rich compounds. These too need to be oxidized, and the nitrogen removed. Ammonium is a toxic poison, especially to freshwater fish, and its degradation in natural water bodies may also pose a threat to consumers of water extracted directly or indirectly from aquifers.

Nitrogen removal in the sewage works exploits those bacteria responsible for nitrification and denitrification (Table 7.4). Urea entering the sewer is rapidly degraded to ammonia before it reaches the treatment plant, and many nitrogen-rich compounds will readily precipitate in the primary sedimentation tanks. Nitrification serves to oxidize the reduced compounds, such as ammonium, further reducing the BOD of the effluent. The nitrogen may be lost to the atmosphere (by denitrification) or be incorporated into the microbial biomass as part of the sludge. The latter becomes the major source of nitrogen for any subsequent anaerobic digestion of the sludge. Where an untreated sludge is spread directly onto land, this nitrogen can make it a worthwhile agricultural fertilizer.

After primary sedimentation, inorganic nitrogen compounds in solution go on to an aerobic treatment, where they will be assimilated by bacteria, and by algae in the trickling filter. The activated sludge process requires a C:N ratio of about 20:1 to encourage bacterial growth and prevent bulking. Nitrogen fixed in either system is lost with the biomass wasted, perhaps up to 30% of that entering the reactor (Mudrack and Kunst, 1986).

Nitrification will also cause a significant loss of nitrogen, but this requires abundant oxygen. Nitrifiers are slow growing and will only become established in an activated sludge tank if their residence time is longer than their generation time. To persist, their rate of population increase has to match that of the heterotrophic bacteria present. An extended period of aeration will be needed for this – for effective nitrification a sludge residence time of around six days is required. One sign of overloading in a conventional activated sludge tank is when this nitrification ceases, as the heterotrophs dominate with the abundant organic matter, and the shorter retention times. In some plants, nitrification may be encouraged in a separate tank.

In biological filtration, nitrification tends to occur toward the bottom of the bed, reflecting the progressive loss of organic substrates (Figure 9.4). The greater segregation of the different genera of bacteria within the bioreactor reduces the competition between species and can make for more efficient nitrification. At very high loads excess ammonium may inhibit *Nitrobacter*, so that nitrification only proceeds as far as nitrite.

The presence of denitrifiers also drives the process faster – *Pseudomonas, Achromobacter, Bacillus* and others which reduce the nitrate and nitrite to molecular nitrogen (Table 7.4). In the absence of oxygen, these bacteria use these nitrogen compounds as oxidizing agents to degrade organic matter, in anaerobic respiration. In the activated sludge process, anoxic zones may be created in the aeration tank to facilitate denitrification before sedimentation. Then the facultative anaerobes, such as *Hyphomicrobium*, will begin to denitrify when the oxygen becomes depleted (Hamer, 1990). If there are insufficient carbon sources to support denitrification, additional organic compounds may be added to a culture to facilitate nitrogen loss (such as raw sewage or occasionally methanol).

In the wrong place, denitrification can be a nuisance in the sewage works. If it occurs in the final settling tank the sludge may float, hindering its sedimentation. This can be minimized if it has a low BOD, or if anaerobic conditions are promoted in a separate tank before the final thickening.

9.2.5 Anaerobic digestion

The sludge produced by both primary and secondary sedimentation still contains large amounts of water and needs further processing to reduce its mass before disposal. A sludge can be de-watered by a variety of physical and chemical methods to increase its solids content from about 6–8% to around 25%, lowering its transportation costs. The sludge may then be dumped in a landfill site or at sea. Direct disposal to farmland is another option if the sludge does not represent a significant source of metal pollution or if grazing is restricted (section 9.3).

One other alternative is further biological treatment. In many sewage treatment plants, the sludge undergoes a final phase of decomposition where the organic matter is degraded by fermentation in a large anaerobic digester (Figure 9.3). A succession of bacteria are used to degrade the sludge, eventually producing a gas composed of methane and carbon dixoide, termed **biogas**. The gas is combustible and is commonly used to generate electricity for the plant.

The process is optimized for the bacteria responsible for methane production, but the fermentation is sequential and other bacteria are necessary to degrade the organic matter. There are four distinct phases to the process – **hydrolysis, acidogeneis, acetogenesis** and **methanogenesis** (Figure 9.8). Hydrolysis of the organic matter reduces its long-chained molecules to substrates that can be used by the acid-forming bacteria. A large variety of fatty acids are produced at this stage, but these have to be converted to acetic acid if they are to be used by the methanogens. These

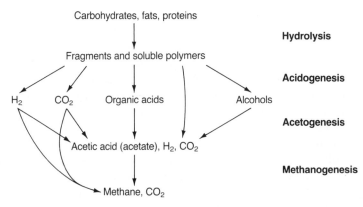

Figure 9.8 The stages in methanogenesis in an anerobic digester (after Mudrack and Kunst, 1986).

bacteria are only able to use acetic and formic acids as substrates – the fermentation of acetic acid produces methane, carbon dioxide and water:

$$4CH_3COOH + 8H \rightarrow 5CH_4 + 3CO_2 + 2H_2O.$$

In addition, autotrophic methanogens are able to use hydrogen and carbon dioxide:

$$CO_2 + 4H_2 \rightarrow CH_4 + 2H_2O.$$

In the absence of methanogens, the organic acids will accumulate and the digestion becomes soured, literally putrefying. Methanogens are inhibited by an abundance of fatty acids, by ammonia or by soluble sulphides, and a correct balance of bacterial activity has to be established from the outset of the digestion. One difficulty is that the acid-producing bacteria have a more rapid rate of population growth, and their activity may have to be controlled if methanogenesis is to develop (Dart and Stretton, 1980). Thus, the hydrolysis of the organic compounds, the formation of the organic acids and the methane generation all have to work in unison or the whole process halts. The digester normally operates close to neutrality, and a significant lowering of the pH indicates that the fatty acid concentration is rising and the methanogens are failing.

All methanogenic bacteria need elevated temperatures to thrive – mesophilic species produce most of their methane around 35°C whereas thermophilic species have an optimum at about 55°C. Ordinarily a digester will be run at one or other temperature. Heat exchangers are commonly used to collect waste heat from any electricity generators to pre-warm the sludge before it enters the digester and the digester itself is insulated to improve its heat retention. The control of the system attempts to stabilize the production of the biogas at the highest level, while at the same time maximizing the reduction of the sludge. Even under optimum conditions the growth in the bacterial biomass is significantly less than under aerobic respiration. A large amount of the carbon in the biomass will be lost as gas, although the amount and quality of the digester gas depends upon the nature of the substrate. The calorific value of the gas depends upon the proportion of CO_2, usually around 70% CH_4 and 30% CO_2.

There are systems where a digesting sludge is used to seed the bioreactor, to accelerate the digestion, especially for wastes that have a high soluble organic content (Berthouex and Rudd, 1977). Bacterial growth in the digester will only be sustained if there is sufficient nitrogen and phosphorus available, although these nutrients are rarely limiting. High levels of ammonia may inhibit the methanogens and the whole process can be easily poisoned by high concentrations of heavy metals or chlorinated hydrocarbons.

An efficient digester can reduce the degradable solids by around 50% over a period of around 20–40 days and will largely remove the grease content of raw sewage (ordinarily about 20% of the total solid mass). After digestion, the sludge can be de-watered by settling in lagoons or by mechanical filtration. The innocuous cake that is produced is a useful soil conditioner, if somewhat lacking in nutrients. The supernatant produced by this separation may be sent back through the plant for treatment with the incoming sewage.

The gas is not the only useful byproduct of the anaerobic digestion. The high temperatures at which the process operates can be very effective in lowering the viability of a number of pathogens and parasitic worms, including the resilient eggs of species such as *Taenia*, a common tapeworm. It is also effective in reducing the odours generated by raw sewage.

Other fermentations may be used with different effluents, using microbial species capable of attacking a particular variety of substrates. For the most part, the practicalities of these processes have yet to be fully worked out. Generally, these fermentations all stem from the acid-forming stage in the initial phase of an anaerobic digestion, but from this a number of pathways are possible, leading to a range of products including alcohols and other fuels. In the same way, the effluents from many industries could be used to recover some of the resources otherwise lost down the drain.

9.3 SLUDGE DISPOSAL PROBLEMS

There are four main methods of sludge disposal – applying to agricultural land, dumping in landfill sites and at sea, or processing to produce a soil conditioner. About 60% of the sewage sludge produced in the UK is applied to land, equivalent to about 0.75 million tonnes of dry matter per year (McGrath, 1987). Thoroughly digested sludge is not a particularly good fertilizer compared with artificial sources of nitrogen, phosphorus and potassium, but its organic content means that, unlike artificial fertilizers, it improves soil texture and water-holding capacity.

The choice of sites where a sludge can be disposed to land may be limited if there are problems of odour or pathogens. These should be of less concern with a treated sludge if the anaerobic digester has been working effectively. However, the digester cannot solve the problem of toxic chemicals, particularly metals, which become concentrated in the sludge. A high metal burden can severely reduce the scope for adding sludge to soils, especially where these pollutants might find their way into the human diet. As a result there are regulations in most industrialized countries that limit the disposal of highly contaminated sludges, or the sites where continued disposal could lead to excessive levels in the soil.

There has been considerable research to establish the impact of contaminated sludges on soil ecology, and to estimate the potential hazard of their toxic metals. As a result we have learnt much about the ecology of metal cycling and of soil mineral dynamics. Most metals are relatively immobile down the soil profile, even over the long term (section 8.3) and relatively small amounts find their way into crop plants (McGrath, 1987). For example, little metal accumulation was found in ryegrass in a permanent pasture which had received frequent surface applications of digested sludge over a period of four years (Davies *et al.*, 1988). In the calcareous and sandy loam soils studied here, around 80% of all metals (cadmium, nickel, zinc, copper, chromium and lead) was confined to the top 5 cm, but some metals, notably zinc and copper, were more mobile and the sludge enhanced their concentrations at a greater depth. The movement of metals down the profile was independent of the soil type, and the principal mechanism appeared to be the action of earthworms.

Adding sludge to a soil also increases the proportion of metals bound in specific fractions of the soil, and most especially the organically bound forms. Cadmium and zinc held in the organic matter was thought to represent the component most available to young barley plants (Chang *et al.*, 1984). Others, such as copper, appear to be less soluble as organic complexes, perhaps because they become incorporated into microbial cells. These processes are far from being understood completely, but the mineralization of the organic matter in sludge appears to lower the availability of some metals to plants. Their availability also changes with their degree of oxidation. In anaerobically digested sludges, most metals are precipitated as sulphides, which will be oxidized on exposure to the atmosphere to produce more soluble sulphates. This occurs mainly through the action of the autotrophic bacteria *Thiobacillus thiooxidans* and *Thiobacillus ferrooxidans*, both of which derive their energy from oxidation of sulphur and reduced sulphur compounds.

The solubility of the various metal fractions is also determined by the pH of the sludge. One way to reduce its metal burden is to acidify the sludge using sulphuric acid and then precipitate the metals from solution using lime. Tyagi *et al.* (1988) review one possible method of reducing the treatment costs, by generating the acid biologically, using the capacity of *Thiobacillus* to produce sulphuric acid from sulphur and metallic sulphides. Inoculating the sludge with *T. ferrooxidans* and *T. thiooxidans* reduces the amount of acid that has to be added and the associated costs by up to 80%. In the bench-scale experiments, the pH fell faster when the two bacteria were working in combination, and together they increased metal yields by 10% compared with the non-biological acid treatment. Maximum rates of removal ranged between 50% for cadmium and 97% for zinc, although this required over 10 days of incubation. More recent work (Tyagi *et al.*, 1991) suggests that the rate of metal leaching can be substantially increased in the sludge if *T. ferrooxidans* is supplied with an energy source, such as ferrous ions (as ferrous sulphate). This employs exactly the same principle as that developed to extract metal ores using microbial leaching (Pooley, 1987).

9.4 ALTERNATIVE METHODS OF TREATING SEWAGE EFFLUENT

A number of natural ecosystems are known to have a massive capacity for nutrient capture and decomposition, particularly those coastal and lowland wetlands that receive large periodic inputs of organic matter. Wetlands throughout the world have been used for effluent treatment, exploiting their established decomposer community. This requires no expensive plant or buildings and only a relatively minor investment in a delivery network. The ecosystems can readily remove nitrogen and phosphorus, nutrients that are difficult to reduce in conventional treatment plants and which may lead to eutrophication in natural water courses.

Swamps in the southern USA and in many tropical areas have been processing sewage for a considerable length of time. The higher temperatures of semi-tropical cypress and mangrove swamps make them particularly effective in treating effluents – decomposition is more rapid, and denitrification is especially effective in their anoxic conditions.

The swamps in Florida are dominated by cypress (*Taxodium* spp.), collected together in wet depressions, and with a water level continuous with the local water table (Deghi *et al.*, 1980). Beneath the water is a thick organic layer where many nutrients are bound, and which promotes anaerobic conditions within the swamp. Adding sewage to swamps

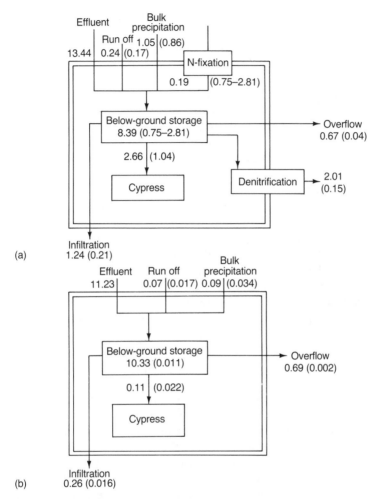

Figure 9.9 A mass balance of (a) nitrogen and (b) phosphorus for two cypress swamps in Florida (g.m^{-2} per year), one receiving partially treated effluent (unbracketed values) and the other with no additions (values in brackets). The nitrogen input is largely nitrate and nitrite: nitrification is insignificant in the swamps because of the anoxic conditions, while for the same reason, denitrification is a major route of loss. This proceeds rapidly also because of the high temperatures and abundant sources of carbon. Nitrogen and phosphorus can also be lost to the ground water or through surface runoff, although in both cases a large proportion is held in the peaty soil and roots of the trees ('below-ground storage') or becomes fixed in the above-ground tissues of the trees. (After Dierberg and Brezonik, 1983.)

increases their anoxic condition, but despite the loss of benthic invertebrates, there is no reduction in their capacity to degrade litter. This is
largely because the decomposer community becomes dominated by
cellulytic fungi (Deghi *et al.*, 1980). Overall there is no increase in litter
accumulation in the enriched swamps, although the trees may increase
their rate of growth substantially.

The routes taken by the principal plant nutrients through the system
also change. Dierberg and Brezonik (1983) used a mass balance to
measure the capacity of these ecosystems to remove nitrogen and
phosphorus from a partially treated effluent, and compared this with a
similar swamp receiving no effluent (Figure 9.9).

Of the nitrogen entering the control swamp, 34–66% was fixed by
microbes within the sediments. Applying a nitrogen-rich effluent slowed
microbial fixation, reducing it to just 1% of the total input. While nitrogen
fixation is switched off by high levels of available nitrite and nitrate in the
effluent, denitrification under the anaerobic conditions becomes more
prominent (Figure 9.9). In the swamp receiving the effluent, denitrification accounts for about 14% of the total nitrogen input, compared
with between 4 and 7% in the control. The enriched swamp also lost more
nitrogen and phosphorus to infiltration into the ground water and to
surface overflow, but both swamps were capable of denitrifying most of
the input, and loss of nitrogen to the surface waters was minimal (1–6%).
Around 1% of the annual phosphorus input and 18% of the nitrogen
input in the enriched swamp becomes incorporated into the above-
ground parts of the cypress trees, associated with significant increases in
their productivity. Overall, 74% of total nitrogen and 92% of total
phosphorus were retained in the enriched swamp, and most of this was
held in the below-ground biomass or the peat of the sediments.

These efficiencies may well be seasonal, changing with temperature,
and low temperatures certainly limit the use of wetlands in temperate
zones for waste treatment. This study lasted three years, but does not
detail the long-term changes this enrichment might bring to the microbial, plant and animal communities of the tropical wetlands. Nevertheless, one swamp that has been receiving effluent for 68 years still shows
low phosphorus losses in its surface water (nitrogen is not mentioned).
Dierberg and Brezonik conclude that the cypress domes may be useful as

---

Figure 9.10 Macrophyte-based wastewater systems. The silhouettes show examples of plants adapted to different depths, each of which may be used in
various stages of a wastewater treatment facility. (a) *Scirpus lacustris;* (b) *Phragmites
australis;* (c) *Typha latifolia;* (d) *Nymphaea alba;* (e) *Potamogeton gramineus;* (f)
Hydrocotyle vulgaris; (g) *Eichornia crassipes;* (h) *Lemna minor;* (i) *Potamogeton crispus;*
(j) *Littorella uniflora.* (With permission from Brix and Schierup, 1989.)

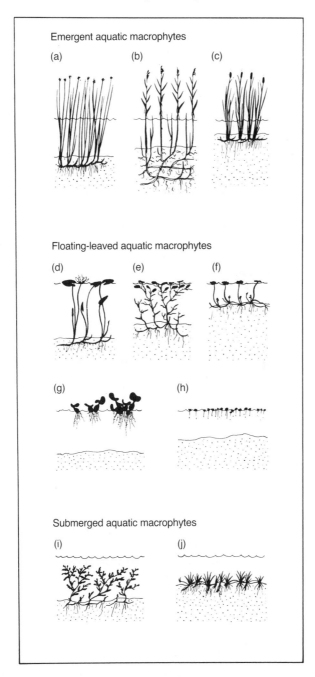

a final treatment plant, handling settled sewage, as a cheap alternative to a large investment in a sewage works.

9.4.1 Macrophyte-based sewage treatment systems

Unfortunately, wetland ecosystems are not always located in the most convenient positions to treat sewage. One answer is to mimic their properties by constructing small, artificial wetland ecosystems near the source of an effluent. A number of systems based on macrophytes have been tried, true sewage treatment plants. These include free-floating forms, such as the water hyacinth *Eichornia crassipes*, water lettuce *Pistia stratiodes*, salvinia *Salvinia rotundifolia* and rooted macrophytes, including reeds *Phragmites australis* and bulrushes *Scirpus lacustris* (Figure 9.10). Many of these species are weeds, opportunists that can accommodate a wide range of conditions, and which will flourish given the abundant nutrients (section 6.6).

Although these systems are cheap and relatively easy to set-up, they do have a limited capacity (Table 9.4). The relative advantages of the different types have been summarized by Reddy and De Busk (1985) and Brix and Schierup (1989).

Macrophytes serve several functions in the purification of wastewater:

1. they reduce water velocity, aiding sedimentation;
2. they create favourable physical conditions for microbial growth and may also ameliorate extreme chemical conditions;
3. rooted macrophytes extend the zone of oxygenation deeper into the sediments, enabling respiration of the organic matter to a greater depth;
4. they absorb nutrients, lowering the load of nitrogen and phosphorus in the effluent.

Floating macrophyte systems have been used largely for removing nitrogen and phosphorus. In this case, the plant biomass has to be harvested frequently to create space for the rapid growth of new plants and tissues. *Eichornia* is an example being used over a wide area in tropical and subtropical areas, which, with harvesting, can achieve rates of nitrogen and phosphorus removal over 0.8 g and 0.15 g respectively for each square metre of the plant per day. In addition, there is significant nitrification and denitrification associated with the microbial community that develops on the roots of the plant. Such systems can be used at the end of a conventional sewage treatment, to 'polish' the water, removing nitrogen and phosphorus before it is finally discharged to a natural water course.

Submerged macrophytes can only be used in systems receiving water with a low BOD, treating water that is largely transparent. Their photosynthesis generates oxygen which aids the oxidation of the residual organic matter, while the corresponding reduction in carbon dioxide raises the pH of the water, increasing the volatilization of ammonia and

Table 9.4 Macrophyte-based wastewater treatment systems compared with conventional sewage works

---

Advantages
- Low operating costs
- Low energy requirements
- Can be established at the source of the wastewater
- More flexible and less susceptible to shock loadings
- Biomass can be harvested for animal feed or soil conditioner (N and P are not lost)
- Biomass can be harvested for fuel

Disadvantages
- Require a large land area for the treatment of large volumes
- Not a viable option for the very large effluent volumes
- Lowered performance during winter (may need to be run in a greenhouse in temperate regions)
- Little capacity for removing pathogens from the effluent
- May be susceptible to high levels of pollutants such as toxic metals

---

the precipitation of phosphates. Emergent macrophytes can serve a similar role, but are more tolerant of turbid water. Both forms of plant have to be harvested if the nutrients they capture are not to be released back to the system when they decay.

Different species may be used in other systems because of the value of the biomass they produce. Duckweed (*Lemna minor*) supports a smaller microbial community than *Eichornia* and is consequently less effective in nutrient removal from wastewater, but it can provide a harvest of animal fodder with a much higher nutritive value (Brix and Schierup, 1988).

In some working systems, these ecosystems are able to achieve comparable and even better standards than conventional treatments. Brix and Schierup describe a unit established in Denmark, based on reeds, which reaches equivalent efficiencies at between 10 and 15% of the cost of a traditional works. Most of the improvement in the BOD, nitrogen and phosphorus is attributable to the microorganisms within the system, with the plants aiding sedimentation and the oxygenation of the sediments. However, this does not appear to be sufficient to complete nitrification in the sediments.

The efficiency of these systems may be improved if these ecosystems are functionally divided, and different macrophytes are used at different stages (Reddy and De Busk, 1985). Similarly, some form of pre-treatment of the effluent, in addition to primary sedimentation, could also improve their performance. As Brix and Schierup note, this is a technology in its infancy, with relatively little known about its capacity, the most appropriate design criteria or the features of the community that should be

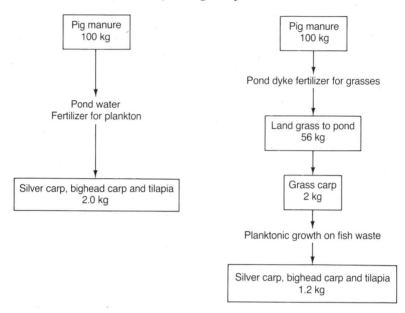

Figure 9.11 Two alternative management strategies used in a Chinese integrated fish pond. The input into each system is the same but, in the second example, the pig manure is used to grow biomass (the grass, *Lolium perenne*) in dykes, which is then added to the pond. Grass carp feed on this, and their waste will support a phytoplankton community that will in turn be fed on by the planktivorous fish (silver carp, bighead carp and tilapia), adding an additional 1.2 kg to the total fish production. (After Zweig, 1985.)

manipulated for its optimization. With further development, it may even be possible to use such systems to supply a fishery, feeding the biomass produced upstream to a herbivorous fish.

This idea is certainly not a new one. In China the technique has a history of over 2500 years. The culture of macrophytes using effluents, which in turn support a fish farm, has been developed into an elaborate technology, one that is still being refined today (Zweig, 1985). Here the ecosystem is managed both by controlling the stock of a small number of fish species, and by making nutrient inputs in addition to the effluent entering the system.

Nine species of fish are commonly grown together, feeding on the macrophytes, phytoplankton, zooplankton and benthic invertebrates, as well as the decomposer community. These are combined in various ratios, according to the nutrient loading of the pond. The different niches occupied by the various fish are created by the other organisms – the macrophytes, herbivores and detritivores that recover the energy from the material added to the pond. There is also some evidence that particular assemblages of fish help to improve yields for individual species (Zweig, 1985).

The system is supported not only by human and animal waste (including silkworm droppings), it is often supplemented by land plants deliberately added to encourage particular species of fish (Figure 9.11). Macrophytes, including *Eichornia* and *Pistia* are grown in adjacent water bodies and are also added to the pond. In addition, plants for human consumption may be grown in or above the fishpond itself. The whole management of these ponds is obviously critical – the stability of the system appears to be a matter of keeping the correct degree of eutrophication. This is down to the experience of the manager, and the method he uses to judge the nutrient loading from the turbidity of the water (Zweig, 1985).

9.5 LANDFILL ECOLOGY

We are not alone in culturing microbial communities to exploit their degradative powers. The sequential degradation of sewage sludge in an anerobic digester mimics the processes found in the rumen of a cow, where the host maintains a community of bacteria and protozoa. Indeed, there have been successful attempts at developing a bioreactor for handling household refuse, based on the digestive system of ruminants.

Within the rumen, bacteria produce cellulases that aid the digestion of the feedstock: following the hydrolysis of the polymers in the vegetation, the products are fermented to produce volatile fatty acids which become the principal energy source for the ruminant. The animal also relies on the microbial biomass for essential parts of its diet. In the sewage plant, the digestion is allowed to run to completion, encouraging the formation of acetic acid, carbon dioxide and hydrogen to provide a substrate for the methanogenic bacteria. To a cow, methane production represents an energy loss, generated from fatty acids that have not been absorbed.

The same processes of acidogenesis, acetogenesis and methanogenesis occur in a landfill site, where methane is produced by the fermentation of domestic refuse. Like the digester, but unlike the cow, we can collect the biogas and burn it as an fuel. This fuel can be a valuable resource and landfill sites have two important advantages as a source of biogas – collecting the gas has a low capital cost and gas production lasts over an extended period. In addition, removing the gas under controlled conditions helps to reduce a hazard at sites where explosive pockets of fuel can develop.

Landfill is the most common method of solid waste disposal in the UK and, although normally associated in the public's mind with domestic waste, it includes building, industrial and mining waste, and low-level radioactive waste. Domestic and trade waste contributes just 8% of the total solid waste in this country – nearly 67% of the total mass is produced by mining and quarrying. The latter is largely collected in spoil heaps, which have their own ecological problems (section 7.3).

Here we consider the use of landfill sites which are rich in organic waste, and which will function as anaerobic digesters, producing methane. In theory, for each kilogram of waste containing 50% organic matter, around $0.4 \, m^3$ of biogas (of which about half is methane) could be generated. In practice, it will be much less – perhaps 0.050–$0.25 \, m^3$ (AFRC Institute of Food Research, 1988) over 10 years. In the UK, it is estimated that landfill gas could produce the energy equivalent of $2.5 \times 10^9 \, kg$ of coal from existing sites over the next 10 years; with attempts to optimize conditions within the tip, this could rise to $6.5 \times 10^9 \, kg$, from the $2.6 \times 10^{10} \, kg$ of biodegradable waste dumped each year (AFRC Institute of Food Research, 1988). The USA is the largest consumer of landfill gas, saving around $1.59 \times 10^9 \, kg$ of coal in 1989 (Richards, 1989). Current estimates suggest that there are about 250 schemes that generate electricity from landfill biogas around the world, burning the methane which would otherwise escape to the atmosphere, and that would otherwise add to the levels of this important 'greenhouse' gas.

One reason that landfill sites do not generate their full potential of biogas is the highly variable nature of the waste. This consists mainly of solids which are poorly mixed. A fluid sludge has a high lipid, high protein content and forms a fairly homogeneous environment. In contrast, domestic refuse is dominated by paper, with a high cellulose, high lignin and high inorganic content. This creates heterogeneous conditions, with an organic component that is far less degradable under anaerobic conditions. The waste may also have high metal levels and organic compounds from plastics which can inhibit the fermentation. An additional problem is that we know so little of the microbiological factors governing the methanogenesis that optimization is more difficult to achieve.

Again, we are faced with managing a microbial community whose stability depends upon the environmental conditions within the bioreactor, and the production of substrates from which its component species can derive energy. Methane generation will develop naturally in landfill sites, and can create a significant explosion hazard where the gas production is not properly managed. Our aim is to control the conditions in the site to promote a large and stable gas production over the longest possible time. Not only do sites have to be properly capped to prevent gas loss and leachate seepage, but the ecosystem itself has to be managed to favour the methanogenic bacteria.

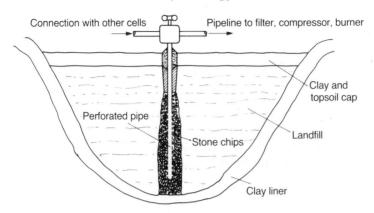

Connection with other cells · Pipeline to filter, compressor, burner

Clay and topsoil cap

Perforated pipe

Landfill

Stone chips

Clay liner

Figure 9.12 The construction of a landfill cell to collect biogas. Modern practice is to divide the site into a series of cells, each operating as a batch culture and proceeding at its own rate. Cells may be isolated from each other by soil or clay, or be interconnected by a series of gravel 'pipelines' that direct the gas to a central collection point. The refuse is normally packed into highly compacted layers to encourage the rapid development of anaerobic conditions. Adding sewage sludge can rapidly accelerate the onset of methane production. Gas passes into the gravel around the perforated base of the collection pipe. The pipe is grouted into the cell with cement to form a seal with the cap. The gas may be collected and bottled for use elsewhere, piped to the customized boilers of a nearby consumer, or used to drive steam turbines and generate electricity on site.

9.5.1 Landfill site management

Rarely is a landfill site dug especially for waste disposal – rather it uses the spent hole left over from excavations or quarrying, although in some cases, natural depressions are filled, or flat areas simply covered. The primary consideration is the permeability of a site and the potential for runoff contaminating local water bodies. A clay liner may be laid to prevent leachate entering the groundwater and, for this reason, sites that are naturally endowed with clay (old brick-pits) are favoured for especially toxic wastes.

For many years, the main strategy in placing the refuse was to pack as much material into the smallest volume possible. More recently, the need to control the pest and pollution problems of the tip required a more considered approach. Legislation demands that the waste is covered at regular intervals to prevent litter being blown off-site, to reduce the access for vermin and limit the infiltration by rain and surface water. Nowadays, the laying and capping of a greater number of sites follows a strategy to maximize methane production. One common method is the cell emplacement strategy, where individual cells of refuse are created, capped with clay, from which gas can be drawn off. Alternatively, beds of gravel or some other porous material can be used to separate cells, acting as pathways down which gas travels to a central collection point. Either way, this creates a series of batch cultures, with cells at different stages of degradation (Figure 9.12).

Very often, refuse is tipped without pre-treatment, but its bulk can be reduced by sorting to remove metals, glass or other recyclable components, and by shredding. Together these raise the proportion of organic material in the waste and improve the amount of gas generated from a volume of the tip. Shredding helps to increase the surface area, accelerating decomposition under the initial aerobic stages, especially by invertebrates (woodlice, worms, millipedes, mites and so on) which further help to reduce particle size.

9.5.2 The microbial succession in a landfill site

As anaerobic conditions develop in the landfill, a sequence of changes in its microbiology mirror the phases we observed in the anaerobic digester. With the onset of the fermentation, an acid-producing stage is followed by acetogenesis and in turn by methanogenesis. However, within this poorly mixed and largely solid feedstock, there are a large number of microenvironments where localized conditions will develop. Water is the main medium transferring nutrients and microorganisms between these habitats, but this imperfect mixing means that very different communities can become established within short distances of each other. This makes the control of the whole ecosystem very difficult; current techniques rely solely on modifying the abiotic conditions of the whole site.

The principal determinant of gas production is the age of the tip and the management of the site responds to these changes (Figure 9.13).

1. In its early stages, aerobic respiration depletes the oxygen supply and raises the temperature of the site. The length of this stage depends upon the degree of compaction, the presence of a cap and the nature of the refuse. The rate at which hydrolysis proceeds governs the other processes in the digestion process – the decomposition of complex organic molecules produces a range of substrates that are used by a variety of microorganisms. More intractable wastes, such as lignin or hemicelluloses, degrade very slowly, whereas simple sugars and starches are hydrolysed very easily. Under wholly anaerobic conditions, lignin is virtually non-degradable; as a result, changes in the cellulose to lignin ratio are measured to follow the decomposition of the tip – this starts out as 4.0 in freshly laid refuse, falling to 0.2 in a well-stabilized landfill.

   Towards the end of this first stage, the C:N ratio of the waste is greater than 55:1, so that a shortage of nitrogen limits the growth of aerobic bacteria. At this time, the temperature may reach 70°C,

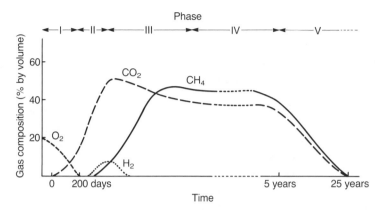

Figure 9.13 Changes in the composition of gas produced by a landfill site as it ages. Hydrolysis of the organic matter during the initial aerobic phase quickly exhausts the oxygen supply of a well-compacted tip. This continues throughout the remaining phases, as anaerobic digestion commences. In phase II, a range of volatile fatty acids are produced, indicated by the rise in hydrogen production, and by the presence of fatty acids in the leachate. As these are converted to acetic acid in phase III, so biogas production commences. There then follows a stable period of gas production (phase IV) which can last for 10 years before gas production declines with the depletion of the substrate. (After AFRC, 1988.)

important for killing pathogens in the refuse, including *Salmonella* and *Typhus*. The high temperature also limits oxygen availability, helping to induce the anoxic conditions of the following stage.

2. As the fermentation begins, hydrogen and carbon dioxide are evolved and a range of volatile fatty acids (VFAs) are produced: acidogenesis. From this stage onwards, the site is comparable to the anaerobic digester in a sewage works or the rumen of a cow. The variety of VFAs produced depend upon the nature of the refuse, but these are mainly acetic, proprionic and butyric acids. Acetogenesis will only continue if the hydrogen and carbon dioxide produced by the acetogenic bacteria are used by methanogenic bacteria.

3. Methanogenesis may take some time to become established in the tip, and along with hydrolysis, this may be the rate-limiting step in the digestion. The methanogenic bacteria are inhibited by acidic conditions so, if methane production is to continue, there needs to be a balance between the VFA production (which will lower pH) and utilization (as a substrate for methane). The leachate from the site remains acidic until the methanogens are established, metabolizing the fatty acids (Figure 9.13). Methanogens are sensitive not only to low pH, but also to toxic metals whose solubility rises with increasing acidity.

4. Stable production of methane is then established. The timing of this is highly variable and will depend on the construction of the tip and the ambient temperatures.
5. A final phase of declining methane production follows. The nature of the refuse and local temperatures will govern the length of time the site remains productive. At high temperatures (50–70°C) thermophilic anaerobes dominate the bacterial community and gas will be generated over a shorter period. In temperate regions, the high temperatures produced during the initial aerobic phase cannot be maintained by the reduced energy release of anaerobic digestion. Typically, the anaerobic phases run at 20–35°C – below the optimum for the mesophilic anaerobes (35–40°C) and gas will then be generated relatively slowly over 10–15 years. In contrast, a site in subtropical Louisiana was exhausted and subsequently reclaimed within just three years (AFRC Institute of Food Research, 1988).

A number of factors regulate methane production, and there is, in each case, some scope for modifying each of these to optimize production (Table 9.5). Temperature governs the range of methanogens present in the tip and their use of the volatile fatty acids needs to balance rates of VFA production if the pH is to stay close to neutral. In addition, the degradability of the refuse is critical – during the early stages, the microorganisms will tend to flourish on those substrates most easily decomposed. This can lead to an early loss of nitrogen and other nutrients in the leachate, corresponding to a rise in protease activity as proteins are broken down. This also accounts for the rapid rise in fatty acids in the leachate before methanogenesis becomes established (Jones and Grainger, 1983). Later microbial communities may be limited by a shortage of nitrogen, slowing the degradation of the more recalcitrant components, such as the cellulose and hemicelluloses in paper.

Biogas production may be reduced by methane-oxidizing bacteria if oxygen is allowed into the site. However, the moisture content of the refuse appears to be the most important factor governing gas production (Lawson, 1990) – the refuse has to be saturated to promote microbial growth and to aid the development of anaerobic conditions. The amount of free water in a tip depends particularly on the proportion of paper in the waste. Applying sewage sludge to the refuse will add moisture, nutrients and a community of microorganisms, all helping to accelerate methane production. Protozoans and fungi are also found in large numbers in landfill sites, but their interactions with the methanogens are unknown.

Waterlogging will lead to a rapid loss of nitrogen and other nutrients. Recycling the leachate will return these nutrients, and bacteria and toxins, back to the tip. This can help to solve the pollution problem that

Table 9.5 Stabilizing gas production from landfill sites

Factor	Role	Method
Temperature	Stable temperatures give constancy of supply Higher temperatures increase rates of CH_4 production	Insulation
pH	Neutrality essential for CH_4 generation – balancing volatile fatty acid production and use	Buffer and recycle leachate
Bioavailability	Differential degradation of refuse components can lead to early loss of important nutrients, e.g. N	Recycle leachate
Moisture	Saturation of refuse needed – dry pockets can slow CH_4 production; flooding can lead to nutrient loss through leachate	Engineering site drainage
Density of refuse	Low density leaves air spaces that prevent anaerobic conditions developing	Increase compaction
Particle size	Reducing particle size can accelerate degradation as long as it does not lead to undue compaction, preventing access of microbes to substrates	Removal of non-compostibles; shredding of wastes
Oxygen content	High oxygen will halt anaerobic digestion and lead to the production of humic materials and CO_2 production	Capping or other practices to stop ingress of O_2

the leachate itself represents, although this may also be collected and treated in separate tanks. At a site that has been properly engineered from the outset to prevent waterlogging, to collect the leachate and to minimize heat loss, the principal method of control is through regulating the nature of the waste placed in a cell and the subsequent input of water.

Little is known about the detailed interactions within the microbial community as the digestion proceeds, although at the moment we have little scope for modifying these interactions either to control gas generation or leachate quality. In most cases, it appears that the site is most rapidly stabilized and its gas production enhanced by recycling the leachate back into the tip. Many of the improvements in this technology will probably come from developments in gas recovery rather than manipulating the biotic components of the community. Even in those sites that have been engineered for the purpose, perhaps only 20% of the biogas currently generated is collected (AFRC Institute of Food Research, 1988). For this reason, the simple hole in the ground may need to be reconsidered.

9.5.3 The problem of landfill leachate

The leachate from landfill sites can represent a considerable toxic hazard to the immediate environment. Optimizing a tip for gas production will also reduce the acids present in the leachate and consequently the soluble metals it carries. A number of toxic metals are known to become more mobile as the fatty acid concentration of the leachate rises (Loch *et al.*, 1981). The leachate will also carry with it microorganisms, especially aerobes, largely those varieties commonly found in the soil (Lawson, 1990).

The quality of the leachate depends on a number of variables:

1. factors external to the tip – such as geology, quality of inflow;
2. the composition of tipped waste, including its microbial community;
3. the emplacement strategy used (topography, permeability, vegetation cover, and after use, etc.);
4. Time-dependent factors, such as age and season.

Water arrives through infiltration and is also generated by the decomposition process within the tip. As it leaves the site it carries with it the solubilized products. In the early stages, before acidogenesis has commenced, the leachate has the highest BOD, COD and total organic carbon, but low levels of metals and sulphates. All metals rise with fatty acid concentration, with the possible exception of copper (Loch *et al.*, 1981). Generally, the concentration of all soluble components in the leachate decline after about seven years.

One method of controlling the hazard from the leachate is to provide a water-impermeable cover, but this can slow down the decomposition rate. Leachate may be pumped out from the cells and recycled back through the tip, to attenuate the pollutants, or passed through a treatment plant, such as a trickling filter. Aeration lagoons have also been used to catch the leachate, but any biological treatment may be poisoned if large amounts of heavy metals are present. In some cases, it may be possible to subject the leachate to anaerobic digestion, producing methane that can add to the productivity of a site (Blakey and Maris, 1990).

9.6 COMPOSTING

The techniques for using aerobic decomposition to reduce the bulk and recover the nutrients in organic wastes will be familiar to most gardeners.

They will know the essential factors that have to be controlled in the process – maintaining high levels of oxygen, occasionally supplying nitrogen and encouraging high temperatures to accelerate the decomposition process. In the compost heap this may be achieved by allowing air to enter the waste and preventing the formation of compressed zones where anaerobic conditions could develop. The garden compost heap relies upon the activity of a number of invertebrates, especially earthworms and woodlice, that serve to reduce particle size, and to inoculate the organic matter with the microorganisms responsible for decomposition.

Composting on an larger scale has been used to treat agricultural and domestic refuse. This tends to be costly in equipment, as some method of maintaining the correct degree of aeration is needed, often using rotating drums. However, this can be used to treat both sewage sludge and domestic refuse, producing a good quality soil conditioner which can be readily sold. A good compost is similar to peat, although usually with more nutrients.

Before treatment, the refuse is usually sorted into compostible and non-compostible materials – the glass and metals collected may themselves be recycled. The compostibles are shredded and in some processes mixed with sewage sludge to raise both the moisture and nitrogen content. This is placed in some form of bioreactor that has a method of adding air to the waste – either through an induced draught, or by mechanical turning. Aerobic respiration of the waste will have begun on the waste before it arrived at the plant, and this now accelerates within the bioreactor. Two days after being laid, the temperature reaches 45°C and thermophilic bacteria become the dominant agents of decomposition. In a commercial plant this stage of rising temperatures lasts for about two weeks (Holmes, 1981). Over about 60°C the thermophilic fungi are lost, and the community is dominated by actinomycetes and spore-forming bacteria. Their respiration causes the temperature to reach around 70°C, high enough to kill most pathogens. If there is excess nitrogen present the high temperatures may drive the nutrient off as ammonia.

The easily degraded material is rapidly lost, after which the production of heat declines. The compost cools, the thermophilic fungi re-establish themselves and decomposition proceeds at a much slower pace. The waste is now dominated by the more intractable lignins and humic acids; over subsequent weeks, it forms a stable humus, making it suitable for applying to the soil. Again, this reflects to some extent the decomposition processes found in the soil (section 7.1), although proceeding at a faster rate.

The rate of decomposition will depend upon the C:N ratio of the waste at the outset. This can range from 20:1 to 70:1 in domestic refuse, but after the production of carbon dioxide this falls to between 15:1 to 20:1. Adding sewage sludge can lower the initial ratio to below 40:1, a level at which composting will proceed at an acceptable speed. The moisture level is also critical to the process: it needs to be around 55% for the process to continue. Above this, waterlogging may occur and the process becomes anaerobic. The high temperatures during the process help to drive off excess water, including that produced by the metabolic activity.

The pH of the compost can be used as an indicator of the stage of the decomposition process: in the first three days it will range between 5.0 and 7.0, and will then fall, but rises again to 8.0–8.5 by the end of the process. A compost that remains consistently below pH 4.5 has become anaerobic, and is undergoing fermentation.

Newer technologies have sought to generate gas from the refuse. Instead of using the landfill site as a bioreactor, these combine domestic refuse and sewage sludge in a tank which, if maintained at high temperature, can rapidly degrade the waste using anaerobic digestion to generate biogas. Following this, the residue is composted to produce a soil conditioner. The largest digester of this sort, at Amiens in France, processes 8.1×10^6 kg (8000 tonnes) of organic waste each year, and produces 3500 m^3 of biogas each day.

More traditional composting simply accelerates the natural processes in the degradation of organic matter. The garden compost heap concentrates the community and processes found in the upper layers of a soil into one location, as a continuous culture. As with all of these technologies, we control the process using those external variables we can manipulate, such as the rate of heat loss or the supply of oxygen, and by regulating the input of nutrients including the main substrate. Although we may monitor several indicative variables within the system – such as temperature, pH or even the abundance of key species, most of our judgements about the performance of the ecosystem comes from its outputs – such as the quality of its final product, the rich humic residue we dig out from the bottom of the heap.

Summary

The degradative and productive capacities of both natural and artificial ecosystems have been exploited for a long time, although new methods continue to be developed to solve particular effluent problems. As there is little prospect of fully describing all the biotic and abiotic interactions within these systems, a reduced form of systems ecology, based upon the simple concept of a mass balance is used. A form of process control is used to manipulate the activity of the system, by altering the inputs into the system, particularly the supply of substrates. In addition a number of external variables, such as temperature, oxygen supply and so on can be used to control the process, and the outputs are used to monitor its progress. Many of these techniques have developed through trial and error and accumulated experience.

Sewage treatment accelerates the processes of oxidation and decomposition, which would naturally occur in a freshwater ecosystem. Following an aerobic phase, designed to lower the BOD of the waste, the biomass of the plant is digested. This follows four distinct phases – hydrolysis, acidogenesis, acetogenesis and methanogenesis, a pattern found in the digestive tract of ruminants and in landfill sites. In addition, microbial communities can be used to remove excess nitrogen from the waste. Alternatives to the conventional sewage works use natural ecosystems, or small mimics of such systems. Some combine these with food or biomass production. In other cases, wastes are combined in an aerobic decomposition process, composting, to produce a nutrient-rich soil conditioner.

Further reading

Mudrack, K. and Kunst, S. (1986) *Biology of Sewage Treatment and Water Pollution Control*, Ellis Horwood, Chichester.

Steel works in Silesia, Poland. Heavy industry in eastern Europe has produced one of the most polluted regions in the world, and is a significant source of acid deposition in Scandinavia.

10

Assessing large-scale ecological change

The most dramatic feature of tropical ecosystems is the sheer variety of species found within them. Tropical forests and the waters around tropical reefs both have an immense diversity of life, unmatched by equivalent communities in the temperate latitudes. Across the planet, moving from the poles to the equator, there is an increase in species richness in nearly all groups of plants and animals.

In terrestrial ecosystems, this pattern would seem to reflect the change in climate with latitude. Indeed, ecologists have long recognized major regional plant communities or **biomes** which broadly follow these climatic zones – such as the coniferous forest belt running around the northern hemisphere or the temperate grassland through much of Europe, Asia and North America. Each biome is associated with particular regimes of temperature and precipitation. However, climate cannot fully explain the latitudinal gradients in species richness as these are also found in marine communities (Figure 10.1). Thus, from the earliest days of the science, ecologists have looked for a more general theory (Fischer, 1960).

The biomes represent a convenient classification of broad community types, but this does little to describe the processes that organize such large-scale assemblages. Most ecological theory has been rooted at the ecosystem level or below, referring to local patterns and processes. Yet many local ecosystem features reflect regional processes – for example, the community supported by a soil is the product of the prevailing climate and regional geology. Its collection of plants and animals depends also upon the soil's history and geography, as well as the existing interactions between species.

Longer-term change affects larger areas – temperate and tropical ecology today has been shaped by the global upheaval of the ice ages during the Pleistocene, finishing around 10 000 years ago. Similarly, the present distribution of marsupials results from the separation of Antarctica and Australia around 45 million years ago, an event with profound

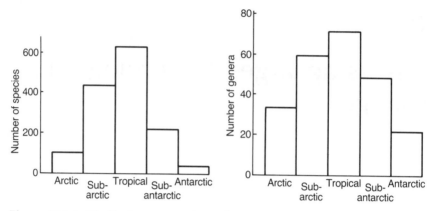

Figure 10.1 Latitudinal patterns of specific and generic diversity in tunicates – these are sessile protochordates, found along coasts from polar to equatorial regions. (After Hartmeyer, 1909, cited in Fischer, 1960.)

consequences for their subsequent evolution. The distribution of species we observe today is a product of large-scale processes in both space and time.

The effects of such processes on local ecology take on a greater significance when the applied ecologist begins to address regional and global environmental problems. Then it becomes a question of how readily our current understanding of ecology can be extended to the larger scales. This chapter reviews a number of attempts to apply existing ecological theory to the landscape and the region, and some of the novel methods being used to monitor change at these levels. This consists of a loose collection of methods and ideas, largely because there is no single coherent theory of large-scale ecology to bind them together. Above the level of the ecosystem, data are scarce and the scope for experimentation is very limited.

The chapter begins by briefly surveying some of the problems in developing a large-scale ecology. It goes on to examine the effect that large-scale ecological processes may have on the ecology of a species and, in particular, how the frequency of disturbance may determine life history strategies in terrestrial and marine ecosystems. The ability of a community to withstand disturbance is then examined from existing population theory, to review the relationship between species richness and stability.

Patterns of disturbance may be central to explaining the global gradients of species diversity. Species richness is closely linked to spatial scale (section 6.3) and diversity can be classified according to the scale over which we sample, and according to the differences between

Table 10.1 Whittaker's classification of species diversity

Inventory diversity	*Differentiation diversity*
1. Point or sample of a homogeneous group	2. Differences between samples
3. α – Within habitat	4. β – Differences between habitats
5. γ – Landscape	6. δ – Differences between landscapes
7. ε – Regional	

habitats. Table 10.1 is a two-fold classification of diversity based on the total species diversity within an area (inventory diversity) or along a gradient (differentiation diversity), and according to the spatial scale over which the diversity has been sampled. Measures of diversity have been suggested as useful summaries of community organization, allowing us to derive quantitative measures of ecosystem 'health'. We go on to examine the use of indices of diversity as indicators of an impact on a community.

Changes in species diversity is one of a variety of measures that has been used for mapping ecological change. Bormann (1985) noted that this was a common feature in stressed ecosystems undergoing a gradual collapse (Table 10.2), accompanied by changes in ecosystem function,

Table 10.2 Stages in the collapse of an ecosystem (after Bormann, 1985)

Stage	*Pollution Stress*	*Impact*
O	Pollutant levels insignificant	Pristine system
I	Low pollution level – some pollutants held in sinks in the system	Species and ecosystem functions generally unaffected
IIA	Levels of pollutant large enough to affect some sensitive species or individuals	Some impact on biological functions, e.g. photosynthesis
IIB	Increased pollution stress leads to loss of individuals or species	Change in community composition
IIIA	Long exposure to high levels of pollution	Size differences become important to survival – larger species lost first Energetic and biogeochemical cycles are severely affected
IIIB	Ecosystem collapse – even removal of pollutant does not allow system to recover in the short to medium term	Highly degraded system with little capacity for self-repair

following repeatable patterns. These follow both the loss of species and the consequent impairment of ecological functions. Similarly, Odum (1985) provided a long list of responses which an ecosystem may show under continuing disturbance, some of which might be used to monitor the impact (Table 1.1). Such methods take a largely holistic view of ecological processes, and we review various efforts to extend these ideas to the larger scales. Because of their close association with theories of ecosystem stability and regulation, many of these ideas bring us back to where we started, in Chapter 1.

10.1 EXTENDING ECOLOGICAL THEORY TO THE LARGER SCALES

Ecologists frequently use natural barriers or discontinuities as convenient boundaries to define ecosystems. In terrestrial ecology, for example, a catchment or watershed is useful for defining the abiotic limits of an ecosystem; a sequence of watersheds together comprise a landscape. A region is composed of a number of landscapes within a defined biogeographical or climatic zone. Increasingly, applied ecology is being required to address problems of environmental degradation at these larger scales, and to predict the consequences of such changes. This begs the question of whether it is possible to have a comprehensive theory of ecological processes at the regional and global level. Much effort has gone into modelling, mapping distributions and rates of change, but by themselves models do not represent a general theory of large-scale ecology.

Quite naturally, we tend to think at a local scale, and much of the ecologist's work is confined to tackling problems in relatively small areas and individual ecosystems. Unfortunately, while such work may solve a local problem it often fails to address the larger problem. In principle, at least, theory could direct our efforts in pollution management or conservation more strategically, to ensure that local action is effective for the landscape or the region. For example, adding limestone to an acidified lake in Scandinavia may solve one local problem, but it fails to address the cause of the regional pollution. The pattern of atmospheric pollution in Europe is a product not only of the distribution of industrial activity, but also of national economic and political policies, of geology and of the predominant wind movements. These regional factors extend beyond physical and ecological zones, and only some of them can be controlled. In these cases, ecological theory might help to direct regional economic and social decision making.

Unfortunately, our capacity to predict the nature of large-scale changes is limited by a lack of data. At best, predictions are based on data collected over one or two decades, short periods in the life of a planet. Even with simpler systems and more abundant records, predictions become less reliable over longer periods and larger areas – forecasting the regional weather a few days hence is becoming less of a lottery, but our capacity to model the global atmosphere for any longer is still relatively poor. For example, climatologists continue to argue whether El Niño, the periodic warming of the southern Pacific ocean, is a predictable regional climatic cycle, or the product of volcanic eruptions pushing dust into the upper atmosphere. We know that each El Niño causes a disruption to the climate in the Americas, and major reductions in agricultural and fishing productivity. The regularity of such events is becoming part of the planning in setting fishing limits for some of the major fisheries in the Pacific (section 4.4).

Even as we refine our models using further data or improved algorithms, nature continues to surprise us. For some, the problem is one of inadequate computer power. More urgent is the need for adequate theory and testable hypotheses that can drive purely descriptive models forward. Our dilemma is whether we trust the models we have now and take action based on their predictions: if climatic regimes change in line with their current forecasts, different management strategies will have to be adopted for agriculture, forestry, habitat and species conservation (Strain, 1987).

10.2 ECOSYSTEM STABILITY

We can observe distinct spatial and temporal patterns from the population level to the biosphere of the whole planet. But to what extent are each of these patterns stable, and what processes generate them? In addition, are the processes of one particular level more important than others, or could the patterns result from the interaction between processes operating at different levels?

For some ecologists, regulation operates from the top down – higher-level, slower-rate processes govern ecological characteristics (section 1.5), even down to the nature of the species found in an ecosystem. Steele (1985) has suggested that the oceans and terrestrial ecosystems have very different communities because of the different frequencies of variation in their abiotic environment. The great mass of the oceans gives this environment a large capacity to buffer temperature and other abiotic factors. Away from the coastal regions oceans show little significant change over the short term (up to about 50 years). Large variation only occurs over the long term, with major shifts in the planet's climate. In contrast, terrestrial communities survive in a much more variable environment, with large-scale fluctuations in both the short (one week) and long term.

Organisms in each habitat have to adapt to the frequency of change, and the oceans will demand different survival strategies to those on land.

Marine animals do not have to maintain their internal environment against highly variable external conditions – for example, the relatively constant temperatures mean these communities are dominated by poikilotherms. Typically, their reproductive strategy is based on the production of large numbers of larvae with well-developed powers of dispersal. Provision for the offspring and parental care is only favoured in more variable environments such as freshwater or terrestrial ecosystems. Marine life relies on massive larval production and rapid dispersal. Steele argues that this is why fish recruitment bears little relation to fish stock size (section 4.4): early life processes are 'de-coupled' from those of the later stages. In contrast to terrestrial vertebrates, simple population models, and particularly the r–K concept, may thus be largely irrelevant to fish population biology. Instead, fish stocks commonly range between multiple equilibrium states, often with a periodicity reflecting the long-term fluctuations in the oceans.

Terrestrial organisms have to accommodate short-term variation, and consequently have physiologies, life histories and ecologies that reflect this. According to Steele, it is the adaptive and regulatory capacities of individual organisms that confer stability on terrestrial ecosystems. Their adaptations and interactions with each other serve to regulate ecological processes. Indeed, we have already considered some evidence for such effects: the nutrient balance in the Hubbard Brook was governed by these interactions (section 8.2). This regulation is not a feature of the more loosely organized communities in the oceans. Population and community ecology here is determined by regional and global variation. Oceanic communities thus follow the long-term, large-scale variation of their abiotic environment, rather than being regulated by their internal organization.

From this, Steele suggests that the concept of the ecosystem is largely meaningless in the oceans. This domain has few boundaries with clear demarcations, and the movement of organisms between patches is effectively unhindered. Food webs merge with each other and nutrients pass freely between different habitats. Variation on both spatial and temporal scales is lower in the oceans, resulting in the essential difference with terrestrial ecosystems: in the oceans stability is imposed by the low frequency of abiotic change, whereas on land, communities are regulated by their internal organization.

If Steele is right, a unifying theory that encompasses the major ecosystems of the planet might be based on the effect of environmental variation on ecosystem and community processes. It may be that our ideas about the effect of disturbance on ecosystem organization need extending to encompass long-term variation, operating over larger areas (section 6.5). Perhaps hierarchy theory, with its functional division of ecosystems based on the frequency of information or materials passing between components (section 1.5), is the one current theory that could encompass the major differences between marine and terrestrial ecology.

10.3 STABILITY AND DIVERSITY

One measure often suggested as a common reference for comparing different commmunities is species diversity. By reducing a complex community to a single number, even just a count of the number of species, we can compare very different ecosystems, or compare the same ecosystem at different times during its development (Table 10.1). Generally, the number of species would be expected to increase during succession, up to some form of steady state, although disturbance may be important for maximizing species richness (section 6.5).

For a long time, the diversity of a community was thought to be a prime determinant of its stability (McIntosh, 1985), with highly diverse communities showing greater inertia. In one of its original forms, the theory argued that food webs with a large number of species provide a number of alternative routes by which energy might flow through the system. So, if one or more species were lost, a predator could switch to alternative prey, and its population would not fluctuate greatly. In effect, a community with a greater number of possible pathways should be buffered against significant change. The theory thus predicts that populations would be more stable in species-rich communities, such as tropical forests or coral reefs. The evidence for population stability in these ecosystems is, however, equivocal (Giller, 1984). Although many of these commmunities appear to be relatively constant, this may have little to do with their species richness or the complexity of their food web.

Much of the analysis of stability has centred on the complexity of the trophic interactions within communities. May (1974) explored this using mathematical models, examining the fluctuations in the population of each species around its equilibrium density, its local stability. The models had different numbers of species (S) and the trophic connections between any two species were ascribed at random. Where both populations were depressed by their interaction there was competition; a predator–prey relationship was indicated when one population gained at the expense of the other. A zero term implied no interaction. Of all the possible interactions, only a proportion had trophic connections and this value was termed the **connectance** (C) of the web. This measures the extent to which the food web is linked and is therefore a direct measure of its complexity. Finally, the average strength of all these interactions (the effect of one species on the population density of another) was also calculated for each model.

The models showed that increasing the number of species, their connectance, or the average magnitude of their interactions above certain critical levels lowered the stability of the constituent populations. In effect, an increase in S above these limits required a reduction in the connectance of the food web for its populations to remain stable. Communities with high S would thus be unlikely to show high C. Put another way, given the same degree of connectance, a stable community could only accommodate a larger number of species if the average strength of their interactions declines.

Any increase in S or C increases the complexity of the food web, so the models predict that stability would actually decline with increasing complexity. From this, May suggested there is only likely to be a very particular set of conditions in nature where complex food webs would be stable and persist – in constant environments where the community would not be subject to significant random variation. Equally, simpler communities would be found where the environment was less predictable.

Modelling food webs using these random allocations can produce some highly unrealistic feeding relations (Lawton, 1989). In other models which avoid circular feeding relationships, or which limit a prey population by its food supply, there are conditions under which stability can increase with complexity. Nevertheless, there is some evidence that connectance and species number are closely related in real ecosystems. The product SC tends to fall within the range 2–6 in different habitats (Begon *et al.*, 1990), indicating that any rise in S must be accompanied by a fall in C, as the model predicts. Connectance appears to be more important for stability than the number of species in the community (O'Neill *et al.*, 1986), suggesting that it is the organization of the species within the community that is critical: rather than the random trophic connections assumed in the model, perhaps it is the structure of the food web that confers stability. If the species are formed into self-regulating subgroups, with strong internal connections, but weak interactions with other groups, stability may follow (May, 1974; O'Neill *et al.*, 1986).

One possible functional group is a guild of species, collected together because they use the same resources in a similar way (section 6.6). We can, for example, distinguish between guilds of plant-sucking insects and leaf-chewing insects. In fact, the insects are one of the few groups in which guilds can be readily identified: in others, defining a guild becomes more problematical. Natural boundaries, differentiation of resources or feeding mechanisms may each be used to indicate guilds, but compartmentalizing food webs in this way is also dependent upon the spatial and temporal scales over which we observe them (Lawton, 1989). Even so, where they can be defined, guilds tend to have a very consistent organization (McNaughton, 1978). Dividing the food web into these functional compartments can allow for greater stability with increased S: within a guild there will be strong competitive interactions between members which, in turn, produces strong regulatory pressures on each population. Here regulation is produced by the internal organization of the community. In hierarchy theory, the stability of the larger system is derived from the interaction of such functional components, and the attenuation of the signal between them. This serves to dampen any displacement of the system (Figure 1.3). Guilds may have this effect in a complex trophic web, even if they are not always readily identifiable by the ecologist.

A further complication is that some species are adjusted to particular frequencies of disturbance (section 1.4). As Steele (1985) suggested for terrestrial ecosystems, the collective effect of the adaptations of its individual species may allow the whole ecosystem to accommodate a disruption and maintain its equilibrium. This has been termed **incorporation** (Urban *et al.*, 1987). Fugitive species, for example, rapidly colonize gaps and facilitate the development of the rest of the community within the gap. In this way, a landscape may incorporate change, forming a mosaic of patches in various stages of recovery. Many natural processes of disturbance, other than the creation of gaps, will be incorporated within an ecosystem or a landscape, because the community or its constituent species are 'adapted' to these changes. The landscape's capacity to accommodate such stresses depends upon their size and frequency relative to the size of the ecosystem, and the extent to which it has incorporated this regime of change; to stay close to an equilibrium, the disturbance should not exceed the elasticity of the community (Figure 1.4). Thus forests subject to different frequencies of disturbance by fire show different equilibrium communities. Urban *et al.* (1987) describe various studies which have measured these equilibrium conditions in communities; one, based on a simulation model measuring the variability in biomass, was used to calculate the ratio of gap size to the area of landscape needed for incorporation of a disturbance. For temperate forests this works out at 50:1.

The relationship of S to C and the strength of species' interactions indicate that complex communities may be less stable in the face of disturbance. From this we would expect that only 'simple' communities would have the adjustment or inertial stability to survive in very variable environments, and there is evidence for lower connectance in communities in fluctuating environments (reviewed in Begon *et al.*, 1990). In contrast, highly connected communities would be confined to environments that experienced little disturbance or, at least, unpredictable changes. For the same reason, complex communities that develop in stable environments will be more susceptible to anthropogenic disturbance and more likely to lose species following displacement. The prime example is the tropical rain forests with their long recovery times from large-scale disturbances.

Much of this work has been dominated by measuring stability in terms of population dynamics within a defined food web. This is not the only possible approach: we could measure other terms, such as species composition or turnover rates, productivity or rates of nutrient transfer. Indeed, some of these processes may be critical for the stability of the whole community. For example, the availability of phosphorus determines the productivity and community structure in many freshwater ecosystems, and also which plants dominate terrestrial ecosystems close to a stable state (section 7.1). The population dynamics of several species may be a useful summary of some of these processes, but the choice of species to be counted is then critical for judging the stability of the whole community.

10.3.1 Indices of diversity as measures of community change

Eutrophication in freshwater ecosystems results from the interactions between certain members of the food web and an increase in the nutrient supply to the primary producers. With abundant nutrients, the balance between the constituent populations shifts and the system moves from one equilibrium state to another. One simple way of measuring this change is to follow the decline in the total number of species, or the increased abundance of one or two species. An index of diversity attempts to combine both of these features – species richness (S) and the distribution of individuals between species (species equitability) – into a single value (Figure 10.2).

An accurate measure of diversity offers the prospect of being able to quantify changes in the structure of a community as it responds to a disturbance. It may also allow us to compare the response of very different communities to the same impact. Some stresses increase diversity by creating gaps: for example, grazing pressure or a reduced nutrient supply have both been shown to increase the value of one commonly used index (the Shannon–Weiner Index) in terrestrial and aquatic ecosystems (Moore, 1983).

A large number of indices have been developed, placing different emphases on species richness and equitability. In several cases, the measure of equitability uses some method of fitting the relative abundance of the different species to a particular pattern or distribution (Figure 10.3). These predict the proportion of abundant and rare species within a sample of the community. A number of different distributions are assumed by various indices, although the biological justification for some of these patterns is not always obvious. One or two patterns are commonly found in very different communities, perhaps indicating some consistency in their organization, at least for the species sampled. The distribution may reflect the competition between the various species for some resource, at least within a particular group. However, competition

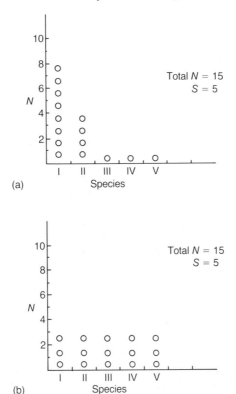

(a)

(b)

Figure 10.2 A simple illustration of the components of diversity. Species richness in both examples is the same ($S = 5$), as is the total number of individuals ($N = 15$). However, the distribution of individuals across the species (the 'species equitability') is very different. In (a) two species dominate whereas in (b) the relative abundance of all species is the same. Intuitively we would rate (b) as being the more diverse community. A simple ratio of S/N for the two samples would be the same. More elaborate measures of diversity attempt to reflect both the species richness and equitability of a community. These can be highly dependent upon sample size and the distribution of individuals between species.

cannot be the sole organizing force within the community, as the abundance of some species is not regulated by these interactions (section 3.2). Indeed, there are relatively few examples of competition determining these patterns (Gray, 1985).

The most commonly observed pattern is the log-normal distribution (Figure 10.3), found in data collected from a variety of communities. Even so, this may be no reflection of the community organization – this distribution can be generated by the combined effect of several factors operating independently on the constituent populations, a feature of the statistics of large numbers (May, 1981a). However, others suggest that a

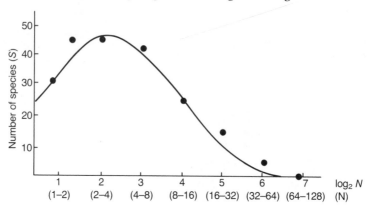

Figure 10.3 One of the commonest patterns of relative abundance observed in nature is the log-normal distribution. Here the number of species with a particular N is plotted against the $\log_2$ of their N. (Log_{10} is sometimes used with very large sets of data.) This produces a curve which approximates to a normal distribution.

partitioning of resources between species would produce the log normal distribution (Magurran, 1988). A departure from this pattern has been used as a measure of an impact and May (1981a) describes how a range of pollution stresses appears to have deflected a community from a stable, log normal distribution in both aquatic and terrestrial habitats.

A range of indices has been used in all major habitats to measure pollution impact at the commmunity level, and a number of these studies are reviewed by Magurran (1988). However, despite their potential value in quantifying community responses, there are several problems in their interpretation. As Green (1979) notes, the index may be responding to a number of factors, other than the disturbance itself. In addition, the indices are usually only applied to a particular group of plants or animals: the difficulties of sampling the larger community means that the response of the whole community is rarely measured. Where a more complete sample is achieved, indices invariably treat individual species as being equivalent. This is a highly questionable assumption given the role of different species, and the significance of particular groups in some ecosystem functions (Karr, 1987). In this case, we need to establish whether some change in an index accurately reflects a change in an ecological process.

Few studies establish what a change in the index represents. An index is commonly used without any attempt to establish a dose–response relationship, making it difficult to decide what magnitude of change is significant for a community. This requires baseline data to judge whether a response differs significantly from the variations observed in the unstressed community (section 2.5). Without evidence that the community responds in a linear fashion to an increasing stress, an index of diversity would give a spurious sense of quantification.

In short, an index can rarely be taken to be a direct measure of some impact on the whole community. Allied to the practical difficulties of sampling and its effect on their interpretation, this may explain why indices of diversity have not been used more widely in pollution ecology.

In some cases, species richness alone has been preferred as a simple measure of a pollutant's impact. Others have suggested a more selective count, concentrating on those species with an intermediate abundance. This may be a more useful approach in the bottom-dwelling commmunities of marine habitats which are dominated by rare species (Gray, 1985). The organic pollution of rivers is often assessed by the loss of sensitive species, or by some groups becoming more common. A number of **biotic indices** have been developed, based on the abundance or presence of such indicator species (section 2.6). Many of these effectively measure the population response of a small number of species, rather than any community-level function. Washington (1984) reviews their use, along with indices of diversity and measures of community similarity. Many indices of diversity are deemed unsuitable for their purpose by Washington, because we have no clear idea about the biological meaning of the index. Equally, selective sampling, counting only a indicator group, can pre-judge the nature of the impact. Organic enrichment can produce a very definite sequence of change in species composition downstream of a sewage outfall, and some biotic indices will indeed reflect the deterioration in water quality, but they are not complete measures of the impact on the community.

Alternative community-level measures of an impact include nutrient fluxes, productivity and community metabolism. One recent suggestion is that scope for growth (*SfG*) (section 2.9) might be scaled up from the individual or population level to measure responses at the community level (Genoni and Pahl-Wostl, 1991). This would again require baseline data against which to compare the performance of the community, although we could speculate from Odum's work (Table 1.1) that the ratio between respiration and biomass ought to increase in such cases. The community *SfG* would then decline as the respiration of individual species increases to 'pump out the disorder' (Odum, 1985). This offers some prospect for identifying the nature of the effect on ecological processes, and making direct comparisons between populations within the community. On its own, however, this might not reflect changes in the community structure, and the practicality of measuring this on more than a small part of any real ecosystem has yet to be addressed.

10.4 REGIONAL DIVERSITY

Not only can we ask if stability is a product of a diverse community, but also whether diverse communities result from a stable environment. May's work on connectance suggests that complex communities are only likely to originate and persist in relatively constant environments. On the other hand, disturbance appears to be necessary to maintain maximum species richness in many communities, at least where it has been incorporated. Without disturbance, the sustained competition found in communities toward the end of their successional development favours K-selective species, leading to an overall reduction in diversity (section 7.1).

The existing complement of species may owe as much to the history of a community as its present ecology. Evidence from pest introductions (section 6.5) suggests that some species would have a much wider distribution were it not for the historical details of their evolution and geographical barriers preventing their dispersal. The diversity of marsupials in Australia reflects their isolation from major competitors for most of their past. A host of large-scale factors including geographical isolation, the regional species pool and rates of evolution are needed to explain local patterns of diversity (Ricklefs, 1987).

These same processes may account for the latitudinal gradient of species diversity across the globe. One obvious feature of the tropics is its greater land area which, according to the species/area hypothesis (section 6.3) should support a larger number of species. However, compared with the northern coniferous forests or temperate woodlands, the forests of the tropical and subtropical regions have more species per unit area. This is true for virtually every taxon, with the exception of aphids (Walker, 1989).

For many years the greater species diversity within the tropics was attributed to two factors linked to the stability of its abiotic environment:

1. Tropical forests grew in a predictable climate with low variability.
2. The tropical regions were not subject to disturbance during the Quaternary glaciations and have thus existed in this form for at least two million years.

It was argued that the predictability of the tropical environment should favour long-lived and competitive, or K-selective species (section 5.1). The relatively minor seasonal changes in the tropics and their predictability were thought to encourage finer niche differentiation because competition and other interactions between species went uninterrupted throughout the year. Fruit production in the forests of Peru is a response to seasonal rains, but lasts for about nine months of the year, supporting over 100 different species of vertebrate (Terborgh, 1986). In contrast, the arctic tundra regions provide food resources for very brief periods and are dominated by r-selective animals that complete their life cycle within a short season.

On the other hand, temperate communities may allow for a greater separation of niches in time, using the distinct seasonal cues, such as day length and temperature. Examples are the snowdrops and bluebells which flower in the early spring, before the canopy of the woodland has formed. Whether this fine synchronization is lacking in tropical forests is debatable; certainly, temperate communities undergo a well-defined sequence of changes as different plants dominate from the spring through to the autumn. It may not be the constancy of a habitat that promotes greater diversity, but rather the predictability of change. For this reason, tropical regions are thought to be more susceptible to disturbances that fall outside the range they have incorporated. Notably, the temperate woodlands with their fewer species have a greater capacity to restore themselves following major disturbance, probably a reflection of the more variable climate under which they grow.

In contrast, Ben-Eliahu *et al.* (1988) attribute the lower species richness of polychaete worms on Israel's temperate coast to competitive exclusions, resulting from a lack of disturbance (section 6.5). The tropical coast benefits from the incorporation of a harsher regime of disruption, with a number of fugitive species able to use the gaps created. In terrestrial habitats, perhaps where disturbance is less dramatic, a constancy of resources might promote greater specialization, and the evolution of new species resulting from niche differentiation (section 3.2).

Over the medium term, such specialization ought to result in a greater diversity of species in each genus. This pattern is not found in the tropical tree flora – generally tropical forests have relatively few congeneric species. Instead, their massive diversity is attributable to different families and genera, suggesting that it results from a long period of evolution, and perhaps a series of competitive exclusions. Ricklefs (1990) concludes that for the trees at least, the greater diversity in the tropics is a product of processes operating over the longer term.

The length of time that the tropical forests have existed was also proposed to account for their greater species richness. The temperate regions have had a short period to acquire species since the end of the Pleistocene, whereas the tropics were believed to have been relatively undisturbed by the advance of the ice sheets. Implicit in this idea is that fewer species go extinct in a constant environment. The problem here is that polar regions have had a longer history with a constant environment, yet have fewer species.

There is also evidence now of major changes in the tropics during the Pleistocene: Amazonia was perhaps 4.5–6°C cooler during the ice ages, with the forest confined to lowland areas (Walker, 1989). In Africa, the tropical forests were perhaps one-quarter of their present extent, while in southeast Asia the lowered sea level probably increased the overall size of the habitat (Wilcox, 1980). Whether this resulted in higher S (and a greater prevalence of K-selected species) in southeast Asia has yet to be decided. Pockets of high endemism for birds and mammals may indicate refugia from this division of Amazonia in the last million years, although again the evidence is not clear (Walker, 1989).

Not only does climate change with latitude, but more specifically, so does the availability of energy. Some have suggested that it is the energy captured by the biota that determines the diversity of the community. The number of insectivorous bird species in various parts of the British Isles correlates with the distribution of available energy when these species are resident (Turner *et al.*, 1988). Indeed climate, and most specifically temperature and sunshine hours, were far more significant factors for this highly mobile group than latitude, longitude or the age of the habitat. In the tropics, the greater input of energy allows for a species to have a larger population, and for each individual to have lower metabolic costs. Turner and his co-workers suggest that this makes extinction less likely in these lower latitudes, so species 'accumulate' with time. There is also a latitudinal gradient of productivity toward the tropics, and perhaps the higher S is a function of the greater resources available. Certainly, most groups tend to increase their biomass per unit area toward the tropics.

A consistent supply of energy may allow finer divisions of the resource spectrum and the greater specialization that produces more species. The variety of trees and other plants in the tropical forests also provides a more heterogeneous environment for the animal community, offering greater opportunities for animals to adapt to specific conditions. The interactions between plant and herbivore may contribute to the plant diversity – one theory suggests that the activity of herbivores will serve to maximize this diversity: any plant species that is common is most likely to support a large commmunity of herbivores, leading to the evolution of specialists adapted to that plant. A common species will then be preferentially consumed, leaving space and resources for other plants with less consumer pressure.

Finally, nutrient-rich ecosystems are often dominated by a small number of highly productive species (section 7.5) and notably, tropical forests have poor soils with the nutrients rapidly cycling through the biomass. The lack of essential nutrients in tropical soils may thus contribute to their high diversity (Ricklefs, 1990), preventing a small number of prolific species becoming dominant.

Each of these hypotheses and speculations need more thorough testing, and at these spatial and temporal scales, this also becomes a test of the ecologist's ingenuity. It seems unlikely that any one theory will explain the latitudinal pattern of species richness for all taxa. Not only does temperature and available energy change with latitude, so may a number of other key environmental factors, such as the stability or predictability of the abiotic environment. The differences in the biology, ecology, mobility and history of the different phyla will mean that each has responded differently to a combination of these variables. They will also be responding to the distribution of other members of their communities. As Ricklefs (1987) points out, the elephant is an example of a long series of historical and ecological processes that governed its evolution; communities can be a product of the same combination of accident, circumstance and process.

Many ecologists argue that community organization can be explained in terms of competition theory and limiting similarities. Understanding why there are more species in the tropics is likely to require a broader perspective. The close relationship between food web complexity, the stability of the community, the predictability of its environment, and the abundance of different resources may begin to explain the global pattern of species diversity, but historical and geographical factors are also important. Identifying these key variables, along with the development of ecosystem theory at the regional and global scales will be crucial for our attempts to apply the science to global pollution and conservation problems.

10.5 LANDSCAPE AND REGIONAL ECOLOGY

One of the central problems, of course, is to describe the significant ecological processes operating at the larger temporal and spatial scales. We then need to decide when a local phenomenon can only be understood by recourse to the regional or global level. Bridging the gap between ecosystem ecology and these larger scales, at least for terrestrial ecology, is landscape ecology. This is the study of the processes that govern spatial and temporal patterns within a broad area defined by a landform or a climatic regime.

Landscape ecology considers the spatial distribution of ecosystems over several square kilometres and regional ecology increases this range further still. A **landscape** consists of a series of interacting ecosystems, subject to the same climate, geomorphology and regime of disturbance (Forman and Godron, 1986). The degree to which a landscape is divided into distinct ecosystems is termed its spatial heterogeneity. The landscape itself may be regarded as a ecosystem, partitioned into patches just as a woodland or heathland may form a mosaic. However, a landscape can include very different ecosystems, such as rivers, lakes and woodlands. Landscape ecology encompasses those large-scale processes that define landscape development, including those resulting from its landform (geomorphological), rock type (geological), history (palaeoecological)

Table 10.3 A classification of landform effects on ecosystem patterns and processes (after Swanson *et al.*, 1988)

Class	Significant features	Nature of effect
1.	Elevation, aspect, slope parent rocks	Air and ground temperatures, availability of moisture, nutrients and pollutants
2.	Wind movement, drainage – boundaries with other communities	Dispersal of plants and animals, seeds, energy and materials
3.	Wind movement, slope and aspect	Frequency of non-geomorphic disturbance – e.g. fire, storm damage
4.	Spatial pattern and processes – slope erosion, glaciation, etc.	Frequency of geomorphic disturbance, e.g. landslip

and climate. Swanson *et al.* (1988) provide a classification of the effects of landform on ecological processes (Table 10.3).

Such factors not only define the pattern but also the process: a key feature of a landform, such as slope, defines a watershed and the range of vegetation types within it, and also the speed with which material, nutrients and energy enter and leave the ecosystem. A range of effects, determined by the degree of spatial heterogeneity, govern ecological processes at this level (Table 10.4).

Table 10.4 Five major principles in landscape ecology (after Forman and Godron, 1986)

1. Landscapes are heterogeneous and differ structurally in the distribution of species and energy. Each individual ecosystem within a landscape is either a patch of significant width, a corridor or part of the background matrix.
2. An increase in heterogeneity in the landscape reduces the area of internal habitats and increases the abundance of species, favouring 'edge' species – those favouring the boundary between two habitats.
3. Landscape heterogeneity has a fundamental effect on the movement of species and may itself be determined by their mobility. For example, the colonization of a gap may require particular species of tree to commence a re-establishment of the woodland community.
4. The movement of plants and animals and the disturbance of patches causes an increase in the movement of nutrients, energy and biomass between patches as landscape heterogeneity increases.
5. The degree and frequency of disturbance serves to maintain and increase diversity and landscape heterogeneity.

Figure 10.4 The network of corridors formed by hedgerows and roadside verges connect areas of woodland, facilitating the movement of animals between them. This reduces the fragmentation of the semi-natural habitats produced by agriculture. Leicestershire, Central England.

A landscape is made up of patches, a network of linear features, and a background matrix (Forman and Godron, 1986). For example, an agricultural landscape may consist of a series of pockets of woodland connected by hedgerows or streams. The matrix, in this case, is the cultivated field separating the patches of woodland. This represents the largest component of the landscape, and is likely to exert the greatest control over its processes (Forman and Godron, 1986). The heterogeneity of landscapes is generally increased by human activity, in fragmenting the habitat, or by providing corridors for movement, such as railway embankments or roadside verges and hedgerows (Figure 10.4). Equally important is patch shape (particularly the length of the boundary relative to its area), its degree of isolation and its age. Not surprisingly, the principles of MacArthur and Wilson's theory of island biogeography have been used to determine the turnover and equilibrium number of species in these fragments (section 6.3).

In addition, the size of fragments or habitat islands could be critical for the population stability of species that form a metapopulation between the patches (section 4.7). The configuration of the patches and the connecting corridors will determine species' movements and their persistence in each fragment. Indeed, the abundance of a particular species within a patch may only be understood by referring to the landscape. Burel (1989) describes the distribution of carabid beetles in the woodlands and connecting hedgerows in western France, and shows that the species found in a particular hedgerow reflects the larger pattern within the landscape, including the distance to the nearest woodland, as well as the local detail, such as the structure of the hedgerow or whether it borders a lane. To maintain forest species of these beetles in rural areas, Burel concludes there has to be a large measure of connectedness between patches. The same will be true of other relatively non-mobile invertebrates. However, evidence that these corridors lead to a permanent increase in species richness within a fragment is generally lacking. Also, some linear features, such as roads and tracks, may be a barrier to the movement of particular species.

10.5.1 Measuring spatial heterogeneity

One critical problem in landscape ecology is deriving suitable methods for quantifying the effects of spatial heterogeneity on ecological processes, and its effect on different groups of organisms with different degrees of mobility. Then it is important to measure the extent to which any patch is isolated. Forman and Godron (1986) make interesting comparisons between the convoluted boundaries within the vertebrate lung and the dendritic patterns formed by the drainage pattern in a watershed. In both cases, the exchange of materials and energy across the boundary is maximized. Finding such patterns in the landscape might indicate that there would be a high mobility of species and nutrients between patches or along these corridors. Indices of diversity have been used to quantify the effects of edges or boundaries between two habitats (Yahner, 1988), and as indirect measures of habitat diversity, these may represent a useful summary of spatial heterogeneity, perhaps as a measure of resource usage. For example, some species are more common at these edges, and predators tend to concentrate their foraging in these areas (Yahner, 1988).

O'Neill *et al.* (1988) have attempted to quantify spatial heterogeneity by using indices of pattern based on three measures – the degree of similarity in land use between adjacent sampling (grid) points, fractal geometry and information theory. In their study of land-use patterns in the eastern USA, they were able to show that a larger fractal dimension (the 'roughness' of a boundary around a land-use patch) was associated with undisturbed forests, but that this dimension was much lower in the regular rectangles and squares associated with agricultural land. In effect, our management of a landscape reduces the length of the boundaries around patches, as well as shrinking their area. This reduction in the fractal dimension could be correlated with a measure of land-use intensity, and offers a simple index of heterogeneity for quantifying spatial patterns, useful in the development of models in landscape ecology.

The variety of these techniques is likely to increase as we draw more heavily on satellite remote-sensing using digitized maps and computer-aided image analysis. Then there is the prospect of a whole range of quantifiable indices or other methods of summarizing spatial patterns (Openshaw, 1991).

10.5.2 Modelling landscape change

The extent to which landscape ecology produces novel theory, or simply extends existing theory is still unclear. The patchiness of an environment has been incorporated into models of predator–prey dynamics and the conservation of endangered species. The movement of energy or materials between patches also falls readily into the remit of hierarchy theory (section 1.5). At the moment it appears that the development of methods will preceed any development of theory, and both are likely to borrow heavily from the spatial pattern analysis techniques used by the geographer. The advent of geographical information systems (GIS) has already had this effect on our capacity to model large-scale processes.

One example is the work of Walker and Walker (1991) in the Alaskan tundra, which classified natural and anthropogenic disturbances associated with the development of the oil field at Prudhoe Bay, and related each to their temporal and spatial scales.

The critical feature of the tundra ecosystem is the thermal stability of the permafrost, the permanently frozen subsoil, and the changes that excessive thawing can bring to the vegetation community growing above it. The mapping of disturbance in this most sensitive area may be of global importance – a significant proportion of the carbon in the biosphere is stored in the peat of the permafrost.

The tundra landscape consists of a mosaic of patches, with the vegetation in various phases reflecting the condition of the permafrost. Disturbance creates a warmer soil: on thawing, this increases the availability of nutrients, and increases decomposition, reducing the organic matter. The change in the overlying vegetation can be used to follow the disturbance or recovery of the soil beneath, and this is being mapped by sampling at various scales, from point sampling in the field all the way up to satellite imaging and GIS. A hierarchical classification of landform and disturbance has been developed to distinguish the scale of the impacts, from those covering a few metres, to those extending over the region (Table 10.5a). These tend to have time scales to match – for example, the deposition of loess, wind-blown sediment, is a regional phenomenon whose impact is only felt over the long term.

Based on their observations of the natural disturbance, Walker and Walker estimated the recovery time from anthropogenic disturbance (Table 10.5b). Excepting long-term atmospheric warming, the most long-lasting human impacts are those that cause significant melting of the permafrost, or an accumulation of waste and debris, including oil spills.

Other studies of the effect of an increase in atmospheric carbon dioxide and temperature on tundra ecology also suggest that their impact would range over various scales. Using enclosures, Strain (1987) was able to demonstrate an increase in photosynthesis by some plants due to both elevated CO_2 and atmospheric temperature. Several species, such as the grass *Eriophorum vaginatum*, grew more rapidly than others, including shrubs such as *Vaccinium* and *Betula*. Thus a long-term elevation of mean atmospheric temperatures would, not surprisingly, lead to a change in the species composition of arctic plant communities. However, the effects may extend beyond this level – a higher concentration of CO_2 allows plants to keep their stomata closed, reducing their transpiration rate which, in turn, may increase rates of runoff from watersheds (Strain, 1987).

10.5.3 Modelling regional ecological problems

In addressing any ecological problem, we need to identify both the source of a stress and the level at which it perturbs the system. Isolating the factors that regulate community and ecosystem processes helps us to follow how effects at one level determine responses at other levels in the system. Rapport (1989) has considered this relationship in his study of pollution at the local and regional levels in the Gulf of Bothnia.

This is the part of the Baltic Sea which divides Sweden and Finland (Figure 10.5). Rapport suggests that the Gulf has a low diversity because it is relatively young, having been established since the retreat of the ice sheet around 10000 years ago. It has a marked gradient of salinity and temperature down its length, producing a range of habitats. Superimposed upon this regional pattern are point sources of pollution (such as streams carrying the waste of a single pulp mill), or more general sources (urban areas) which introduce pollutants over a broader area. Many of these sources have been created in the last few decades.

Table 10.5a A hierarchy of natural disturbance for landscape and vegetation of the Alaskan tundra (adapted from Walker and Walker, 1991)

Disturbance	Geomorphological effect	Vegetation effect	Spatial scale (m^2)	Frequency of event (years)
Continental drift and mountain building	Rock formation, formation of landscape and drainage patterns	Evolution of arctic flora	$>10^{12}$ (Continental–global)	$>10^6$
Climatic fluctuations over ice ages	Glacial landscape features	Speciation and extinction	10^4–10^{11} (Regional)	10^4–10^6
Climatic fluctuations during the Holocene	Development of permafrost	Species migrations, landscape mosaics	10^2–10^{10} (Regional)	10^3–10^5
Tundra fire	Local erosion	Gap creation, loss of nutrients	10^4–10^8 (Site/ecosystem)	10^3–10^4
Storms, ice-wedge formation Annual cycles of disturbance	Erosion, sedimentation	Change in microenvironment and mosaic – burial and removal of vegetation – addition and loss of nutrients	10^{-1}–10^6 (Ecosystem)	10^0–10^4

Table 10.5b A hierarchy of anthropogenic disturbances on the landscape and vegetation of the Alaskan tundra (adapted from Walker and Walker, 1991)

Disturbance	Geomorphic effect	Vegetation effect	Spatial scale (m^2)	Frequency of event (years)
Climatic change due to greenhouse gases	Alteration of depth of permafrost movement of steep slopes	Species migration, changes in landscape mosaic, movement of tree line	10^2–10^{11} (Regional)	10^3–10^5
Oil fields and transportation	Changes in microclimatic flooding, dust, waste heaps, compression of soils	Introduction of weed species – plant communities – new end point for succession Burial of natural vegetation Creation of barren surfaces Killing of vegetation – local pollution incidents	10^3–10^8 (Regional)	10^3–10^4
Trails, buildings, wastes	Elimination of landforms, hummocks, increased erosion	Trampling, compaction of vegetation Eutrophication and killing of vegetation	10^{-1}–10^4 (Ecosystem)	10^0–10^4

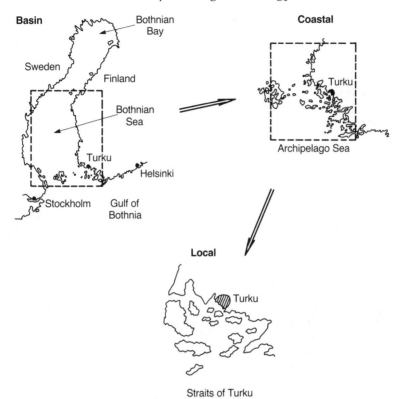

Figure 10.5 The scales used by Rapport (1989) to examine levels of ecosystem stress in the Gulf of Bothnia – the local, coastal and basin levels. This is a young sea, with a marked gradient of salinity, running from the fresher waters in the north (where the sea is dominated by runoff) down to the more saline zones in the south. There is also a large temperature gradient down its length. These changes in salinity mean that some communities are naturally stressed by the low salt content, with low species diversity and low productivity.

This study tried to detect the impact on ecoystem-level criteria of various stresses originating at three levels (Figure 10.5): 'local' areas – the protected waters around bays, inlets and so on, which are close to major sources of pollution; the 'coastal' areas outside sheltered bays and the offshore waters of the two major basins of the Gulf. Some of these effects were confined to the local areas and had little impact at the larger scale, but others had implications from the local to the regional level.

At the local scale, change could be detected in all of the parameters measured, including increases in nitrogen and phosphorus, reductions in species diversity and reductions in the average size of species within the benthic communities (Table 10.6). Zones devoid of life only occur very

Table 10.6 Ecosystem effects at three spatial scales in the Gulf of Bothnia (after Rapport, 1989)

	Scale		
Ecosystem response	*Local*	*Coastal*	*Basin*
Nutrients	↑	↑	↑ ?
Productivity	↑	↑	?
Abiotic zones	↑	?	–
Species diversity	↓	–	–
Genetic diversity	↓	↓ ?	↓
Size distribution	↓	↓ ?	↓ ?
Prevalence of disease	↑ ?	–	↑ ?
Proportion of opportunist species	↑	↑ ?	–

– = No data; ↑ = increase, ↓ = decrease.

locally, around point sources of pollution. A switch to smaller, opportunistic species (*r*-selective) is found at local and probably at coastal levels. Within the whole gulf, Rapport suggests that a reduction in stocks of commercial fish and of large mammals (seals) would indicate a loss of genetic diversity that applied to all levels.

These impacts need not originate simply in local areas and spread to the coastal or basin region. The effect of some stresses can be transmitted in both directions: the pressure of salmon fishing in the basins has led to a reduction in spawning activity in rivers and inlets, a local effect. Indeed, effects at all levels are likely to be multifactorial and involve interchanges between each level (Rapport, 1989).

Extrapolating from the local level has also been used to assess the regional problem of **acid deposition**, and to predict its effect on the lakes of Scandinavia. Sulphur and nitrogen oxides arise as localized pollution in industrial and urban areas, but become a regional problem through their long-range transport in the atmosphere (Figure 10.6). A number of factors determine which areas suffer the greatest stress from their deposition, such as the predominant wind direction and the sensitivity of the ecosystems receiving the insult. Because of their geology, geomorphology, prevailing climate and history, some ecosystems lack the capacity to buffer their environment against an increase in acidity. For example, mountainous and well-forested areas induce precipitation, and will tend to purge the atmosphere of its pollutant burden.

Scandinavia has been particularly susceptible to such effects, removing much of the industrial pollution in the winds arriving from south western Europe. It also suffers occasional massive inputs from central and eastern

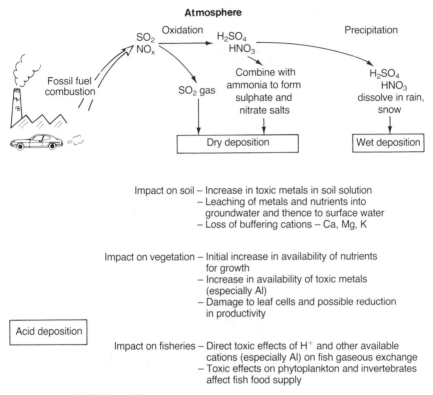

Figure 10.6 Acid deposition.

Europe (Elsom, 1987). The thin granitic soils that overlie much of the region reflect both the parent rock and the extensive glaciation in the last ice age. These soils have little capacity for binding the free protons that raise the acidity, and are unable to buffer the soil solution or surface waters against low pH. In addition, the cool, wet climate of the region favours the formation of peatlands or the cultivation of conifers, each of which produce naturally acidic conditions in the soil.

Overall, this collection of regional factors make Scandinavia especially susceptible to the impact of acid deposition. At the local level, it is possible to work out the sensitivity of an individual ecosystem if its particular features are known. One of the key variables in a lake, for example, is the level of organic matter – generally clear-water lakes, with small amounts of organic matter, are more susceptible to acidification (Kämäri *et al.*, 1991). However, this also varies with some regional factors. In some parts of Finland, much of the water drains through acidic, organic soils, to produce acidic, 'brownwater' lakes. Their acidity is dominated by natural organic acids, and sulphate additions from the

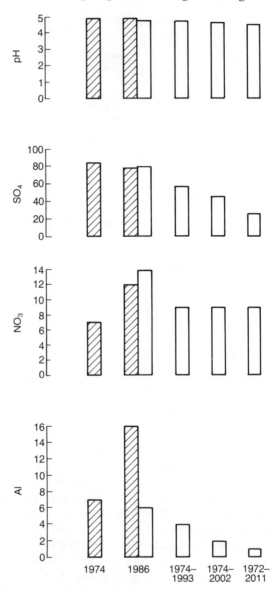

atmosphere represent a small proportion of the total acidity. Nevertheless, brownwater lakes have been found to be especially sensitive to anthropogenic inputs of acidity. Most of the sulphur dioxide pollution is found in the south and it is the fisheries in these lakes that have suffered the largest impact (Kämäri *et al.*, 1991).

Some ecologists have sought to extend ecosystem models to the regional level to predict these effects. One example is MAGIC (Model of

Acidification of Groundwater In Catchments) which has been used to make predictions about future trends in lake and soil chemistry in southern Norway (Wright *et al.*, 1991). This is based on soil solution equilibria equations and a series of budgets, or mass balances, for the major ions in a single catchment. Wright *et al.* use this model to simulate the pattern of chemical variables in lakes throughout the region. The model was calibrated against a known set of data, in this case, 464 lakes surveyed in 1974–75. The inputs of the model were adjusted to give accurate values for a range of chemical factors for that year, in a process akin to sensitivity analysis (section 8.1). From the calibrated model, predictions were made for 1986, and compared against data collected for 180 lakes in that year (Figure 10.7). The simulation allowed for a 5% reduction in sulphate deposition in the region between 1974–86, and its predictions showed a high level of accuracy for most ions. The exceptions were nitrates and aluminium – the rate of nitrate deposition nearly doubled over this period, keeping the pH of the lakes very low. This in turn maintained high levels of free aluminium in the waters, significant because aluminium toxicity is the main cause of fish death in acidified freshwaters. Overall, the general success of the model suggests that the procedure of scaling up the model from the watershed to the region has been valid for these lakes.

Based on this model, the authors make predictions about the effects of reducing acid deposition in the immediate future (Figure 10.7). It would require reductions of over 50% for a significant improvement in the proportion of Norwegian lakes restored to a non-acid state – markedly above the levels of 30% currently set for the region under the United Nations environment programme. These greater reductions are predicted to give a sustained improvement in the water quality of the lakes. The model also predicts that these reductions would significantly improve soil acidification in the whole region.

Figure 10.7 The water chemistry of lakes in southern Norway in 1974 and 1986 (shaded boxes) and the predictions of a regional model (open boxes). The model was calibrated against the 1974 data and then used to predict the values in 1986, assuming a 5% reduction in acid deposition over this period. The model was effective for sulphate and a number of other ions (values in μeq.l^{-1}). However it did not allow for an increase in nitrate deposition and the effects this would have on acidification and the availability of cations, including aluminium. This probably accounts for the discrepancy between the observed and predicted values for 1986. Further predictions have been made assuming reductions of 30% over 1974–1993, 50% for 1974–2002 and 70% for 1974–2011. Only at the higher levels would a significant proportion of the lakes in the region recover (35 and 50% respectively). (After Wright *et al.*, 1991.)

10.6 EVALUATING GLOBAL ECOLOGICAL CHANGE

These simple extrapolations from the local to the regional scale may not always be successful, failing in some cases because processes operating at the global scale regulate regional or lower level processes. Hierarchy theory suggests that the attenuation of a 'signal' between these different levels can govern ecological processes at each stage (section 1.5). An example is provided by O'Neill *et al.* (1991) who discuss the application of hierarchy theory to the problem of increased atmospheric carbon and global atmospheric warming, working from the community level. They consider the specific role of mycorrhizal associations in carbon fixation (section 7.3).

Applied ecology has been instrumental in establishing the true significance of mycorrhizal associations – Miller (1987) points out that the nature of this association only became apparent from studies of planting success on reclaimed soils. Our ignorance of the details of this association, and how the fungi aids phosphate uptake by trees, means that most mycorrhizal research is concentrated at the cellular and subcellular level (O'Neill *et al.*, 1991). However, as phosphorus may be the limiting factor in many terrestrial communities, the survival of the mycorrhizal spores, their distribution and dispersal, could be critical factors determining the productivity of these ecosystems. The mycorrhizae represent 'keystone mutualists', establishing the link between biochemical and geochemical processes that are crucial to ecosystem functioning. The mycorrhizae demonstrate what they term **upward constraint** – the control of a function at a higher level is limited by the rate process at some lower level. The presence or absence of mycorrhizae is one factor controlling the productivity of a woodland.

O'Neill *et al.* suggest that hierarchy theory helps us to decide research priorities in this area: an understanding of the rate of dispersal and survival of particular mycorrhizae might drastically improve our ability to restore the productive capacity of degraded communities. Burning down the tropical forests not only adds carbon dioxide to the atmosphere, it reduces the capacity of the biosphere to remove it again. Our dependence upon forests and woodlands to store carbon and absorb atmospheric CO_2 suggests that research could usefully examine the ecology of these associations. In particular, we need to establish how well microbial symbionts survive in disturbed or degraded soils and the ease with which new associations become established. It may not be enough to simply re-plant local areas of tropical forests, but may also demand a proper regard for the microbial associations essential to sustain the forest community. We thus need to know how easily mycorrhizae and other members of the community can colonize a site, perhaps from studying their ecology in the landscape.

Figure 10.8 As both consumer and disposer we each have an impact over very large temporal and spatial scales. The quality and sustainability of the environment we exploit depends upon the decisions we, as individuals, make. To limit our impact, we need to think beyond the immediate effects of our actions to include a broader perspective.

In the past 100 years, human perspectives have shifted from the local, to the regional and increasingly to the global scale (Clark, 1989). In the past 10 years, regional pollution priorities have been replaced by global pollution problems. 'Any significant improvements in our ability to manage planet earth will require that we learn how to relate local development action to a global environmental perspective' (Clark, 1989). It will also require individuals to understand the consequences of their actions and choices, and their cumulative effect on global ecological processes (Figure 10.8). Some authors have attempted to make a direct connection between the size of the human population and the nature of the global environment (Figure 10.9).

Others have tried to provide a more comprehensive picture of the global problem by incorporating some measure of human economic activity. Glasby (1988) cites the laws of thermodynamics to suggest that in maintaining his social and economic structures, man increases the entropy of the biosphere. One expression of this is the simple correlation

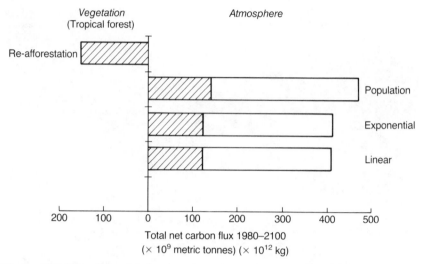

Figure 10.9 The effect of deforestation in the tropical forests on the flux of carbon in the atmosphere by the year 2100. The burning of tropical forest contributes around one-third of the carbon to the atmosphere as the burning of fossil fuels. It also reduces one of the more significant sinks for carbon from the atmosphere. Models have been used to estimate the effect of continuing deforestation on the carbon balance of the atmosphere. Three possible scenarios are considered here – a linear projection that allows for a rate of tropical deforestation based on the present rate of increase; the exponential projection extrapolates the rate of deforestation between 1980–85 exponentially until 2100; the population projection relates the rate of deforestation to the growth of the world population between now and 2100. For each a high and low estimate is given – the high estimate (open boxes) includes the clearance of fallow and open forest recovering from agricultural use. The low estimate (shaded boxes) assumes that only dense and closed canopy forest is cleared. A final projection measures the net flux of carbon into the tropical forests if deforestation stopped in the tropical areas in 1991 and the 'most-possible area' (around 865 x 10^6 ha) was replanted. (After Houghton, 1990.)

of *per capita* energy consumption with *per capita* income across different societies. As we become more wealthy, so we use more energy. Each society relies upon energy transformations to maintain its communications, its agriculture and its manufacturing. In contrast, natural ecosystems tend, during a succession, to reduce the energy they need to maintain their structures and organization. So mature ecosystems are characterized by a decrease in the net energy flow per unit of biomass (Table 1.1). The energy and materials that we release through our economic activity lead to a degradation of these natural systems. This is very much a global and mechanistic view of us in our environment.

With a rapidly expanding population, we should naturally ask how long will the biosphere sustain this growth. And then, what factors will halt it? In the past 50 years, the biosphere has had to withstand the greatest ever increase in the amounts of energy, radioactivity, metals and organic chemicals passing through it. We have increased the rate of loss of species perhaps a thousand times above the natural rate. Rivers now carry three times their estimated natural load of silt, the soil that was once the major nutrient store of the land and the foundation of our agricultural productivity.

Glasby summarizes the arguments of several authors: rather than adapting to our environment, we use energy to adapt our environment to suit us, inevitably leading to environmental disturbance and disorder. Applied ecology is just one part of our effort to manage the problem, but it cannot be the whole answer. With a rapidly swelling human population sitting on a shrinking bubble of resources, we also need to ask whether economics is the appropriate mechanism for managing human affairs.

Summary

Increasing attention to regional and global environmental issues has led to several attempts to extend ecological thinking to larger scales. Many of these rely on existing community level measures.

Species richness and species equitability are two important attributes that measure the organization of communities. Previous speculation that high diversity implies high stability seems largely unfounded, although the latitudinal increase in diversity toward the tropics may reflect the relative stability of these environments. However, a range of other factors change with latitude and may contribute to this pattern, including available energy, shortage of nutrients and the degree and predictability of disturbance. Similarly, the frequency of disturbance may account for the very different communities that we can observe between the land and the oceans. According to one view, two different ecologies apply – the first in which regulation comes from the internal organization of the community, and the second from the constancy of the abiotic environment.

Measures of diversity have been used to detect community change under stress, although their application is problematical. An ecology that attempts to embrace the landscape and the larger regional and global scales is being developed, although most quantification has centred around measuring spatial heterogeneity, connectivity and the frequency of disturbance.

Several attempts have been made to link ecological processes at the ecosystem, local and regional levels. Regulation can operate in both directions – from the local level upwards, and from the regional level downwards. This has been shown with the pollution in the Gulf of Bothnia in Scandinavia, and in the acidification of Scandinavian lakes. Efforts are also being made to predict responses at the global level, and to address the relationship between local action and large-scale change.

Further reading

Forman, R.T.T. and Godron, M. (1986) *Landscape Ecology*, John Wiley, Chichester.
Macquire, D.J., Goodchild, M.F. and Rhind, D.F. (eds) (1991) *Geographical Information Systems: Principles and Applications*, Longman, Harlow, Essex.

References

Abbott, I. (1989) The influence of fauna on soil structure, in *Animals in Primary Succession* (ed. J. D. Majer), Cambridge University Press, Cambridge, pp. 39–50.

Admiraal, W., De Ruyter van Steveninck, E.D. and De Kruijf, H.A.M. (1989) Environmental stress in five aquatic ecosystems in the floodplain of the river Rhine. *Sci. Total Environ.*, **78**, 59–75.

AFRC Institute of Food Research (1988) *A Basic Study of Landfill Microbiology and Biochemistry.* Contractor Report to the Department of Energy, UK.

Aiking, H., Kok, K., van Heerikhuizen, H. and van T Reit, J. (1982) Adaptation to cadmium by *Klebsiella aerogenes* growing in continuous culture proceeds mainly by formation of cadmium sulfide. *Appl. Environ. Microb.*, **44**, 938–944.

Allen, T.F.H. and Hoekstra, T.W. (1987) Problems of scaling in restoration ecology: a practical application, in *Restoration Ecology* (eds W.R. Jordan, M.E. Gilpin and J.D. Aber), Cambridge University Press, Cambridge, pp. 289–299.

Allendorf, F.W. and Leary R.F. (1986) Heterozygosity and fitness in natural populations of animals, in *Conservation Biology: The Science of Scarcity and Diversity* (ed. M.E. Soulé), Sinauer, Sunderland, MA, pp. 57–76.

Allendorf, F. W. and Servheen, C. (1986) Genetics and the conservation of grizzly bears. *Trends Ecol. Evol.*, **1**, 88–89.

Andrews, S.M. and Cooke, J.A. (1984) Cadmium within a contaminated grassland ecosystem established on metalliferous mine waste, in *Metals in Animals* (ed. D. Osborn), Natural Environment Research Council/ Institute of Terrestrial Ecology, Abbots Ripton, Huntingdon, pp. 11–15.

Anselin, A., Meire, P.M. and Anselin, L. (1989) Multicriteria techniques in ecological evaluation: an example using the analytical hierachy process. *Biol. Conserv.*, **49**, 215–229.

Atlas, R.M. (1988) Biodegradation of hydrocarbons in the environment, in *Environmental Biotechnology* (ed. G.S. Omenn), Plenum Press, New York, pp. 211–222.

Atlas, R.M. and Sayler, G.S. (1988) Tracking microorganisms and genes in the environment, in *Environmental Biotechnology* (ed. G.S. Omenn), Plenum Press, New York, pp. 31–45.

Ausmus, B.S. (1984) An argument for ecosystem level monitoring. *Environ. Monitor. Assessm.*, **4**, 275–293.

Barrett, G. W. (1985) A problem-solving approach to reserve management. *BioScience*, **35**, 423–427.

Bayne, B.L. (1989) Measuring the biological effects of pollution: the mussel watch approach. *Water Sci. Tech.*, **21**, 1089–1100.

Bayne, B.L., Livingstone, D.R., Moore, M.N. and Widdows, J. (1976) A cytochemical and a biochemical index of stress in *Mytilus edulis* L. *Mar. Pollut. Bull.*, **7**, 221–224.

Bazzaz, F.A. (1987) Experimental studies on the evolution of niche in successional plant populations, in *Colonization, Succession and Stability* (eds A.J. Gray, M.J. Crawley and P.J. Edwards), Blackwell, Oxford, pp. 245–272.

Begon, M., Harper, J. L. and Townsend, C.R. (1990) *Ecology*, 2nd edn, Blackwell, Oxford.

Belovsky, G.E. (1987) Extinction models and mammalian persistence, in *Viable Populations for Conservation* (ed. M. E. Soulé), Cambridge University Press, Cambridge, pp. 35–57.

Ben-Eliahu, M.N., Safriel, U.N. and Ben-Tuvin, S. (1988) Environmental stability is low where polychaete diversity is high: quantifying tropical vs temperate within-habitat features. *Oikos*, **52**, 255–273.

Ben-Shlomo, R. and Nevo, E. (1988) Isozyme polymorphism as monitoring of marine environments: the interactive effects of cadmium and mercury pollution on the shrimp *Palaemon elegans*. *Mar. Pollut. Bull.*, **19**, 314–317.

Bender, E.A.. Case. T.J. and Gilpin, M.E. (1984) Perturbation experiments in community ecology: theory and practice. *Ecology*, **65**, 1–13.

Benndorf, J. and Henning, M. (1989) Daphnia and toxic blooms of *Microcytis aeruginosa* in Bantzen reservoir (GDR). *Int. Rev. ges. Hydrobiol.*, **74**, 233–248.

Berendse, F., Oomes, M.J.M., Altema, H.J. and Elberse, W.Th. (1991) Restoration of species-rich grassland communities, in *Terrestrial and Aquatic Ecosystems; Perturbation and Recovery* (ed. O. Ravera), Ellis Horwood, Chichester, pp. 456–463.

Berger, J. (1990) Persistence of different-sized populations: an empirical assessment of rapid extinctions in Bighorn sheep. *Conserv. Biol.*, **4**, 91–98.

Berthouex, P.M. and Rudd, D.F. (1977) *Strategy of Pollution Control*, John Wiley, New York.

Beukema, J.J. (1988) An evaluation of the ABC (abundance/biomass comparison) as applied to macrozoobenthic communities living on tidal flats in the Dutch Wadden Sea. *Mar. Biol.*, **99**, 425–433.

Blair, J.M. and Crossley, D.A. (1988) Litter decomposition, nitrogen dynamics and litter microarthropods in a southern Appalachian hardwood forest. *J. appl. Ecol.*, **25**, 683–698.

Blakey, N.C. and Maris, P.J. (1990) *Methane Recovery from Anaerobic Digestion of Landfill Leachates*. Contractor Report to the Water Research Centre, UK.

Boatman, N.D., Dover, J.W., Wilson, P.J., Thomas, M.B. and Cowgill, S.E. (1989) Modification of farming practice at field margins to encourage wildlife, in *Biological Habitat Reconstruction* (ed. G.P. Buckley), Belhaven, London, pp. 299–311.

Boerema, L.K. and Gulland, J.A. (1973) Stock assessment of the Peruvian Anchovy *(Engraulis ringens)* and management of the fishery. *J. Fish Res. Board Can.*, **30**, 2226–2235.

Borchardt, T., Burchert, S., Hablizel, H., Karbe, L. and Zeitner, R. (1988) Trace metal concentrations in mussels: comparisons between estuarine, coastal and offshore regions in the southeastern North Sea from 1983 to 1986. *Mar. Ecol. Prog. Ser.*, **42**, 17–31.

Borgegård, S.-O. and Rydin, H. (1989) Utilization of waste products and inorganic fertilizers in the restoration of iron-mine tailings. *J. appl. Ecol.*, **26**, 1083–1088.

Borgman, U., Millard, E.S. and Charlton, C.C. (1988) Dynamics of a stable, large volume laboratory ecosystem containing *Daphnia* and phytoplankton. *J. Plankton Res.*, **10**, 691–713.

Bormann, F.H. (1985) Air pollution and forests: an ecosystem perspective. *BioScience*, **35**, 434–441.

Bormann, F.H, and Likens, G.E. (1979) Catastrophic disturbance and the steady state in northern hardwood forests. *Amer. Scientist*, **67**, 660–669.

Bormann, F.H., Likens, G.E. and Melillo, J.M. (1977) Nitrogen budget for an aggrading northern hardwood forest ecosystem. *Science*, **196**, 981–983.

Bormann, F.H., Likens, G.E., Siccama, T.G., Pierce, R.S. and Eaton, J.S. (1974) The export of nutrients and recovery of stable conditions following deforestation at Hubbard Brook. *Ecol. Monogr.*, **44**, 255–277.

Bourgeron, P.S. (1988) Advantages and limitations of ecological classification for the protection of ecosystems. *Conserv. Biol.*, **2**, 218–220.

Boyden, C.R. (1974) Trace element content and body size in molluscs. *Nature*, **251**, 311–314.

Bradshaw, A.D. (1983) The reconstruction of ecosystems. *J appl. Ecol.*, **20**, 1–17.

Bradshaw, A.D. (1987a) Restoration: an acid test for ecology, in *Restoration Ecology* (eds W.R. Jordan, M.E. Gilpin and J.D. Aber), Cambridge University Press, Cambridge, pp. 24–29.

Bradshaw, A.D. (1987b) The reclamation of derelict land and the ecology of ecosystems, in *Restoration Ecology* (eds W.R. Jordan, M.E. Gilpin and J.D. Aber), Cambridge University Press, Cambridge, pp. 53–74.

Bradshaw, A.D. (1991) What the ecologist has to offer, in *Terrestrial and Aquatic Ecosystems; Perturbation and Recovery* (ed. O. Ravera), Ellis Horwood, Chichester, pp. 27–40.

Bradshaw, A.D. and Chadwick, M.J. (1980) *The Restoration of Land*, Blackwell, Oxford.

Bradshaw, A.D. and McNeilly, T. (1981) *Evolution and pollution*, Arnold, London.

Briand, F. and McCauley, E. (1978) Cybernetic mechanisms in lake plankton systems: how to control undesirable algae. *Nature*, **273**, 228–230.

Brisbin, I.L., Breshears, D.D., Brown, K.L. *et al.* (1989) Relationships between levels of radiocaesium in components of terrestrial and aquatic food webs of a contaminated streambed and floodplain community. *J. appl. Ecol.*, **26**, 173–182.

Brix, H. and Schierup, H.H. (1989) The use of aquatic macrophytes in water-pollution control. *Ambio*, **18**, 100–107.

Brookes, P.C., Newcombe, A.D. and Jenkinson, D.S. (1987) Adenylate energy charge measurements in soil. *Soil. Biol. Biochem.*, **19**, 211–217.

Brown, V.K. and Southwood, T.R.E. (1987) Secondary succession: pattern and strategies, in *Colonization, Succession and Stability* (eds A.J. Gray, M.J. Crawley and P.J. Edwards), Blackwell, Oxford, pp. 315–337.

Brown, V.M. and Dinsmore, J.J. (1988) Habitat islands and the equilibrium theory of island biogeography: testing some predictions. *Oecologia*, **75**, 426–429.

Bryan, G.W., Langston, W.J., Hummerstone, L.G. and Burt, G.R. (1985) *A Guide to the Assessment of Heavy Metal Contamination in Estuaries using Biological Indicators*. Occasional publications of the Marine Biological Association of the United Kingdom., No. 4.

Buckley, G.P. and Knight, D. G. (1989) The feasibility of woodland reconstruction, in *Biological Habitat Reconstruction* (ed. G.P. Buckley), Belhaven, London, pp. 171–188.

Burel, F. (1989) Landscape structure effects in Carabid beetle spatial patterns in western France. *Landscape Ecol.*, **2**, 215–226.

Cable, T.T., Brack, V. and Holmes, V.R. (1989) Simplified method for wetland habitat assessment. *Environ. Mgmt.*, **13**, 207–213.

Cairns, J. (1987) Disturbed ecosystems as opportunities for research in restoration ecology, in *Restoration Ecology* (eds W.R. Jordan, M.E. Gilpin and J.D. Aber), Cambridge University Press, Cambridge, pp. 301–320.

Callaghan, T.V., Abdelnow, H. and Lindley, D.K. (1988) The environmental crisis in the Sudan: the effect of water absorbing synthetic polymers on tree germination and early survival. *J. Arid Environ.*, **14**, 301–317.

Callaghan, T.V., Lindley, D.K., Ali, O.M., Abd El Nour, H. and Bacon, P.J. (1989) The effect of water-absorbing synthetic polymers on the stomatal conductance, growth and survival of transplanted *Eucalyptus microtheca* seedlings in the Sudan. *J. appl. Ecol.*, **26**, 663–672.

Calow, P. and Berry, R.J. (eds) (1989) *Evolution, ecology and environmental stress*, Academic Press, London.

Calow, P. and Sibley, R.M. (1990) A physiological basis of population processes: ecotoxicological implications. *Funct. Ecol.*, **4**, 283–288.

Campbell, R. (1990) Current status of biological control of soil-borne diseases. *Soil Use Mgmt*, **6**, 173–178.

Case, T.J. (1990) Invasion resistance arises in strongly interacting species-rich model competition communities. *Proc. Natl. Acad. Sci. USA*, **87**, 9610–9614.

Chaney, W.R., Kelly, J.M. and Strickland, P.C. (1978) Influence of cadmium and zinc on carbon dioxide evolution from litter and soil from a Black Oak forest. *J. Environ. Qual.*, **7**, 115–119.

Chang, A.C., Page, A.L., Warneke, J.E. and Grgurevic, E. (1984) Sequential extraction of soil heavy metals following a sludge application. *J. Environ. Qual.*, **13**, 33–38.

Chapman, P.J. (1988) Constructing microbial strains for degradation of halogenated aromatic hydrocarbons, in *Environmental Biotechnology* (ed. G.S. Omenn), Plenum Press, New York, pp. 81–95.

Chapman, S.B., Rose, R.J. and Basanta, M. (1989) Phosphorus adsorption by soils from heathland in southern England in relation to successional change. *J. appl. Ecol.*, **26**, 673–680.

Cherrett, J.M. (1989) Key concepts: the results of a survey of our members opinions, in *Ecological Concepts* (ed. J.M. Cherrett), Blackwell, Oxford, pp. 1–16.

Cherrett, J.M. (ed.) (1989) *Ecological Concepts*, Blackwell, Oxford.

Christiansen, F.B. and Feldman, M.W. (1986) *Population Genetics*, Blackwell, Oxford.

Clark, W.C. (1989) Managing planet earth. *Sci. Amer.*, **261**, 47–54.

Clements, W.H., Cherry, D.S. and Cairns, J. (1988) Structural alterations in aquatic insect communities exposed to copper in laboratory systems. *Environ. Toxicol. Chem.*, **7**, 735–747.

Colinvaux, P. (1986) *Ecology*, John Wiley, New York.

Connell, J.H. (1978) Diversity in tropical rain forests and coral reefs. *Science*, **199**, 1302–1310.

Connell, J.H. and Slatyer, R.D. (1977) Mechanisms of successions in natural communities and their role in community stability and organization. *Amer. Nat.*, **111**, 1119–1144.

Connell, J.H. and Sousa, W.P. (1983) On the evidence needed to judge ecological stability and persistence. *Amer. Nat.*, **121**, 789–824.

Connor, E.F. and McCoy, E.D. (1979) The statistics and biology of the species–area relationship. *Amer. Nat.*, **113**, 791–833.

Conquest, L.L. and Taub, F.B. (1989) Repeatability and reproducibility of the standardized aquatic microcosm: statistical properties, in *Aquatic toxicology and Hazard Assessment*, Vol. 12 (eds U.M. Cowgill and L.R. Williams), American Society for Testing and Materials, Philadelphia, pp. 159–177.

Conrad, S.E. and Radosevich, S.R. (1979) Ecological fitness of *Senecio vulgaris* and *Amaranthus retroflexus* biotypes susceptible or resistant to atrazine. *J. appl. Ecol.*, **16**, 171–177.

Cook, R.J. (1988) Management of the environment for the control of pathogens, in

Biological Control of Pests, Pathogens and Weeds (eds R.K.S. Wood and M.J. Way), Royal Society, London, pp. 171–182.

Crawley, M.J. (1987) What makes a community invasible? in *Colonization, Succession and Stability* (eds A.J. Gray, M.J. Crawley and P.J. Edwards), Blackwell, Oxford, pp. 429–453.

Cullen, J.M. and Hasan, S. (1988) Pathogens for the control of weeds, in *Biological Control of Pests, Pathogens and Weeds* (eds R.K.S. Wood and M.J. Way), Royal Society, London, pp. 213–224.

Cushing, D.H. (1981) *Fisheries Biology*, 2nd edn, University of Wisconsin Press, Madison, WI.

Cushing, D.H. (1987) Population biology and the management of fisheries. *Trends Ecol. Evol.*, **2**, 138–139.

Danell, K., Belin, P. and Wickerman, G. (1989) [137]Caesium in northern Swedish moose: the first year after the Chernobyl accident. *Ambio*, **18**, 108–111.

Danielson, R.M. (1985) Mycorrhizae and reclamation of stressed terrestrial environments, in *Soil Reclamation Processes* (eds R.L.Tate and D.A. Klein), Marcel Dekker, New York, pp. 173–201.

Dart, R.K. and Stretton, R.J. (1980) *Microbiological Aspects of Pollution Control*, 2nd edn, Elsevier, Amsterdam.

Davies, R.D., Carlton-Smith, C.H., Stark, J.H. and Campbell, J.A. (1988) Distribution of metals in grassland soils following surface application of sewage sludge. *Environ. Pollut.*, **49**, 99–115.

Day, C.A. and Lisanky, S.G. (1987) Agricultural alternatives, in *Environmental Biotechnology* (eds C.F. Forster and D.A.J. Wase), Ellis Horwood, Chichester, pp. 234–294.

Deghi, G.S., Ewel, K.C. and Mitsch, W.J. (1980) Effects of sewage effluent application on litter fall and litter decomposition in cypress swamps. *J. appl. Ecol.*, **17**, 397–408.

Department of the Environment (1990) *This Common Inheritance: Britain's Environmental Strategy*, HMSO, London.

Diamond, J.M. (1976) Island biogeography and conservation: strategy and limitations. *Science*, **193**, 1027–1028.

Diamond, J.M. (1986) The design of a nature reserve system for Indonesia, in *Conservation Biology: The Science of Scarcity and Diversity* (ed. M.E. Soulé), Sinauer, Sunderland, MA, pp. 485–503.

Diamond, J.M. and May, R.M. (1981) Island biogeography and the design of nature reserves, in *Theoretical Ecology* (ed. R.M. May), Blackwell, Oxford, pp. 228–252.

Diamond, S.A., Newman, M.C., Mulvey, M., Dixon, P.M. and Martinson, D. (1989) Allozyme genotype and time to death of mosquitofish, *Gambusia affinis* (Baird and Girard) during acute exposure to inorganic mercury. *Environ. Toxicol. Chem.*, **8**, 613–622.

Dickinson, N.M., Turner, A.P. and Lepp, N.W. (1991) How do trees and other long-lived plants survive in polluted environments? *Funct. Ecol.*, **5**, 5–11.

Dierberg, F.C. and Brezonik, P.L. (1983) Nitrogen and phosphorus mass balances in natural and sewage enriched cypress domes. *J. appl. Ecol.*, **20**, 323–327.

Domsch, K.H. (1984) Effects of pesticides and heavy metals on biological processes in soil. *Plant Soil*, **76**, 367–378.

Donkin, P. and Widdows, J. (1986) Scope for growth as a measure of environmental pollution and its interpretation using structure-activity relationships. *Chemy. Ind.*, **21**, 732–735.

Dony, J.G. and Denholm, I. (1985) Some quantitative methods of assessing the conservation value of ecologically similar sites. *J. appl. Ecol.*, **22**, 229–238.

Down, G.S. and Morton, A.J. (1989) A case study of whole woodland transplanting, in *Biological Habitat Reconstruction* (ed. G.P. Buckley), Belhaven, London, pp. 251–257.

Dunger, W. (1989) The return of soil fauna to coal-mined areas in the German Democratic Republic, in *Animals in Primary Succession* (ed. J.D. Majer), Cambridge University Press, Cambridge, pp. 301–337.

Duxbury, T. (1985) Ecological aspects of heavy metal responses in microorganisms. *Adv. Microb. Ecol.*, **8**, 185–235.

Ehler, L.E. (1991) Planned introductions in biological control, in *Assessing Ecological Risks of Biotechnology* (ed. L.R. Ginsburg), Butterworth–Heinemann, Boston, pp. 22–39.

Elkins, N.Z., Parker, C.W., Aldon, E. and Whitford, W.G. (1984) Responses of soil biota to organic amendments in stripmine spoils in northwestern New Mexico. *J. Environ. Qual.*, **13**, 215–219.

Ellis, G.M. and Fisher, A.C. (1987) Valuing the environment as input. *J. Environ. Mgmt.*, **25**, 149–156.

Elsom, D.M. (1987) *Atmospheric Pollution*, Blackwell, Oxford.

Ernst, W.H.O. (1986) Longterm pollution and stress. *Proc. International Conf. Heavy Metal Pollution*, Amsterdam, CEP, Edinburgh, pp. 10–15.

Everett, R.D. (1979) The monetary value of the recreational benefits of wildlife. *J. Environ. Mgmt.*, **8**, 203–213.

Ewens, W.J., Brockwell, P.J., Gani, J.M. and Resnick, S.I. (1987) Minimum viable population size in the presence of catastrophes, in *Viable Populations for Conservation* (ed. M. E. Soulé), Cambridge University Press, Cambridge, pp. 59–68.

Fagerstrom, T. (1977) Body-weight, metabolic rate and trace substance turnover in animals. *Oecologia*, **29**, 99–104.

Farris, J.L., Van Hassel, J.H., Belanger, S.E., Cherry, D.S. and Cairns, J. (1988) Application of cellulolytic activity of Asiatic Clams (*Corbicularia* sp.) to in-stream monitoring of power-plant effluents. *Environ. Toxicol. Chem.*, **7**, 707–713.

Felkner, I.C., Worthy, B., Christison, T. *et al.* (1989) Laser/microbe bioassay systems, in *Aquatic Toxicology and Hazard Assessment* (eds U.M. Cowgill and L.R. Williams), American Society for Testing and Materials, Philadelphia, pp. 95–103.

Fenner, M. (1987) Seed characteristics in relation to succession, in *Colonization, Succession and Stability* (eds A.J. Gray, M.J. Crawley and P.J. Edwards), Blackwell, Oxford, pp. 103–114.

Ferard, J.F., Jouany, J.M., Truhaut, R. and Vasseur, P. (1983) Accumulation of cadmium in a freshwater food chain experimental model. *Ecotoxicol. Environ. Safety*, **7**, 43–52.

Fischer, A.G. (1960) Latitudinal variations in organic diversity. *Evolution*, **14**, 64–81.

Focht, D.D. (1988) Performance of biodegradative microorganisms in soil: xenobiotic chemicals as unexploited metabolic niches, in *Environmental Biotechnology* (ed. G.S. Omenn), Plenum, New York, pp. 15–29.

Focht, D.D. and Brunner, W. (1985) Kinetics of biphenyl and polychlorinated biphenyl metabolism in soil. *Appl. Environ. Microbiol.*, **50**, 1058–1063.

Forcella, F. and Harvey, S.J. (1983) Relative abundance in an alien weed flora. *Oecologia*, **59**, 292–295.

Forman, R.T.T. and Godron, M. (1986) *Landscape Ecology*, John Wiley, New York.

Forster, C.F. and Wase, D.A.J. (eds) (1987) *Environmental Biotechnology*, Ellis Horwood, Chichester.

Francis, A.J. (1985) Low-level radioactive wastes in sub-surface soils, in *Soil*

Reclamation Processes (eds R.L. Tate and D.A. Klein), Marcel Dekker, New York, pp. 279–331.

Franklin, I.R. (1980) Evolutionary change in small populations, in *Conservation Biology: An Evolutionary–Ecological Perspective* (eds M.E. Soulé and B.A. Wilcox), Sinauer Associates, Sunderland, MA, pp. 135–149.

Friedland, A.J. and Johnson, A.H. (1985) Lead distribution and fluxes in a high elevation forest in northern Vermont. *J. Environ. Qual.*, **14**, 332–336.

Futuyama, D.J. and Moreno, G. (1988) The evolution of ecological specialization. *Ann. Rev. Ecol. Syst.*, **19**, 207–233.

Gadd, G.M. and Griffiths, A.J. (1978) Microorganisms and heavy metal toxicity. *Microb. Ecol.*, **4**, 303–317.

Gaertner, F. and Kim, L. (1988) Current applied recombinant DNA projects. *Trends Ecol. Evol.*, **3**, 54–57.

Gardenfors, U., Westermark. T., Emanuelsson, U., Mutuei, H. and Wladen, H. (1988) Use of land-snails as environmental archives. *Ambio*, **17**, 347–349.

Gemmell, R.P. (1977) *Colonization of Industrial Wasteland*, Arnold, London.

Genoni, P. and Pahl-Wostl, C. (1991) Scope for change in ascendency, a new concept in community ecotoxicology for environmental management, in *Terrestrial and Aquatic Ecosystems; Perturbation and Recovery* (ed. O. Ravera), Ellis Horwood, Chichester, pp. 69–75.

Getz, W.M. and Haight, R.G. (1989) *Population Harvesting: Demographic Models of Fish, Forest and Animal Resources*, Princeton University Press, Princeton, NJ.

Giller, P.S. (1984) *Community Structure and the Niche*, Chapman & Hall, London.

Gilpin, M.E. (1987a) Spatial structure and population vulnerability, in *Viable Populations for Conservation* (ed. M. E. Soulé), Cambridge University Press, Cambridge, pp. 125–139.

Gilpin, M.E. (1987b) Experimental community assembly: competition, community structure and the order of species introductions, in *Restoration Ecology* (eds W.R. Jordan, M.E. Gilpin and J.D. Aber), Cambridge University Press, Cambridge, pp. 151–161.

Gilpin, M.E. and Soulé, M.E. (1986) Minimum viable populations – processes of species extinction, in *Conservation Biology: The Science of Scarcity and Diversity* (ed. M.E. Soulé), Sinauer, Sunderland, MA, pp. 19–34.

Ginzburg, L.R. (ed.) (1991) *Assessing Ecological Risks of Biotechnology*, Butterworth–Heinemann, Boston.

Ginzburg, L.R., Ferson, S. and Akakaya, H.R. (1990) Reconstructibility of density dependence and the conservative assessment of extinction risks. *Conserv. Biol.*, **4**, 63–70.

Glasby, G.P. (1988) Entropy, pollution and environmental degradation. *Ambio*, **17**, 330–335.

Goeden, G.B. (1979) Biogeographic theory as a management tool. *Environ. Conserv.*, **6**, 27–32.

Goldsmith, F.B. (1983) Evaluating nature, in *Conservation in Perspective* (eds A. Warren and F.B. Goldsmith), John Wiley, Chichester, pp. 233–246.

Goldsmith, F.B. (ed.) (1991) *Monitoring for Conservation and Ecology*, Chapman & Hall, London.

Goldsmith, F.B. (1991) The selection of protected areas, in *The Scientific Management of Temperate Communities for Conservation* (eds I. F. Spellerberg, F.B. Goldsmith and M.G. Morris), Blackwell, Oxford, pp. 273–291.

Goodman, D. (1980) Demographic intervention for closely managed populations, in *Conservation Biology: An Evolutionary–Ecological Perspective* (eds M.E. Soulé and B.A. Wilcox), Sinauer Associates, Sunderland, MA, pp. 171–195.

Goodman, D. (1987) The demography of chance extinction, in *Viable Populations for Conservation* (ed. M. E. Soulé), Cambridge University Press, Cambridge, pp. 11–34.

Gordon, M.S., Bartholomew, G.A., Grinnell, A.D. *et al.* (1982) *Animal Physiology*, 4th edn, Macmillan, New York.

Gore, J.A. (1985) Mechanism of colonization and habitat enhancement for benthic macroinvertebrates in restored river channels, in *The Restoration of Rivers and Streams* (ed. J.A. Gore), Butterworths, Boston, pp. 81–101.

Gorman, M.L. (1979) *Island Ecology*, Chapman & Hall, London.

Grant, A., Hateley, J.G. and Jones, N.V. (1989) Mapping the ecological impact of heavy metals on the estuarine polychaete *Nereis diversicolor* using inherited metal tolerance. *Mar. Pollut. Bull.*, **20**, 139–142.

Gray, A.J. (1991) Management of coastal communities, in *The Scientific Management of Temperate Communities for Conservation* (eds I. F. Spellerberg, F.B. Goldsmith, M.G. Morris), Blackwell, Oxford, pp. 227–243.

Gray, J.S. (1985) Ecological theory and marine pollution monitoring. *Mar. Pollut. Bull.*, **16**, 224–227.

Gray, J.S. (1989) Effect of environmental stress on species-rich assemblages, in *Evolution, ecology and environmental stress* (eds P. Calow and R.J. Berry), Academic Press, London, pp. 19–32.

Greaves, M.P., Poole, N.J., Domsch, K.H., Jagnow, G. and Verstraete, N. (1980) Recommended tests for assessing the side-effects of pesticides on the soil microflora. *Weed Research Organisation, Technical Report No. 59.*

Green, B.H. (1989) Agricultural impacts on the rural environment. *J. appl. Ecol.*, **26**, 793–802.

Green, B.H. and Burnham, C.P. (1989) Environmental opportunities offered by surplus agricultural production, in *Biological Habitat Reconstruction* (ed. G.P. Buckley), Belhaven, London, pp. 92–101.

Green, R. (1979) *Sampling Design and Statistical Methods for Environmental Biologists*, John Wiley, New York.

Griesbach, S., Peters, R.H. and Youakim, S. (1982) An allometric model for pesticide bioaccumulation. *Can. J. Fish Aquat. Sci.*, **39**, 727–735.

Grime, J.P. (1974) Vegetation classification by reference to strategies. *Nature*, **250**, 26–31.

Grime, J.P. (1987) Dominant and subordinate components of plant communities: implications for succession stability and diversity, in *Colonization, Succession and Stability* (eds A.J. Gray, M.J. Crawley and P.J. Edwards), Blackwell, Oxford, pp. 413–428.

Grime, J.P. (1989) The stress debate: symptom of impending synthesis? in *Evolution, ecology and environmental stress* (eds P. Calow and R.J. Berry), Academic Press, London, pp. 3–17.

Gross, K.L. (1987) Mechanisms of colonization and species persistence in plant communities, in *Restoration Ecology* (eds W.R. Jordan, M.E. Gilpin and J.D. Aber), Cambridge University Press, Cambridge, pp. 173–188.

Grubb, P.J. (1977) The maintenance of species richness in plant communities: the importance of the regeneration niche. *Biol. Rev.*, **52**, 107–145.

Grubb, P.J. (1987) Some generalizing ideas about colonization, succession in green plants and fungi, in *Colonization, Succession and Stability* (eds A.J. Gray, M.J. Crawley and P.J. Edwards), Blackwell, Oxford, pp. 81–102.

Gulati, R.D., van Liere, L. and Siewertsen, K. (1991) The Loosdrecht lake system: man's role in its creation, perturbation and rehabilitation, in *Terrestrial and Aquatic Ecosystems: Perturbation and Recovery* (ed. O. Ravera), Ellis Horwood, Chichester, pp. 593–606.

Gulland, J.A. (1983) *Fish Stock Assessment*, John Wiley, Chichester.

Gutierrez, A.P., Wermelinger, B., Schulthess, F. *et al.* (1988a) Analysis of biological control of cassava pests in Africa. I. Simulation of carbon, nitrogen and water dynamics in cassava. *J. appl. Ecol.*, **25**, 901–920.

Gutierrez, A.P., Neuenschwander, P., Schulthess, F. *et al.* (1988b) Analysis of cassava pests in Africa. II. Cassava mealybug *Phenacoccus manihoti. J. appl. Ecol.*, **25**, 921–940.

Gutierrez, A.P., Yaninek, J.S., Wermelinger, B., Herren, H.R. and Ellis, C.K. (1988c) Analysis of cassava pests in Africa. III. Cassava green mite *Mononychellus tanajoa. J. appl. Ecol.*, **25**, 941–950.

Haanstra, L., Doelman, P. and Voshaar, J.H.O. (1985) The use of sigmoidal dose response curves in soil ecotoxicological research. *Plant Soil*, **84**, 293–297.

Hall, C.A.S. (1988) An assessment of the historically most influential theoretical models in ecology and of the data provided in their support. *Ecol. Modelling*, **43**, 5–31.

Hamer, G. (1990) Aerobic biotreatment: the performance limits of microbes and the potential for exploitation, in *Effluent Treatment and Waste Disposal*, Institute of Chemical Engineers, Symposium No. 116. Rugby, UK, pp. 57–68.

Hamilton, S.J., Mehrle, P.M. and Jones, J.R. (1987) Evaluation of metallothionein measurement as a biological indicator of stress from cadmium in Brook Trout. *Trans. Am. Fish Soc.*, **116**, 551–560.

Hansen, H.S. and Hove, K. (1991) Radiocesium bioavailability: transfer of Chernobyl and radiotracer radiocesium to goat milk. *Health Phys.*, **60**, 665–673.

Harborne, J.B. (1988) *Introduction Ecological Biochemistry*, 3rd edn, Academic Press, London.

Hardman, D.J. (1987) Microbial control of environmental pollution: the use of genetic techniques to engineer organisms with novel catalytic capabilities, in *Environmental Biotechnology* (eds C.F. Forster and D.A.J. Wase), Ellis Horwood, Chichester, pp. 295–317.

Hardy, A.C. (1959) *The Open Sea*, Collins, London.

Harper, J.L. (1981) The concept of population in modular organisms, in *Theoretical Ecology* (ed. R.M. May), Blackwell, Oxford, pp. 53–77.

Harper, J.L. (1987) The heuristic value of ecological restoration, in *Restoration Ecology* (eds W.R. Jordan, M.E. Gilpin and J.D. Aber), Cambridge University Press, Cambridge, pp. 35–45.

Harris, P. (1986) Biological control of weeds. *Forstch. Zool.*, **32**, 123–128.

Harris, R.P. (1987) *Satellite Remote Sensing*, Routledge Kegan Paul, London.

Hassell, M.P. (1981) Arthropod predator–prey systems, in *Theoretical Ecology* (ed. R.M. May), Blackwell, Oxford, pp. 105–131.

Hassell, M.P. and Anderson, R.M. (1989) Predator–prey and host–pathogen interactions, in *Ecological Concepts* (ed. J.M. Cherrett), Blackwell, Oxford, pp. 147–196.

Hawksworth, D.L. and Rose, F. (1976) *Lichens as Pollution Monitors*, Arnold, London.

Hellawell, J.M. (1991) Development of a rationale for monitoring, in *Monitoring for Conservation and Ecology* (ed. F.B. Goldsmith), Chapman & Hall, London, pp. 1–14.

Hendrix, P.F., Langner, C.L. and Odum, E.P. (1982) Cadmium in aquatic microcosms: implications for screening the ecological effects of toxic substances. *Environ. Mgmt.*, **6**, 543–553.

Hickey, D.A. and McNeilly, T. (1975) Competition between metal tolerant and normal plant populations: a field experiment on normal soil. *Evolution*, **29**, 458–464.

Higgs, A.J. and Usher, M.B. (1980) Should nature reserves be large or small? *Nature*, **285**, 568–569.

Hilder, V.A., Gatehouse, A.M.R., Sheerman, S.E., Barker, R.F. and Boulter, D. (1987) A novel mechanism of insect resistance engineering into tobacco. *Nature*, **330**, 160–163.

Hodgson, J.G. (1989) Selecting and managing plant materials used in habitat construction, in *Biological Habitat Reconstruction* (ed. G.P. Buckley), Belhaven, London, pp. 45–67.

Hoffman, A.A. and Parsons, P.A. (1989) An integrated approach to environmental stress tolerance and life history variation: dessication tolerance in *Drosophila*, in *Evolution, ecology and environmental stress* (eds P. Calow and R.J. Berry), Academic Press, London, pp. 117–136.

Holmes, J.R. (1981) *Refuse Recycling and Recovery*, John Wiley, Chichester.

Hopkin, S.P. (1989) *Ecophysiology of Metals in Terrestrial Invertebrates*, Elsevier Applied Science, London.

Horn, H.J. (1981) Succession, in *Theoretical Ecology* (ed. R.M. May), Blackwell, Oxford, pp. 253–271.

Horn, D.J. (1988) *Ecological Approach to Pest Management*, Guilford Press, New York.

Houghton, R.A. (1990) The future role of tropical forests in affecting the carbon dioxide concentration of the atmosphere. *Ambio*, **19**, 204–209.

Howard, B.J. (1987) ^{137}Cs uptake by sheep grazing tidally-inundated and inland pastures near the Sellafield reprocessing plant, in *Pollutant Transport and Fate in Ecosystems* (eds P.J.Coughtrey, M.H. Martin and M.H. Unsworth), Blackwells, Oxford, pp. 371–383.

Howard, B.J. and Beresford, N.A. (1989) Chernobyl radiocaesium in an upland sheep farm ecosystem. *Br. vet. J.*, **145**, 212–219.

Hubbell, S.P. and Foster, R.B. (1986) Commonness and rarity in a neotropical forest. Implications for tropical tree conservation, in *Conservation Biology: The Science of Scarcity and Diversity* (ed. M.E. Soulé), Sinauer, Sunderland, MA, pp. 205–231.

Hunter, B.A., Johnson, M.S. and Thompson, D.J. (1984) Food chain relationships of copper and cadmium in herbivorous and insectivorous small mammals, in *Metals in Animals* (ed. D. Osborn), Natural Environment Research Council/Institute of Terrestrial Ecology, Abbots Ripton, Huntingdon, pp. 5–10.

Ingham, E.R., Trofymow, J.A., Ames, R.N. *et al.* (1986a) Trophic interactions and nitrogen cycling in a semi-arid grassland soil. I. Seasonal dynamics of the natural populations, their interactions and effects on nitrogen cycling. *J. appl. Ecol.*, **23**, 597–614.

Ingham, E.R., Trofymow, J.A., Ames, R.N. *et al.* (1986b) Trophic interactions and nitrogen cycling in a semi-arid grassland soil. II. System responses to removal of different groups of soil microbes or fauna. *J. appl. Ecol.*, **23**, 615–630.

Istock, C.A. (1991) Genetic exchange and genetic stability in bacterial populations, in *Assessing Ecological Risks of Biotechnology* (ed. L.R. Ginzburg), Butterworth–Heinemann, Boston, pp. 123–149.

Jackson, D. and Smith, A.D. (1987) Generalized models for the transfer and distribution of stable elements and their radionuclides in agricultural systems, in *Pollutant Transport and Fate in Ecosystems* (eds P.J.Coughtrey, M.H. Martin and M.H. Unsworth), Blackwell, Oxford, pp. 385–402.

Jackson, D.R., Selvidge, W.J. and Ausmus, B.S. (1978) Behaviour of heavy metals in forest microcosms: II. Effects on nutrient cycling processes. *Water Air Soil Pollut.*, **10**, 13–18.

Janssen, M.P.M., Bruins, A., De Vries, T.H. and van Straalen, N.M. (1991)

Comparison of cadmium kinetics in four soil arthropod species. *Arch. Environ. Contam. Toxicol.*, **20**, 305–312.

Jefferies, R.A., Bradshaw, A.D. and Putwain, P.D. (1981) Growth, nitrogen accumulation and nitrogen transfer by legume species established on mine spoils. *J. appl. Ecol.*, **18**, 945–956.

Jeffers, J.N.R. (1978) *An Introduction to Systems Analysis with Ecological Applications*, Arnold, London.

Jenkinson, D.S. and Oades, J.M. (1979) A method for measuring adenosine triphosphate in soil. *Comp. Biol. Biochem.*, **11**, 193–199.

Johnson, M.S., McNeilly, T. and Putwain, P.D. (1977) Revegetation of metalliferous mine spoil contaminated by lead and zinc. *Environ. Pollut.*, **12A**, 261–277.

Jones, K.L. and Grainger, J.M. (1983) The application of enzyme activity measurements to a study of factors affecting protein, starch and cellulose fermentation in domestic refuse. *Eur. J. Appl. Biotechnol.*, **18**, 181–185.

Jordan, C.F. (1986) Local effects of tropical deforestation, in *Conservation Biology: The Science of Scarcity and Diversity* (ed. M.E. Soulé), Sinauer, Sunderland, MA, pp. 410–426.

Jordan, W.R., Gilpin, M.E. and Aber, J.D. (1987) Restoration ecology: ecological restoration as a technique for basic research, in *Restoration Ecology* (eds W.R. Jordan, M.E. Gilpin and J.D. Aber), Cambridge University Press, Cambridge, pp. 3–21.

Jørgensen, S.E. (1986) *Fundamentals of Ecological Modelling*, Elsevier, Amsterdam.

Jutsum, A.R. (1988) Commercial application of biological control: status and prospects, in *Biological Control of Pests, Pathogens and Weeds* (eds R.K.S. Wood and M.J. Way), Royal Society, London, pp. 357–373.

Kämäri, J., Forsius, M., Kortelainen, P., Mannio, J. and Verta, M. (1991) Finnish lake survey: present status of acidification. *Ambio*, **20**, 23–27.

Karr, J.R. (1987) Biological monitoring and environmental assessment: a conceptual framework. *Environ. Mgmt.*, **11**, 249–256.

Kaule, G. and Krebs, S. (1989) Creating new habitats in intensively used farmland, in *Biological Habitat Reconstruction* (ed. G.P. Buckley), Belhaven, London, pp. 161–170.

Keddy, P.A. (1991) Biological monitoring and ecological prediction: from nature reserve management to national state of the environment indicators, in *Monitoring for Conservation and Ecology* (ed. F.B. Goldsmith), Chapman & Hall, London, pp. 249–267.

Khan, D.H. and Frankland, B. (1984) Cellulolytic activity and root biomass production in some metal contaminated soils. *Environ. Pollut.*, **33A**, 63–74.

Kim, J., Ginzburg, L.R. and Dykhuizen, D.E. (1991) Quantifying the risks of invasion by genetically engineered organisms, in *Assessing Ecological Risks of Biotechnology* (ed. L.R. Ginsburg), Butterworth–Heinemann, Boston, pp. 193–214.

Kitching, R.L. (1983) *Systems Ecology*, University of Queensland Press, St Lucia.

Killham, K. (1987) Assessment of tress to microbial biomass, in *Chemical Analysis in Environmental Research* (ed. A.P. Rowland), Institute of Terrestrial Ecology/Natural Environment Research Council, Symposium No. 18, pp. 79–83.

Klein, D.A., Sorensen, D.L. and Redente, E.F. (1985) Soil enzymes: a predictor of reclamation potential and progress, in *Soil Reclamation Processes* (eds R.L. Tate and D.A. Klein), Marcel Dekker, New York, pp. 141–171.

Kline, V.M. and Howell, E.A. (1987) Prairies, in *Restoration Ecology* (eds W.R. Jordan, M.E. Gilpin and J.D. Aber), Cambridge University Press, Cambridge, pp. 75–83.

Koehn, R.K. and Bayne, B.L. (1989) Towards a physiological and genetical understanding of the genetics of the stress response, in *Evolution, Ecology and Environmental Stress* (eds P. Calow and R.J. Berry), Academic Press, London, pp. 151–171.

Lande, R. and Barrowclough, G.F. (1987) Effective population size, genetic variation and their use in population management, in *Viable Populations for Conservation* (ed. M. E. Soulé), Cambridge University Press, Cambridge, pp. 87–123.

Lavie, B. and Nevo, E. (1986) Genetic selection of homozygote allozyme genotypes in marine gastropods exposed to cadmium pollution. *Sci. Total Environ*, **57**, 91–98.

Law, R. (1991) Fishing in evolutionary waters. *New Scient.*, **129**, 35–37.

Law, R. and Watkinson, A.R. (1989) Competition, in *Ecological Concepts* (ed. J.M. Cherrett), Blackwell, Oxford, pp. 243–284.

Lawson, P.S. (1990) *Summary of UK Research on Landfill Microbiology: Mid 1970s to 1990*, Harwell Waste Management Symposium. Department of Energy, UK.

Lawton, J.H. (1987a) Fluctuations in a patchy world. *Nature*, **326**, 328–329.

Lawton, J.H. (1987b) Are there assembly rules for successional communities, in *Colonization, Succession and Stability* (eds A.J. Gray, M.J. Crawley and P.J. Edwards), Blackwell, Oxford, pp. 225–244.

Lawton, J.H. (1988) Biological control of bracken in Britain: contrasts and opportunities, in *Biological Control of Pests, Pathogens and Weeds* (eds R.K.S. Wood and M.J. Way), Royal Society, London, pp. 335–355.

Lawton, J.H. (1989) Food webs, in *Ecological Concepts* (ed. J.M. Cherrett), Blackwell, Oxford, pp. 43–78.

Le Houérou, H.N. and Gillet, H. (1986) Conservation versus desertization in African arid lands, in *Conservation Biology: The Science of Scarcity and Diversity* (ed. M.E. Soulé), Sinauer, Sunderland, MA, pp. 444–461.

Leahy, J.G. and Colwell, R.R. (1990) Microbial degradation of hydrocarbons in the environment. *Microb. Rev.*, **54**, 305–313.

Ledig, F.T. (1986) Heterozygosity, heterosis and fitness in outbreeding plants, in *Conservation Biology: The Science of Scarcity and Diversity* (ed. M.E. Soulé), Sinauer, Sunderland, MA, pp. 77–104.

Leigh, E.G. (1981) The average lifetime of a population in a varying environment. *J. theor. Biol.*, **90**, 213–239.

Lenksi, R.E. and Nguyen, T.T. (1988) Stability of recombinant DNA and its effect on fitness. *Trends Ecol. Evol.*, **3**, 518–522.

Lincoln, R.J., Boxshall, G.A. and Clark, P.F. (1982) *A Dictionary of Ecology, Evolution and Systematics*, Cambridge University Press, Cambridge.

Livens, F.R., Horrill, A.D. and Singleton, D.L. (1991) Distribution of radiocesium in the soil-plant systems of upland areas of europe. *Health Phys.*, **60**, 539–545.

Loch, J.P.G., Lagas, P. and Haring, B.J.A.M. (1981) Behaviour of heavy metals in soil beneath a landfill: results of model experiments. *Sci. Total Environ.*, **21**, 203–213.

Loucks, O.L. (1985) Looking for surprise in managing stressed ecosystems. *BioScience*, **35**, 428–432.

Luoma, S.N. (1977) Detection of trace contaminant effects in aquatic ecosystems. *J. Fish. Res. Board Can.*, **34**, 436–439.

McDowell, C., Sparks, R., Grindley, J. and Moll, E. (1989) Persuading the landowner to conserve natural ecosystems through effective communication. *J. Environ. Mgmt.*, **28**, 211–225.

McGaughey, W.H. (1985) Insect resistance to the biological insecticide *Bacillus thuringiensis. Science*, **229**, 193–195.

McGrath, S.P. (1987) Long term studies of metal transfers following application of sewage sludge, in *Pollutant Transport and Fate in Ecosystems* (eds P.J. Coughtrey, M.H. Martin and M.H. Unsworth), Blackwells, Oxford, pp. 301–317.

McIntosh, R.P. (1985) *The Background of Ecology*, Cambridge University Press, Cambridge.

McNaughton, S.J. (1978) Stability and diversity of ecological communties. *Nature*, **274**, 251–253.

McNeilly, T. (1987) Evolutionary lessons and establishment for degraded ecosystems, in *Restoration Ecology* (eds W.R. Jordan, M.E. Gilpin and J.D. Aber), Cambridge University Press, Cambridge, pp. 271–286.

MacArthur, R.H. and Wilson, E.O. (1967) *The Theory of Island Biogeography*, Princeton University Press, Princeton, New Jersey.

Macnair, M.R. (1987) Heavy metal tolerance in plants: a model evolutionary system. *Trends Ecol. Evol.*, **2**, 354–359.

MacPhee, D.G. (1985) Indications that mutagenesis in *Salmonella* may be subject to catabolic repression. *Mutat. Res.*, **151**, 35–41.

Ma, W.-C. and Eijsackers, H. (1989) The influence of substrate toxicity on soil macrofauna return in reclaimed land, in *Animals in Primary Succession* (ed. J. D. Majer), Cambridge University Press, Cambridge, pp. 223–244.

Maciorowski, A.F. (1988) Populations and communities: linking toxicology and ecology in a new synthesis. *Environ. Toxicol. Chem.*, **7**, 677–678.

Magurran, A.E. (1988) *Ecological Diversity and its Measurement*, Croom Helm, London.

Majer, J.D. (ed.) (1989a) *Animals in Primary Succession: the Role of Fauna in Reclaimed Land*, Cambridge University Press, Cambridge.

Majer, J.D. (1989b) Fauna studies and land reclamation technology – a review of the history and need for such studies, in *Animals in Primary Succession* (ed. J. D. Majer), Cambridge University Press, Cambridge, pp. 5–33.

Manasse, R. and Kareiva, P. (1991) Quantifying the spread of recombinant genes and organisms, in *Assessing Ecological Risks of Biotechnology* (ed. L.R. Ginsburg), Butterworth–Heinemann, Boston, pp. 215–231.

Marrs, R.H. and Bradshaw, A.D. (1982) Nitrogen accumulation, cycling and the reclamation of china clay waste. *J. Environ. Mgmt.*, **15**, 139–157.

Marrs, R.H. and Gough, M.W. (1989) Soil fertility – a potential problem for habitat restoration, in *Biological Habitat Reconstruction* (ed. G.P. Buckley), Belhaven, London, pp. 29–44.

Martin, M.H. and Coughtrey, P.J. (1982) *Biological Monitoring of Heavy Metal Pollution: Land and Air*, Applied Science, London.

Martin, M.H. and Coughtrey, P.J. (1987) Cycling and fate of heavy metals in a contaminated woodland, in *Pollutant Transport and Fate in Ecosystems* (eds P.J. Coughtrey, M.H. Martin and M.H. Unsworth), Blackwells, Oxford, pp. 319–336.

Matis, J.H., Miller, T.H. and Allen, D.M. (1991) Stochastic models of bioaccumulation, in *Metal Ecotoxicology* (eds M.C. Newman and A.W. McIntosh), Lewis, Michigan, pp. 171–206.

May, R.M. (1974) *Stability and Complexity in Model Ecosystems*, 2nd edn, Princeton University Press, Princeton, NJ.

May, R.M. (1981a) Patterns in multi-species communities, in *Theoretical Ecology* (ed. R.M. May), Blackwell, Oxford, pp. 197–227.

May, R.M. (1981b) Models for single populations, in *Theoretical Ecology* (ed. R.M.May), Blackwell, Oxford, pp. 5–29.

May, R.M. (1981c) Models for two interacting populations, in *Theoretical Ecology* (ed. R.M. May), Blackwell, Oxford, pp. 78–104.

May, R.M. (1986) The search for patterns in the balance of nature: advances and retreats. *Ecology*, **67**, 1115–1126.

May, R.M. and Hassell, M.P. (1988) Population dynamics and biological control, in *Biological Control of Pests, Pathogens and Weeds* (eds R.K.S. Wood and M.J. Way), Royal Society, London, pp. 129–169.

Meeusen, R.L. and Warren, G. (1989) Insect control with genetically engineered crops. *Ann. Rev. Entomol.*, **34**, 373–381.

Meire, P.M. and Dereu, J. (1990) Use of the abundance/biomass comparison method for detecting environmental stress: some considerations based on intertidal macrozoobenthos and bird communities. *J. appl. Ecol.*, **27**, 210–233.

Meredith, G.N. and Campbell, R.R. (1988) Status of the Fin Whale *Balaenoptera physalus* in Canada. *Can. Field Naturalist*, **102**, 351–368.

Messenger, P.S. (1975) Parasites, predators and population dynamics, in *Insects, Science and Society* (ed. D. Pimentel), Academic Press, New York, pp. 201–223.

Metcalfe, J.L. (1989) Biological water quality assessment of running waters based on macroinvertebrate communities: history and present status in Europe. *Environ. Pollut.*, **60**, 101–139.

Miles, J. (1987) Vegetation succession: past and present perceptions, in *Colonization, Succession and Stability* (eds A.J. Gray, M.J. Crawley and P.J. Edwards), Blackwell, Oxford, pp. 1–29.

Miller, R.M. (1987) Mycorrhizae and succession, in *Restoration Ecology* (eds W.R. Jordan, M.E. Gilpin and J.D. Aber), Cambridge University Press, Cambridge, pp. 205–219.

Miller, R.M. (1988) Potential for transfer and establishment of engineered and genetic sequences, *Trends Ecol. Evol.*, **3**, S23–S27.

Mitchley, J. (1988) Restoration of species-rich calcicolous grassland on ex-arable land in Britain. *Trends Ecol. Evol.*, **3**, 125–127.

Moore, D.R.J., Keddy, P.A., Gaudet, C.L. and Wisheu, I.C. (1989) Conservation of wetlands: do infertile wetlands deserve a higher priority? *Biol. Conserv.*, **47**, 203–217.

Moore, M.N., Livingstone, D.R., Widdows, J., Lowe, D.M. and Pipe, R.K. (1987) Molecular, cellular and physiological effects of oil derived hydrocarbons on molluscs and their use in impact assessment. *Phil. Trans. R. Soc. Lond.*, B. **316**, 603–623.

Moore, P.D. (1983) Ecological diversity and stress. *Nature*, **306**, 17.

Moriarty, F. (1988) *Ecotoxicology*, 2nd edn, Academic Press, London.

Moriarty, F. and Walker, C.H. (1987) Bioaccumulation in food chains – a rational approach. *Ecotoxicol. Environ. Safety*, **13**, 208–215.

Morris, M.G. (1991) The management of reserves and protected areas, in *The Scientific Management of Temperate Communities for Conservation* (eds I. F. Spellerberg, F.B. Goldsmith and M.G. Morris), Blackwell, Oxford, pp. 323–347.

Moss, B. (1988) *Ecology of Fresh Waters*, 2nd edn, Blackwells, Oxford.

Moss, B. (1989) Water pollution and the management of ecosystems: a case study of science and scientist, in *Toward a More Exact Ecology* (eds P.J. Grubb and J.B. Whittaker, J.B.), Blackwell, Oxford, 401–422.

Moss, B., Balls, H., Irvine, K. and Stansfield, J. (1986) Restoration of two lowland lakes by isolation from nutrient-rich water sources with and without removal of sediment. *J. appl. Ecol.*, **23**, 391–414.

Moss, B., Stansfield, J. and Irvine, K. (1991) Development of Daphnid communities in diatom- and cyanophyte-dominated lakes and their relevance to lake restoration by biomanipulation. *J. appl. Ecol.*, **28**, 586–602.

Mudrack, K. and Kunst, S. (1986) *Biology of Sewage Treatment and Water Pollution Control*, Ellis Horwood, Chichester.

Murdoch, W.W., Chesson, J. and Chesson, P.L. (1985) Biological control in theory and practice. *Am. Nat.*, **125**, 344–366.

Neuenschwander, P. and Herren, H.R. (1988) Biological control of the cassava mealybug *Phenacoccus manihoti* by the exotic parasitoid *Epidinocarsis lopezi* in Africa, in *Biological Control of Pests, Pathogens and Weeds* (eds R.K.S. Wood and M.J. Way), Royal Society, London, pp. 319–333.

Nevo, E., Noy, R., Lavie, B., Beiles, A. and Muchtar, S. (1986) Genetic diversity and resistance to marine pollution. *Biol. J. Linn. Soc.*, **29**, 139–144.

Newbold, C. (1989) Semi-natural habitats or habitat re-creation: conflict or partnership, in *Biological Habitat Reconstruction* (ed. G.P. Buckley), Belhaven, London, pp. 9–17.

Newman, M.C. and Heagler, M.G. (1991) Allometry of metal bioaccumulation and toxicity, in *Metal Ecotoxicology* (eds M.C. Newman and A.W. McIntosh), Lewis, MI, pp. 91–130.

Newman, M.C., Diamond, S.A., Mulvey, M. and Dixon, P. (1989) Allozyme genotype and time to death of mosquitofish *Gambusia affinis* (Baird and Girard) during acute toxicant exposure: a comparison of arsenate and inorganic mercury. *Aquat. Toxicol.*, **15**, 141–156.

Newsom, L.D. (1975) Pest management: concept to practice, in *Insects, Science and Society* (ed. D. Pimentel), Academic Press, New York, pp. 257–277.

Nicholls, M.K. and McNeilly, T. (1985) The performance of *Agrostis capillaris* L. genotypes, differing in copper tolerance, in ryegrass swards on normal soil. *New Phytol.*, **101**, 207–217.

Niering, W.A. (1987) Vegetation dynamics (succession and climax) in relation to plant community management. *Conserv. Biol.*, **1**, 287–295.

Nordgren, A., Kauri, T., Bååarth, E. and Söderstrom, B. (1986) Soil microbial acitivity, mycelial lengths and physiological groups of bacteria in a heavy metal polluted area. *Environ. Pollut.*, **41A**, 89–100.

O'Brien, S.J., Roelke, M.E., Marker, L. *et al.* (1985) Genetic basis for species vulnerability in the Cheetah. *Science*, **227**, 1428–1434.

O'Neill, E.G., O'Neill, R.V. and Norby, R.J. (1991) Hierarchy theory as a guide to mycorrhizal research on large-scale problems. *Environ. Pollut.*, **73**, 271–284.

O'Neill, R.V., DeAngelis, D.L., Waide, J.B. and Allen, T.F.H. (1986) *A Hierarchical Concept of Ecosystems*, Princeton University Press, Princeton, NJ.

O'Neill, R.V., DeAngelis, D.L., Milne, B.T. *et al.* (1988) Indices of landscape pattern. *Landscape Ecol.*, **1**, 153–162.

Odum, E.P. (1985) Trends expected in stressed ecosystems. *BioScience*, **35**, 419–422.

Omenn, G.S. (ed.)(1988) *Environmental Biotechnology*, Plenum Press, New York.

Openshaw, S. (1991) Developing appropriate spatial analysis methods for GIS, in *Geographical Information Systems* (eds D.J. Macquire, M.F. Goodchild and D.F. Rhind), Longman, Harlow, pp. 389–402.

Packham, J.R. and Harding, D.J.L. (1982) *Ecology of Woodland Processes*, Arnold, London.

Palmer, J.P. and Iverson, L.R. (1983) Factors affecting nitrogen fixation by white clover (*Trifolium repens*) on colliery spoils. *J. appl. Ecol.*, **20**, 287–301.

Parker, C.A. (1989) Soil biota and plants in the rehabilitation of degraded agricultural soils, in *Animals in Primary Succession* (ed. J. D. Majer), Cambridge University Press, Cambridge, pp. 423–438.

Parry, G.D. (1981) The meaning of r- and K-selection. *Oecologia*, **48**, 260–264.

Parsons, P.A. (1987a) Features of colonizing animals: phenotypes and genotypes, in *Colonization, Succession and Stability* (eds A.J. Gray, M.J. Crawley and P.J. Edwards), Blackwell, Oxford, pp. 133–154.

Parsons, P.A. (1987b) Evolutionary rates under environmental stress. *Evol. Biol.*, **21**, 311–347.

Pascoe, D. (1983) *Toxicology*, Arnold, London.

Paul, E.A. (1989) Soils as components and controllers of ecosystem processes, in *Toward a more exact ecology* (eds P.J. Grubb and J.B. Whittaker), Blackwell, Oxford, 353–374.

Paul, E.A. and Ladd, J. N. (eds) (1981) *Soil Biochemistry*, Vol. 5, Marcel Dekker, New York.

Payne, C.C. (1988) Pathogens for the control of insects: where next?, in *Biological Control of Pests, Pathogens and Weeds* (eds R.K.S. Wood and M.J. Way), Royal Society, London, pp. 225–248

Payne, J.F., Fancy, L.L., Rahimtula, A.D. and Porter, E.L. (1987) Review and perspective on the use of mixed-function oxygenase enzymes in biological monitoring. *Comp. Biochem. Physiol.*, **86C**, 233–245.

Pearsall, S.H., Durham, D. and Eager, D.E. (1986) Evaluation methods in the United States, in *Wildlife Conservation Evaluation* (ed. M.B. Usher), Chapman & Hall, London, pp. 111–133.

Peterman, R.M. and Bradford, M.J. (1987) Windspeed and mortality rate of a marine fish, the Northern Anchovy (*Engraulis mordax*). *Science*, **235**, 354–356.

Phillips, D.J.H. (1980) *Quantitative Aquatic Biological Indicators*, Applied Science, London.

Phillips, D.J.H. and Rainbow, P.S. (1988) Barnacles and mussels as biomonitors of trace elements: a comparative study. *Mar. Ecol. Prog. Ser.*, **49**, 83–93.

Pianka, E.R. (1983) *Evolutionary Ecology*, 3rd edn, Harper Row, New York.

Pickett, J.A. (1988) Integrating the use of beneficial organisms with chemical crop protection, in *Biological Control of Pests, Pathogens and Weeds* (eds R.K.S. Wood and M.J. Way), Royal Society, London, pp. 203–211.

Pimentel, D., Glenister, C., Fast, S. and Gallahan, D. (1984) Environmental risks of biological pest control. *Oikos*, **42**, 283–290.

Pimm, S.L. (1984) The complexity and stability of ecosystems. *Nature*, **307**, 321–326.

Pimm, S.L. (1986) Community stability and structure, in *Conservation Biology: The Science of Scarcity and Diversity* (ed. M.E. Soulé), Sinauer, Sunderland, MA, pp. 309–329.

Pitcher, A. and Hart, P.J.B. (1982) *Fisheries Ecology*, Croom Helm, London.

Platt, H.M., Shaw, K.M. and Lambshead, P.J.D. (1984) Nematode species abundance patterns and their use in the detection of environmental perturbations. *Hydrobiol.*, **118**, 59–66.

Pooley, F.D. (1987) Mineral leaching with bacteria, in *Environmental Biotechnology* (eds C.F. Forster and D.A.J. Wase), Ellis Horwood, Chichester, pp. 114–134.

Prance, G.T. (1991) Rates of loss of biological diversity: a global view, in *The Scientific Management of Temperate Communities for Conservation* (eds I. F. Spellerberg, F.B. Goldsmith and M.G. Morris), Blackwell, Oxford, pp. 27–44.

Pratt, J.R., Bowers, N.J., Niederlehner, B.R. and Cairns, J. (1988) Effects of chlorine on microbial communities in naturally derived microcosms. *Environ. Toxicol. Chem.*, **7**, 679–687.

Puschnig, M., Schettler-Wiegel, J. and Schulz-Berendt, V. (1991) Can soil animals be used to improve decontamination of oil residues in polluted soils?, in *Terrestrial and Aquatic Ecosystems: Perturbation and Recovery* (ed. O. Ravera), Ellis Horwood, Chichester, pp. 485–492.

Pyle, R.M. (1980) Management of nature reserves, in *Conservation Biology: An Evolutionary–Ecological Perspective* (eds M.E. Soulé and B.A. Wilcox), Sinauer Associates, Sunderland, MA, pp. 318–327.

Quinn, J.F. and Harrison, S.P. (1988) Effects of habitat fragmentation and isolation on species richness: evidence from biogeographic patterns. *Oecologia*, **75**, 132–140.

Rabinowitz, D., Cairns, S. and Dillon, T. (1986) Seven forms of rarity and their frequency in the flora of the British Isles, in *Conservation Biology: The Science of Scarcity and Diversity* (ed. M.E. Soulé), Sinauer, Sunderland, MA. pp. 182–204.

Ramade, F. (1984) *Ecology of Natural Resources*, John Wiley, Chichester.

Rapport, D.J. (1989) Symptoms of pathology in the Gulf of Bothnia (Baltic Sea): ecosystem response to stress from human activity, in *Evolution, Ecology and Environmental Stress* (eds P. Calow and R.J. Berry). Academic Press, London, pp. 33–49.

Rapport, D.J., Regier, H.A. and Hutchinson, T.C. (1985) Ecosystem behaviour under stress. *Am. Nat.*, **125**, 617–640.

Ratcliffe, D.A. (1986) Selection of important areas for wildlife conservation in Great Britain: The Nature Conservancy Council's approach, in *Wildlife Conservation Evaluation* (ed. M.B. Usher), Chapman & Hall, London, pp. 135–159.

Rauser, W.E. and Curvetto, N.R. (1980) Metallothionein occurs in roots of *Agrostis* tolerant to excess copper. *Nature*, **287**, 563–564.

Reddy, K.R. and DeBusk, W.F. (1985) Nutrient removal potential of selected aquatic macrophytes. *J. Environ. Qual.*, **14**, 459–462.

Reed, J.M., Doerr, P.D. and Walters, J.R. (1988) Minimum viable population size of the Red-Cockaded Woodpecker. *J. Wildl. Mgmt.*, **52**, 385–391.

Reiners, W.A. (1981) Nitrogen cycling in relation to ecosystem succession. *Ecol. Bull.*, **33**, 507–528.

Richards, B.N. (1987) *The Microbiology of Terrestrial Ecosystems*, Longmans, Harlow, Essex.

Richards, K.M. (1989) Landfill gas: working with Gaia. *Biodeterioration Abstr.*, **3**, 317–331.

Ricklefs, R.E. (1987) Community diversity: relative roles of local and regional processes. *Science*, **235**, 167–171.

Ricklefs, R.E. (1990) *Ecology*, 3rd edn, W.H. Freeman, New York.

Ritz, D.A., Lewis, M.E. and Shen, M. (1989) Response to organic enrichment of infaunal macrobenthic communities under salmonid sea cages. *Mar. Biol.*, **103**, 211–214.

Rosenzweig, M.L. (1987) Restoration ecology: a tool to study population interactions?, in *Restoration Ecology* (eds W.R. Jordan, M.E. Gilpin and J.D. Aber), Cambridge University Press, Cambridge, pp. 189–203.

Rupela, O.P. and Tauro, P. (1973) Utilization of *Thiobaccillus* to reclaim alkali soil. *Soil Biol. Biochem.*, **5**, 899–901.

Salt, G.W. (1979) A comment on the use of the term emergent properties. *Am. Nat.*, **113**, 145–161.

Samways, M.J. (1981) *Biological Control of Pest and Weeds*, Arnold, London.

Sayler, G.S., Shields, M.S., Tedford, E.T. *et al.* (1985) Application of DNA–DNA colony hybridization to the detection of catabolic genotypes in environmental samples. *Appl. Environ. Microbiol.*, **49**, 1295–1303.

Schafer, W.M., Nielsen, G.A. and Nettleton, W.D. (1980) Minesoil genesis and morphology in a spoil chronosequence in Montana. *Am. J. Soil Sci.*, **44**, 802–807.

Scheffer, M. (1989) Alternative stable states in eutrophic, shallow freshwater systems: a minimal model. *Hydrobiol. Bull.*, **23**, 73–83.

Schmidt-Nielsen, K. (1979) *Animal Physiology: Adaptation and Environment*, Cambridge University Press, Cambridge.

Schoener, T.W. (1989) The ecological niche, in *Ecological Concepts* (ed. J.M. Cherrett), Blackwell, Oxford, pp. 79–113.

Shaffer, M. (1987) Minimum viable poplations: coping with uncertainty, in *Viable Populations for Conservation* (ed. M. E. Soulé), Cambridge University Press, Cambridge, pp. 69–86.

Sheehan, P.J. (1984) Effects on individuals and populations, in *Effects of Pollutants at the Ecosystem Level* (eds P.J.Sheehan, D.R. Miller, G.C. Butler and Ph. Bourdeau), SCOPE/John Wiley, New York, pp. 23–50.

Sheehan, P.J. (1989) Statistical and non-statistical considerations in quantifying pollutant-induced changes in microcosms, in *Aquatic toxicology and Hazard Assessment*. Vol. 12, (eds U.M. Cowgill and L.R. Williams), American Society for Testing and Materials, Philadelphia, pp. 178–188.

Simberloff, D. (1986) Design of nature reserves, in *Wildlife Conservation Evaluation* (ed. M.B. Usher), Chapman & Hall, London, pp. 315–337.

Simberloff, D. (1991) Keystone species and community effects of biological introductions, in *Assessing Ecological Risks of Biotechnology* (ed. L.R. Ginzburg), Butterworth–Heinemann, Boston, pp. 1–19.

Simkiss, K. and Taylor, M. (1981) Cellular mechanisms of metal ion detoxification and some new indices of pollution. *Aquatic Toxicol.*, 1, 279–290.

Simonsen, L. and Levin, B.R. (1988) Evaluating the risk of releasing genetically engineered organisms. *Trends Ecol. Evol.*, 3, S27–S30.

Sinclair, A.R.E. (1989) Population regulation in animals, in *Ecological Concepts* (ed. J.M. Cherrett), Blackwell, Oxford, pp. 197–241.

Slobodkin, L.B. (1988) Intellectual problems of applied ecology. *BioScience*, 38, 337–342.

Södergren, A. (1973) Transport, distribution and degradation of chlorinated hydrocarbon residues in aquatic model ecosystems. *Oikos*, 24, 30–41.

Soulé, M.E. and Simberloff, D. (1986) What do genetics and ecology tell us about the design of nature reserves? *Biol. Conserv.*, 35, 19–40.

Soulé, M.E., Wilcox, B.A. and Holtby, C. (1979) Benign neglect: a model for faunal collapse in the game reserves of East Africa. *Biol. Conserv.*, 15, 259–272.

Southwood, T.R.E. (1981) Bionomic strategies and population parameters, in *Theoretical Ecology* (ed. R.M.May), Blackwell, Oxford, pp. 30–52.

Southwood, T.R.E. (1988) Tactics, strategies and templets. *Oikos*, 52, 3–18.

Southwood, T.R.E. and Comins, H.N. (1976) A synoptic population model. *J. Anim. Ecol.*, 45, 949–965.

Spellerberg, I.F. (1991) Biogeographical basis of conservation, in *The Scientific Management of Temperate Communities for Conservation* (eds I. F. Spellerberg, F.B. Goldsmith and M.G. Morris), Blackwell, Oxford, pp. 293–322.

Spellerberg, I.F., Goldsmith, F.B. and Morris, M.G. (1991) *The Scientific Management of Temperate Communities for Conservation*, Blackwell, Oxford.

St John, W.D. and Sikes, D.J. (1988) Complex industrial waste sites, in *Environmental Biotechnology* (ed. G.S. Omenn), Plenum, New York, pp. 237–252.

Stay, F.S., Flum, T.E., Shannon, L.J, and Yount, J.D. (1989) An assessment of the precision and accuracy of SAM and MFC microcosms exposed to toxicants, in *Aquatic toxicology and Hazard Assessment*. Vol. 12. (eds U.M. Cowgill and L.R. Williams), American Society for Testing and Materials, Philadelphia, pp. 189–203.

Steele, J.H. (1985) A comparison of terrestrial and marine ecological systems. *Nature*, 313, 355–358.

Stotzky, G., Zeph, L.R. and Devanas, M.A. (1991) Factors affecting the transfer of genetic information among microorganisms in soil, in *Assessing Ecological Risks of Biotechnology* (ed. L.R. Ginzburg), Butterworth–Heinemann, Boston, pp. 95–122.

Strain, B.R. (1987) Direct effects of increasing atmospheric CO_2 on plants and ecosystems. *Trends Ecol. Evol.*, 2, 18–21.

Strong, D.R. (1986) Density-vague population changes. *Trends Ecol. Evol.*, 1, 39–42.

Suteau, P., Daubeze, M., Migaud, M.L. and Narbonne, J.F. (1988) PAH-metabolizing enzymes in whole mussels as biochemical tests for chemical pollution monitoring. *Mar. Ecol. Prog. Ser.*, **46**, 45–49.

Swanson, F.J., Kratz, T.K., Caine, N. and Woodmansee, R.G. (1988) Landform effects on ecosystem patterns and processes. *BioScience*, **38**, 92–98.

Tate, R.L. and Klein, D.A. (1985) *Soil Reclamation Processes*, Marcel Dekker, New York.

Taylor, E.M. and Schuman, G.E. (1988) Fly-ash and lime amendment of acidic coal spoil to aid revegetation. *J. Environ. Qual.*, **17**, 120–124.

Terborgh, J. (1974) Preservation of natural diversity: the problem of extinction prone species. *BioScience*, **24**, 715–722.

Terborgh, J. (1976) Island biogeography and conservation: strategy and limitations. *Science*, **193**, 1029–1030.

Terborgh, J. (1986) Keystone plant resources in the tropical forest, in *Conservation Biology: The Science of Scarcity and Diversity* (ed. M.E. Soulé), Sinauer, Sunderland, MA, pp. 330–344.

Terborgh, J. (1988) The big things that rule the world – a sequel to E.O. Wilson. *Conserv. Biol.*, **2**, 402–403.

Terborgh, J. and Winter, B. (1980) Some causes of extinction, in *Conservation Biology: An Evolutionary–Ecological Perspective* (eds M.E. Soulé and B.A. Wilcox), Sinauer Associates, Sunderland, MA, pp. 119–133.

Thomas, J.A. (1991) Rare species conservation: case studies of European butterflies, in *The Scientific Management of Temperate Communities for Conservation* (eds I. F. Spellerberg, F.B. Goldsmith and M.G. Morris), Blackwell, Oxford, pp. 149–197.

Thomas, M.B., Wratten, S.D. and Sotherton, N.W. (1991) Creation of 'island' habitats to manipulate populations of beneficial arthropods: predator densities and emigration. *J. appl. Ecol.*, **28**, 906–917.

Thompson, N.B. (1988) The status of Loggerhead *Caretta caretta*, Kemp's Ridley, *Lepidochelys kempi*, and Green, *Chelonia mydas*, sea turtles in US waters. *Mar. Fish Rev.*, **50**, 16–23.

Thompson, S. (1988) Range expansion by alien weeds in the coastal farmlands of Guyana. *J. Biogeogr.*, **15**, 109–118.

Thorpe, G.J. and Costlow, J.D. (1989) The relation of the acute (96h) uptake and subcellular distribution of cadmium and zinc to cadmium toxicity in larvae of *Rhithropanopeus harrisii* and *Palaemonetes pugio*, in *Aquatic toxicology and Hazard Assessment*, Vol 12, (eds U.M. Cowgill and L.R. Williams), American Society for Testing and Materials, Philadelphia, pp. 82–94.

Timmis, K.N., Rojo, F. and Ramos, J.L. (1988) Prospects for laboratory engineering of bacteria to degrade pollutants, in *Environmental Biotechnology* (ed. G.S. Omenn), Plenum, New York, pp. 61–79.

Turner, J.R.G., Lennon, J.J. and Lawrenson, J.A. (1988) British bird species distributions and the energy theory. *Nature*, **335**, 539–541.

Tyagi, R.D., Couillard, D. and Tran, F. (1988) Heavy metals removal from anaerobically digested sludge by chemical and microbiological methods. *Environ. Pollut.*, **50**, 295–316.

Tyagi, R.D., Couillard, D. and Grenier, Y. (1991) Effects of medium composition on the bacterial leaching of metals from digested sludge. *Environ. Pollut.*, **71**, 57–67.

Tyler, G. (1981) Heavy metals in soil biology and biochemistry, in *Soil Biochemistry*, Vol. 5 (eds E.A. Paul and J.N. Ladd), Marcel Dekker, New York, pp. 371–414.

Tyler, G. (1984) The impact of heavy metal pollution on forests: a case study of Gusum, Sweden. *Ambio*, **13**, 18–24.

Uhlmann, D. (1991) Anthropogenic perturbation of ecological systems: a need for transfer from principles to applications, in *Terrestrial and Aquatic Ecosystems: Perturbation and Recovery* (ed. O. Ravera), Ellis Horwood, Chichester, pp. 47–61.

Underwood, A.J. (1989) The analysis of stress in natural populations, in *Evolution, Ecology and Environmental Stress* (eds P. Calow and R.J. Berry), Academic Press, London, pp. 51–78.

Unterman, R., Bedard, D.L., Brennan, M.J. *et al.* (1988) Biological approaches for polychlorinated biphenyl degradation, in *Environmental Biotechnology* (ed. G.S. Omenn), Plenum, New York, pp. 253–269.

Urban, D.L., O'Neill, R.V. and Shugart, H.H. (1987) Landscape ecology. *BioScience*, **37**, 119–127.

Urbanek, R.P. (1989) The influence of fauna on plant productivity, in *Animals in Primary Succession* (ed. J. D. Majer), Cambridge University Press, Cambridge, pp. 71–106.

Usher, M.B. (1986a) Attributes, criteria and values, in *Wildlife Conservation Evaluation* (ed. M.B. Usher), Chapman & Hall, London, pp. 3–44.

Usher, M.B. (ed.) (1986b) *Wildlife Conservation Evaluation*, Chapman & Hall, London.

Usher, M.B. (1987) Modelling successional processes in ecosystems, in *Colonization, Succession and Stability* (eds A.J. Gray, M.J. Crawley and P.J. Edwards), Blackwell, Oxford, pp. 31–55.

Usher, M.B. (1989) Scientific aspects of nature conservation in the United Kingdom. *J.appl. Ecol.*, **26**, 813–824.

Usher, M.B. (1991) Scientific requirements of a monitoring programme, in *Monitoring for Conservation and Ecology* (ed. F.B. Goldsmith), Chapman & Hall, London, pp. 15–32.

Vaeck, M., Reynaerts, A., Höfte, H. *et al.* (1987) Transgenic plants protected from insect attack. *Nature*, **328**, 33–37.

Van den Bosch, R., Messenger, P.S. and Gutierrez, A.P. (1982) *An Introduction to Biological Control*, Plenum, New York.

Van Brummelen, T.C., Verweij, R.A. and van Straalen, N.M. (1991) Determination of Benzo (a) pyrene in isopods (*Porcellio scaber* Latr.) exposed to contaminated food. *Comp. Biochem. Physiol.*, **100C**, 21–24.

Van Donk, G. and Gulati, R.D. (1991) Ecological management of aquatic ecosystems: a complementary technique to reduce eutrophication-related perturbations, in *Terrestrial and Aquatic Ecosystems: Perturbation and Recovery* (ed. O. Ravera), Ellis Horwood, Chichester, pp. 566–575.

Van Emden, H.F. (1988) The potential for managing indigenous natural enemies of aphids on field crops, in *Biological Control of Pests, Pathogens and Weeds* (eds R.K.S. Wood and M.J. Way), Royal Society, London, pp. 183–201.

Van Hattum, B., de Voogt, P., van den Bosch, L. *et al.* (1989) Bioaccumulation of cadmium by the freshwater isopod *Asellus aquaticus* (L.) from aqueous and dietary sources. *Environ. Pollut.*, **62**, 129–151.

Van Hook, R.I., Harris, W.F. and Henderson, G.S. (1977) Cadmium, lead and zinc distributions and cycling in a mixed deciduous forest. *Ambio*, **6**, 281–286.

Vance, E.D., Brookes, P.C. and Jenkinson, P.C. (1987) An extraction method for measuring soil microbial biomass C. *Soil Biol. Biochem.*, **19**, 703–707.

Veith, G.D., Call, D.J. and Brooke, L.T. (1983) Structure–toxicity relationships for the Fathead Minnow, *Pimephales promelas*: narcotic industrial chemicals. *Can. J. Fish Aquat. Sci.*, **40**, 743–748.

Visser, S. (1985) Management of microbial processes in surface mined land reclamation in Western Canada, in *Soil Reclamation Processes* (eds R.L. Tate and D.A. Klein), Marcel Dekker, New York, pp. 203–241.

Vitousek, P.M. and Walker, L.R. (1987) Colonization, succession and resource availability: ecosystem-level interactions, in *Colonization, Succession and Stability* (eds A.J. Gray, M.J. Crawley and P.J. Edwards), Blackwell, Oxford, pp. 207–233.

Waage, J.K. and Greathead, D.J. (1988) Biological control: challenges and opportunities, in *Biological Control of Pests, Pathogens and Weeds* (eds R.K.S. Wood and M.J. Way), Royal Society, London, pp. 111–128.

Waage, J.K. and Hassell, M.P. (1982) Parasitoids as biological control agents – a fundamental approach. *Parasitology*, **84**, 241–268.

Walker, C.H. (1990) Kinetic models to predict bioaccumulation of pollutants. *Funct. Ecol.*, **4**, 295–301.

Walker, D. (1989) Diversity and stability, in *Ecological Concepts* (ed. J.M. Cherrett), Blackwell, Oxford, pp. 115–145.

Walker, D.A. and Walker, M.B. (1991) History and pattern of disturbance in Alaskan artic terrestrial ecosystems: a hierarchical approach to analysing landscape change. *J. appl. Ecol.*, **28**, 244–276.

Wapshere, A.J. (1985) Effectiveness of biological control agents for weeds: present quandaries. *Agric. Ecosystems Environ.*, **13**, 261–280.

Ward, L.K. and Jennings, R.D. (1990) Succession of disturbed and undisturbed chalk grassland at Aston Rowant national nature reserve: dynamics of species changes. *J. appl. Ecol.*, **27**, 897–912.

Wardlaw, A.E. (1985) *Practical Statistics for Experimental Biologists*, John Wiley, Chichester.

Waring, R.H. (1989) Ecosystems: fluxes of matter and energy, in *Ecological Concepts* (ed. J.M. Cherrett), Blackwell, Oxford, pp. 17–41.

Warren, A. and Goldsmith, F.B. (eds) (1983) *Conservation in Perspective*, John Wiley, Chichester.

Warwick, R.M., Pearson, T.H. and Ruswahyuni (1987) Detection of pollution effects on marine macrobenthos: further evaluation of the species abundance/biomass method. *Mar. Biol.*, **95**, 193–200.

Washington, H.G. (1984) Diversity, biotic and similarity indices. A review with special reference to aquatic ecosystems. *Water Res.*, **18**, 653–694.

Webb, N.R. and Hopkins, P.J. (1984) Invertebrate diversity on fragmented *Calluna* heathland. *J. appl. Ecol.*, **21**, 921–933.

Welch, E.B. and Cooke, G.D. (1987) Lakes, in *Restoration Ecology* (eds W.R. Jordan, M.E. Gilpin and J.D. Aber), Cambridge University Press, Cambridge, pp. 109–129.

Wells, T.C.E. (1983) The creation of species-rich grassland, in *Conservation in Perspective* (eds A. Warren and F.B. Goldsmith), John Wiley, Chichester, pp. 215–232.

Westman, W.E. (1985) *Ecology, Impact Assessment and Environmental Planning*, John Wiley, New York.

Wheeler, B.D. (1988) Species richness, species rarity and conservation evaluation of rich-fen vegetation in lowland England and Wales. *J. appl. Ecol.*, **25**, 331–353.

Whelan, R.J. (1989) The influence of fauna on plant species competition, in *Animals in Primary Succession* (ed. J. D. Majer), Cambridge University Press, Cambridge, pp. 107–142.

Whittaker, R.H. (1975) *Communities and Ecosystems*, 2nd edn, MacMillan, New York.

Wiegart, R.G. (1988) Holism and reductionism in ecology: hypotheses, scale and systems models. *Oikos*, **53**, 267–269.

Wilcove, D.S., McLellan, C.H. and Dobson, A.P. (1986) Habitat fragmentation in the temperate zone, in *Conservation Biology: The Science of Scarcity and Diversity* (ed. M.E. Soulé), Sinauer, Sunderland, MA, pp. 237–256.

Index

Numbers in **bold** denote figures; those in *italic* refer to tables.

Abundance Biomass Comparison
(ABC) 54–6, **56**
Acacia senegal 226, **267**
Acclimation 33–4, 78
and heterozygosity 78
in plants 91–2
Accumulation factor 38
Acetogenesis 346, 357–62
Achromobacter 98, 346
Acid deposition 394–7, **395**
in lakes 395–7, **396**
in soils 395
Acidity in spoil 273
Acidogenesis 346, 357–62
Acinetobacter **99**
Acroptilon repens 189
Actinomycetes 94, 265, 269, 277, 365
Activated sludge **340**, 341–5
floc formation in 342
mean cell retention time 342
process control of 344
Adair run, Virginia 313
Adaptation 30, 32–4
and the distribution of species 32
and Fisher's theorem of natural
selection 77
genotypic 73
and homeostasis 30, 33
phenotypic 73
Adenylate energy charge (AEC) 61
Adversity selection 165, 252–4
Aerobic respiration 330
Age distribution
and functional responses 171–2
and population growth 118, 140
stable 118
Agouti, *Dasyprocta* 205, 224
Agrilus hyperici 181

Agrobacterium tumefasciens 106, 186
Agrostis canina 89
A. capillaris 88–90, 92, 332
A. gigantea 92
Alcaligenes 95, 98, 105, 338
Alder, *Alnus* 269
Algae 313
in effluent treatment 338–9
as freshwater phytoplankton 301–14
see also Individual genera
Alleles 73
Allozymes 85
as a measure of genetic variation
85–6
as pollution indicators 86–7
Aluminium 256, 397
Amazonia
and bird diversity 205, 384
and mammalian diversity 384
native agriculture in 225
Amiens 366
Ammonia, ammonium
in soil 268, 296
in effluent treatment 340, 342, 345,
348
Ammonification 258, **261**
Amphithrix 338
Amplitude, *see* Stability
Amoco Cadiz 97
Anabaena planctonica 305
Anacytis 338
Anaerobic digestion
of halogenated wastes 105
in landfill sites **361**
process control of 347
of sewage sludge 346–8
see also Fermentation
Anagrus epos 194

Anchovy, *Engraulis ringens* **127**
Antagonism 32
Anthoxanthum odoratum **90**
Ants in soil reclamation 227
Apanteles flavipes **175**
Aphids 163, 168, 182
  biological control of 183
  chemical control of 194–5
  'honeydew' 190
  kairomones 190
  and reclamation 277
  Walnut, *Chromaphis juglandicola* 177
Aphytis maculicornis 179
Arsenate 86, *196*
Ascomycetes 254
Ascophyllum nodosum 50–1
Asellus aquaticus 320
Aspergillus 98
Assimilation efficiency 36
ATP in soils and spoils 60
Atrazine 92, **93**
Aureobasidium 338
Autotrophic microorganisms *258*, 336
  in effluent treatment 339

Bacillus 98, *258*, 346
  B. popilliae 188
  B. subtilis 68
  B. thuringiensis 106, 186–8, **188**
Bacteria
  coli-form 334
  conjugation in **74**
  heterotrophic 336, 343
    and metal-binding 275
  mesophilic 347, 362
  metal-tolerant 95–6
  in reclamation 277
  soil 255
  sulphur-oxidizing 275
  thermophilic 347, 362, 365
  transduction in **75**
  transformation in 74
  transposition in **76**
  see also Individual genera
Baja California **213**
Balantiophorus minutus 344
Baltic Sea 390–4
Barley, *Hordeum vulgare* 274, 349
Barnacles 52
Barro Colorado island 204, **205**, 214, 224
Basal metabolic rate 35
Basidiomycetes 254–5

Beetles
  Carabidae 40, 388
  Cerambycidae 277
  Coccinellidae 180, 194, 197
Benzo(a)pyrene 62
  and molluscs 62
  and woodlice 323
Bighorn sheep, *Ovis canadensis* 145, **146**
Bioaccumulation 38, 314–23
  and body size 40–1, 317
  compartment models of 40
  and metabolic rate 41
  microcosm models of 318–21
  of radionuclides 321–3
Biochemical oxygen demand (BOD) *333*, 333–44, 354–5, 365
Biochemical measures of stress 61–8
Bioconcentration, *see* Bioaccumulation
Biogas
  and the digestion of sludge 346–8
  and landfill 358–60
  scope for generation 358
Biological availability 315
  see also Individual pollutants
Biological control
  and the action threshold **159**
  classical 173
  by native Brazilian farmers 158
  and profitability 160
  see also Pests
Biological filtration **337**, 336–40
  biological films in 339
  communities in
    invertebrate 338
    microbial 338
  regulation of 339
Biological monitoring 46–68
  at community level 54–61
  at the ecosystem level 297–300
  metallothioneins in 63–4
  soil enzymic activity 59–60
  scope for growth and 64–8
  see also Indicator; Monitor; Sentinel species
Biomagnification, *see* Bioaccumulation
Biomanipulation 305–8
Biomes 369
Bioreactors **101**
  in effluent treatment 335–48
Bioremediation 99
  degradation of hydrocarbons in 99–102

and genetically engineered
microorganisms (GEMs) 102–6
microbial seeding in 100
soil 94–102
Biotechnological Sciences Coordinating Committee (BSSC) (USA)
110
Biotic indices 54, 381
Birch, *Betula* 91, 247, 272, 390
Birds
Barro Colorado island 214
Hawaiian 182, **217**, 223
New Guinea 220
Passerine 207
species equilibrium of 211
tropical 205
wetland 235
see also Individual genera
Birth rate (b) 117
Blackberry 194
Blackfly (Simulidae) 188
Black locust tress, *Robinia pseudoacacia*
277
Black scale, *Saissetia oleae* 181
Body size
and bioaccumulation 40–1,
317–18
in K-selected species 164
and persistence time 145
and reproduction rate 145
in r-selected species 162
Bracken, *Pteridium aquilinum* **156**, 169,
170, 190
Brazil nut, *Bertholletia excelsa* 205
Bream, *Abramis brama* 304–5, 308
Brussels sprout 194
Budget, *see* Mass balance

Cadmium
bioaccumulation of 319–21
in freshwater microcosms 313, 319
in mammals 320
and microorganisms 95–6
in molluscs 41, 52, **87**
sewage sludge 350
soil 298–300, 349
in soil arthropods 39
Caesium
bioaccumulation of 21–2
Calcium
and legumes 272
in molluscs 41
soil 298

Capitella capitata 54
Captive breeding programmes 141,
240
Carbon
global atmospheric 398–401, **400**
soil cycles of **257**
Cardiaspina albitexture **168**
Carp
bighead, *Aristichthys nobilis* **356**
grass, *Ctenopharyngodon idella* **356**
silver, *Hypophthalmichthys molitrix*
356
Carrying capacity (K) 119
as a criteria in reserve selection 234
density-dependence and 139
Cassava green mite, *Mononychellus
tanajoa* 180
Cassava, *Manihot esculenta* 180
Cassava mealy bug, *Phenacoccus
manihoti* 172, 180, 183
Catchability (q) 123–4
Catchable stock (N_c) 122
estimation of 123–4
Catch per unit effort (U) **126**, **127**
Cellulose decomposition 255, 362
Cellulose:lignin ratio 360
Ceratophyllum demersum 305
Cerithium 87
Cheetah, *Acinonyx jubatus* **154**
Cherry, wild, *Prunus pennsylvanica* 7
Chickweed, *Stellaria media* 253
Chlorella 338, 318–19
Chlorine **314**
Chlorophyll α **314**
Chemical oxygen demand (COD) 333,
364
Chemostat 330
China clay wastes 269, 270
Chondrilla, *Chondrilla juncea* 181
Chromium 349
Chromosomes 73, **74–6**
Chrysolina hyperici 185
 C. *quadrigemina* 185
Chrysomela aenicollis 190
Citrobacter 95
Clam, *Corbicula* 61
Clear-cutting 136, 295
Clinch River, Tennessee 61
Clover, *Trifolium* 269, **270**, 271
C:N ratios
and composting 365–6
in effluent treatment 345
in landfill sites 360

C:N ratios (*cont'd*)
  in soils 257–8, 276, 298
Coccophagoides utilis 179
Cod, *Gadus morhua* **131, 134**
Cohorts 129
Collembola
  and biological filtration 339
  cadmium accumulation by 39
  in soil reclamation 277
Colombia 279
Co-metabolism
  in activated sludge 344
  and dehalogenation 104
Community 5
  climax 244
    monoclimax theory of 244
    plagioclimax 280
  integration 13, 243–4
    'rules of assembly' 244
  responses to stress 56–61
Competition
  and the colonization of islands 211
  and community organization 244,
    385
  in genetically engineered
    microorganisms (GEMs) 108
  and introduced species 223–4
  in late successional communities 248
  between temperate trees 248
  see also Intraspecific and Interspecific
    competition
Competitive dominants 252–4
Composting 264–6
  process control of 366
Concentration factors 38, 316, 320, 322
  and compartment models **39,
    40**
  in soil invertebrates 40
  see also Bioaccumulation
Conjugation **74**, 109
Conservation
  evaluating communities for 228–39
  of populations 138–53
Conservula cinisigna 170
Copper
  in crabs 45, 63
  and metallothioneins (Mt) 63
  in microorganisms 96
  in molluscs 50, 52, 61
  soil 59, 298–300
    mobility in 349
    nitrogen mineralization and **58**
    phosphatase activity and 58, 59

scope for growth (SfG) in mussels
    66, **67**
tolerance to
  in *Agrostis* 88, 91
  in *Mimulus* 91
  in *Nereis* 85
Corophium volutator 55
Corn *192*
Cotton boll weevil, *Anthonomus grandis*
    191, *196*
Crassostrea virginica 52
Cucumber *192*
Culling 240
Cyclades islands 207
Cyanophyta
  and eutrophication 300–9
  filamentous forms of 306–7
  toxic strains of 306
see also Individual genera
Cypress domes **351**
Cypress, *Taxodium* 350
Cytochrome P-450 62

2, 4, D **104**, 105
Dactylella oviparastitica 185
Dactylopius 197
Dandelion, *Taraxacum* 164
Daphnia 44, 306–9, 319
Darenth wood, Kent 284
Daucus carota 250
DDT
  bioaccumulation of 316–18, **319**
  breakdown products of **319**
  pest control and *196*, 197
Death rate, *see* Mortality rate
Deforestation
  Himalayas 279
  Temperate 279
  Tropical 278–9, **400**
Dehalogenation, microbial 104–5, 344
Deltamethrin 195
Dendrobaena octaedra 277
Denitrification *258*, 345
  in effluent treatment 345–6
  in swamps 352
Density-dependence 119
  and birth rate (b) 166
  and fishery stocks 129, **133–4**
  and predation 166
Desertization 279
Detergents and molluscs **87**
Diaretiella rapae 190
Diartraea **175**

Diflubenzuron 191
Dilution rate (D) 242
Disturbance 21
  and community structure 255, 377
  ecosystem responses to 393–4
  frequency of 7, 23, **24**, 225
  induced 226
  intermediate disturbance
      hypothesis 225
  gap size 7–8, 377
  hierarchical classification of
    in the Alaskan tundra 390
    in the Gulf of Bothnia 390–1
  landscape ecology and 227
  management of 228, 281, **282**
  perturbation 22
  predictability of 377
  reserve management and 227–8
  in tropical forests 226, 377
  see also Stress
Diversity
  in artificial ecosystems 343
  of British birds 384
  conservation value and 230
  disturbance and 225–7, 393
  global pattern of (latitudinal
      gradient) 369–70, 382
    distribution of energy and 384–5
    extinction rates and 384
    habitat constancy and 383
  indices of 378–81
    log-normal **380**
    as measures of community stress
        378, 380–1
  microcosm 313
  regional 382–5
  scale and 370–1
  S:N ratio and **379**
  species
    equitability 378
    richness 246, 378, 388
  Whittaker's classification of **371**
DNA 73
  extra-chromosomal **74**, **76**, 77, 79, 95
  mobility in the environment 109
  technology **103**
Dose 43–4
  acute 43
  chronic 43
Dose-response relationships 29, 298,
    311–12, 380
  a graded response 44
  quantal response 44

soil bacteria and metal concen-
    trations 57
Dreissena 307
Drosophila
  climatic races of 79
  heat-shock proteins in 77
  interspecific competition in 244
Ducks 235
Duckweed, *Lemna minor* **353**, 355
Dynamic pool models 129–35

Earthworms
  in composting 365
  in landfill sites 360
  metal mobility in 340
  and soil reclamation 265, 277
  see also Individual genera
Echinochloa colonum 163–4
Ecological dose range (EDR) and soil
    respiration 46
Ecology
  community 9–10
  holistic 10–13
  methodology of 8–17
  modelling 2
  population 9–10
  systems 10
  theory 2, 372–3
Ecosystems 5
  management *26*
  monitoring
    nutrient loss in 297–300
  nutrient cycling in 288
  oceanic 373–4
  responses to stress *11*
  terrestrial 373–4
Economic activity and pollution
    399–401
Ecotypes 34
  Amaranthus retroflexus 93
  commercial 92–4
  grasses 89–94
  microorganisms 95–6
  reclamation and 92–4
  Senecio vulgaris **93**
  see also Tolerance
Ectotherms 36, 220
Ecuador 204
Edge communities 206, 218, 228
Effective population size (N_E) 149–53
Effluent treatment
  in conventional sewage works
      335–48, **334**

Effluent treatment (*cont'd*)
  in macrophyte systems **353**, 354–5
  see also Activated sludge; Anaerobic
    digestion; Biological filtration
Eichornia crassipes **353**, 354–5, 357
Elephant, *Loxodonta* 240
Elasticity, *see* Stability
Elodea 320
El Niño 373
Emergent properties 12
Encarsia formosa 192
Enclosures, *see* Mesocosms
Endemism 204
Endotherms 36, 220
δ- endotoxin
  and genetically engineered
    microorganisms (GEMs) 106,
    108
  in pest control 186–8
  resistance to 192
Energetics
  allocation of energy 34–6
  budgets **35**
  community 36–8
  efficiencies 36–7
  stress and 37
  and trophic levels 37
Energy consumption, human 400
Enzymes
  allozymes 85–8
  Benzo(a)pyrene hydroxylase
    (B(a)PH) 62
  cellulases 61
  fish
    glucose phosphate isomerase 86
    isocitrate dehydrogenase 86
    malate dehydrogenase 86
  luciferase 60
  molluscs
    and cellulolytic activity 61
    mixed function oxidases (MFOs)
      62–3
  relaxation of 104
  soil 59
    dehydrogenase 60
    phosphatase **58**
    urease 59
Epidinocarsis lopezi 180, 183, 186
Equilibrium
  community
    in eutrophic lakes 310
  of fish stocks 374
  steady-state

  in the activated sludge process
    342
  of the monod relationship 330,
    331
  of pollutants in compartment
    models 38–9, 315
  yield to fishery 125
  see also Stability
Erica, *Erica* 222, 272, 322
Eriophorum vaginatum 390
Escherichia coli 79, 108
Eucalyptus 266
Euglena 338
Eurytopic species 80
Eutrophication 300–10
  management of
    in agricultural land 283
    in freshwaters 309–10
  models of 300–9
  and sewage treatment 334
  soil 280
Exploitation efficiency 37
Extinction
  and genetic variation 147–9
  and habitat fragmentation 151–3
  and population stability 149–51
  rates 201
  vortex 149, **152**
Exxon Valdez 31

Fecundity 120
Fermentation 330, 336, 346–8,
  360–4
Festuca ovina 253
Fig, *Ficus* 224
Fish, *see* Individual genera
Fish farms, organic pollution beneath
  cages 55
Fish ponds **356**, 356–7
Fisheries
  by-catches of 130
  management of 8
  see also Yield (Y)
Fishing effort (E) 123, **126**
Fishing mortality rate (F) 123
  age-specific 129
Flavobacterium 95, 98, 338
Food chains, *see* Trophic levels
Food web 374
  complexity and stability of 375–6
  connectance in 375–6
Forest fires 8

Forest management 8, 135–7
  and recruitment 136
  and yield 137
Fractal geometry 388
Fragmentation, *see* Insularization
Frankia 269
Fugitive species 7, 247, 377
Fungi
  in biological control 181, 185, **187**
  in composting 365
  in effluent treatment 338–9
  and reclamation 277
  soil 254–5, 275, 352
  see also Individual genera
Fungi imperfecti 254, 338
Fusarium 338
Fusarium root rot, *Fusarium culmorum*
  194

Gambusia affinis 85–6, 181
Game management 128
Garden snail, *Helix aspersa* 53
Generalist species 80
Generation time 116
Gene 73
  markers and probes 110
Genetic drift 148
Genetic fixation 91, 149
Genetic variation
  in finding natural enemies 181
  loss of 147–53, 219
  and population size 148
Genetically engineered micro-
    organisms (GEMs)
  in bioremediation 102–6
  in pest control 186, 191–2
  release 106–10
Genotype 72
Geographical Informative Systems
  (GIS) 240
Geotrichum 344
Global pollution 399–401
Glyphosate 106
Goldenrod, *Solidago* 164
Gorse, *Ulex europaeus* 247, 283
Grape leafhopper, *Erythroneura*
  elegantula 194
Grasses
  and mycorrhizae 272
  in reclamation **267**, 269, **271**
  in sand dune succession 252
  see also Individual genera
Great Barrier Reef 214

Grizzly bear, *Ursus arctos* 153
Guilds
  in birds 205
  and the stability of communities
    376–7
Gulf of Bothnia **393**
Gypsy Moth, *Lymantria* 191

Habitat
  diversity 214
  fragmentation in Dorset heathland
    221, **222**
  insularization 215
Hawaii **217**, 223
Health and Safety Executive (UK) 110
Heather, *Calluna* 222, 247, 272
Heat stress
  Drosophila 77
  Salmonella 79
Hedgerows
  as corridors 387, 388
  headlands 284
  pests and 284
Heterorhabditis 188–9
Heterosis, *see* Heterozygosity
Heterozygosity 73
  and scope for growth (SfG) 78
  vigour of 107, 148
Hierarchy theory 13–17, 398
  and ecosystem regulation 15, 377
  functional groups in 14–17
  stress and 17, 23
  upward constraint and 398
Homeostasis
  and adaptation 33
  and metal uptake 41
Homeotherms 36
Homeorhesis, *see* Stability
Homozygosity 73
Holism, *see* Methodology
Hong Kong 52
Hubbard Brook 295–7, 300
Humates 255
Hyacynthoides 281
Hybanthus prunifolus 204
Hydrobia ulvae 55–6
Hydrocotyle vulgaris **353**
Hydrocarbon pollution
  microbial degradation of 99–102
  molluscs and **65**
  partition coefficients of 39
Hydrolysis 346, 357, 360
Hyphomicrobium 346

Hypotheses, testing 2–3, 288

Ice ages, *see* Pleistocene
Inbreeding depression 148
Incorporation 377
Incidence functions **221**, 221–3
Indicator species 47, 53–4, 229
    in activated sludge *343*
    in biotic indices 54
    marine polychaetes as 54
Inertia, *see* Stability
Islands
    colonization rates (λ) of 209–12
    habitat 207, 216
        beetles in heathland 217
        higher plants in quarries 217
    extinction rates (μ) 210–12
    island biogeography theory 209,
        210, 387–8
    oceanic 207, 209, 216
    relaxation in **213**
    size and species number 216–18, **218**
    vertebrate fauna 221
    see also Species–area relationships
Insecticides
    organochlorines 316–17
    see also Pesticides
Insularization 206–7
    and reserve design 216–20
Integrated Pest Management (IPM)
    194–7
Integration, *see* Conjugation
Interspecific competition 84
    and pests 167–8
Intraspecific competition 84, 119
    and pests 161–2
Intrinsic rate of population increase
    (r), *see* r
Introduced species
    as pests 169
    see also Natural enemies
Iron 256, 306, 309
    pyrites 96, 273
    wastes **267**
Israel 225

Japanese beetle, *Popillia japonica* 188

K, *see* Carrying capacity
Kairomones 190
Kaolinite 278
Keystone mutualists 398
Keystone species 206–7, 219, 229

Klebsiella 95, 96
K-selected species 120, 164–5,
    382
    birth rate of 164
    earthworms 277
    extinction and 206
    habitat characteristics of 165
    mortality rate and 164
    as pests 168
    characteristics of 163

Lacewings 194
Lakes
    acidification of 396–7
    Douglas 313, **314**
    eutrophic 301
    Norwegian **396**
    oligotrophic 301
    Washington 302, 306
    Wingra **303**, 303–4, 306
    Zwemlust 308
    see also Eutrophication
Landfill ecology 357–64
    cell emplacement strategy 359–60
    and invertebrates 360
    leachate 364
    microbiology of 360–3
    nature of waste 358
    optimizing gas production from 358,
        362–5, *363*
    see also Anaerobic digestion
Landscape ecology 385–390, *386*
    pattern in
        background matrix 387
        corridors in 387
        hierarchical classification of 390
        patches in 387–8
    processes in *386*
    scale and 385–6
    watersheds 390
Lantana camara 169, 184
Large Blue butterfly, *Maculinea arion*
    203, 206, 240
Laser detection of pollution 68
Laterization 278
LD_{50} **44**
    derivation of **43**
    variations on *45*
    see also Toxicity test
Lead
    in molluscs 41, 52
    soil 298–300, 349
    tolerance to 88

Legumes
    in reclamation of mine spoils
        269–70, **271**
    sensitivity to metals 94
Leicestershire 387
Leucaspius 319–20, 319
Levels of organisation 5, 5–6
Lichens **28**
    and sulphur dioxide 32
Life history strategies 253
    of animals
        in terrestrial and marine habitats
            374
    of plants **253**
    see also r- and K-selected species
Lignin 255, 360
Liming mine spoil 273
Littorella uniflora **353**
Littorina 87
Lixophaga diaetraeae **175**
Locus 73
London **263**
Lotus **269**
Lumbricus terrestris 277
Lupin, *Lupinus perennis* 269
Lysosome latency and hydrocarbon
        toxicity 62

Macroptilium lathyroides 163–4
Macrophytes
    and eutrophication 302–9
    in effluent treatment 353, 354–7
Madagascar 204
Magnesium 298
Malathion 191
Management
    of grassland 227, 280, 281–3
    of eutrophic lakes 303, 309–10
    of fisheries 122–35
    of forests 135–7
    of game 128
    of prairie 227, 281
    of salt marshes 280
    of semi-natural habitats 280
    of soil nutrient levels *283*
    of woodland 281
Manduca sexta 192
Marsupials 369, 382
Mass balances 289, 327, **329**, 330, **351**
    Hubbard Brook 295–7
Maximum population size (N_m) 142
Maximum sustainable average yield
        (MSAY) 129

Maximum sustainable yield (MSY)
        122–3
    in fisheries 122–35
    in forestry 135–7
Mediterranean sea 225
Megacyllene robiniae 277
Melilotus alba 281
Meloidogyne 185
Mercuralis 281
Mercury
    allozymes and 86
    methylation of 96
    microorganisms and 95
    in molluscs 87
Mesocosms 288, 310, 313, 390
Metal tolerance
    in *Alyssum* 88
    in grasses 88–92
    in microorganisms 95–6
    in *Thlaspi* 88
Metallothioneins (Mt)
    in fish 63–4
    in invertebrates 66
    in plants 91
Metapopulations 151, 388
Metaphycus helvolus 181
Methanogenesis 346, **347**, 357–62
Microorganisms
    soil 254–5
        biomass of 60–1
        development of microbial
            community in 275
    see also Individual genera
Microcosms 288, 310, 311
    using artificial substrates 313
    and bioaccumulation 318–21
    freshwater 308–14
    nutrient loss from 298, 314
    standardized 312
Microcytis aeruginosa 306
Microthrix **344**
Millipedes 360
Mimulus guttatus 91
Mine spoil, *see* Reclamation
Minimum viable population (MVP)
        140–1, 218
Mites 360
    Oribatid 39
    in soil development 255
Mixed-function oxidases 62–3
Models
    of acidification of groundwater
        (MAGIC) 396–7

Beverton and Holt, *see* Dynamic
pool
  bioaccumulation 317–23
  of biological control 179–81
    deterministic 179
    integrated pest management 197
    multiple species 178–9
    multivariate 180
    synoptic population growth **166**
  Birth and death process 140–5
  ecosystem **292**, **303**
  fishery
    dynamic pool 129–35
    surplus yield 124–9
  of eutrophication 302–9
  food webs 375–6
  forestry 137
  large-scale ecology 373
  methodology *290*, 291–5
  objectives 289
  of pollutant uptake **39**, 317–18
  population
    logistic 116–21
    predator–prey 173–6
    synoptic 165–9
  Schaefer, *see* Surplus yield
  sensitivity analysis 293, 397
  systems ecology
  types 289–91
    analytical 289–90
    compartment **39**, 317
    deterministic 4, 180, 290–1
    dynamic 294
    multi-compartment *39*
    multivariate 180
    stochastic 4, 290–1
    simulation 289–90
  validation 293
Molluscs
  basal metabolic rate of 40
  body weight and pollutant uptake
    by 41
  and detergents **87**
  and hydrocarbon pollution 41, 62–3,
    87
  niche width of **87**
  scope for growth (SfG) in 64–6
  shell in pollution monitoring 53
  toxic metals in 41
  see also Individual genera
Mongoose, *Herpestes* 182
Monitor species 47, 50–3
  see also Sentinel and Indicator species

Monod constant **330**
Monodonta **87**
Moose, *Alces alces* 322
Mortality rate 117
Mosquitoes 188, 189
Moulting hormone 190
Mussels, *Mytilus edulis*
  metal uptake by 52
  as monitor species 52–3
  'mussel watch' 53
  and protein production 78
  scope for growth (SfG) in **64**, 65
  as sentinel species **51**
Mutation 77
  of genetically engineered
    microorganisms (GEMs) 108
  rate 89
Mycorrhiza 272
  and the global climate 398
  and protection against disease 185
  in reclamation 272
    and tree establishment 272
  and toxic metals 272
Myriophyllum spicatum 303–4
Myrmica sabuleti 203, 206
Myxmatosis 186

Natural enemies
  finding 181–2
  'ravine' 168
  release of 183–5
  selection of 182–3
  specificity of 182–3
Natural mortality rate (M) 123
Natural selection, Fisher's theorem 77
Nature reserves, *see* Reserves
Negative feedback 293, **294**
Nematodes
  in effluent treatment 339, 344
  in pest control 188–9
  see also Individual genera
Nereis diversicolor 85
Net productive rate (R_N) 116
Nettle, *Urtica dioica* 253
New Guinea Harpy Eagle, *Harpyopsis
  novaeguinae* 203
Niche 71
  and environmental gradients **81**
  fundamental 82
  overlap 82
    in lizards 83
  realized 82
  regeneration 204, 226, 281

and pioneer plants 251
theory 80–4
width 83
  and competition 84
  of ecotypes 84
  of specialist and generalist species
    83
  and succession **252**
Nickel 46, 349
Nicotine 190
Nitrate 260, 268, 296, 397
Nitrification *258*, 260, **261**, 345–6
  in activated sludge 346
  bacterial
    Nitrobacter 258
    Nitrosomonas 258
  in biological filtration 339,
    345–6
  in mine spoils 275
Nitrite 260
Nitrogen
  in freshwaters 302–10
  soil 268–70
    budget **296**
    capital 270, 275, 295
    cycling **257**
  in swamps **351**
Nitrogen fixation *258*, 269
  by legumes 269
  and phosphates 271
  by *Rhizobium* 269
  in swamps 352
Nocardia 98
n-Octanol, *see* Partition coefficient
Norfolk broads 306–7
North Sea **51**
Nosema locustae **187**
Nuclear reactors
  Chernobyl 322
  Hinkley Point **286**
  Savannah River 321
  Sellafield 322
Nutrient cycling
  accelerated loss *283*, 284, 297–8
  in Hubbard Brook 296
  and pollution 298–300
Nymphaea alba **353**

Oenothera biennis 250
Oil pollution
  in Alaska 389–90
  in Brittany 97
  crude oil

degradation and fertilizers 97
microbial degradation of 96–102
mousse *97*
properties of *97*
see also Hydrocarbon pollution
Oligotrophic freshwaters 301, 305
Olive scale, *Parlatoria oleae* 179
Opportunist species, *see* r-selected
  species
Orchesella cincta 40
Oriental fruit fly, *Dacus dorsalis* 191
Oscillatoria 306
Outbreeding depression 153
Overfishing 122, 137
Oxalis 281
Oxygenation ditches **326**

Panotima 170
Paramecium caudatum **344**
Parasitoids 176, 181, *182*
Pathogens
  in landfill 361
  as natural enemies 178, 181, 183, *187*
  in sewage 334–5
Partition coefficient 39
  log K_{ow} and 41
Parys Mountain, Wales **70**
Pea, *Pisum sativum* 277
Peccary, *Dicotyles* 224
Penicillium 98
Perch, *Perca fluviatilis* **319**
Permafrost 389
Persistence of pollutants 31, 315
  see also Individual pollutants
Persistence time of populations 140–7
  and body size 145
  and density dependence 143, **144**
  in herbivorous mammals 146, **147**
  and population size 142
  and variability in r 141–3
Perturbation, *see* Disturbance
Peru 224
Peruvian anchovy, *Engraulis ringens*
  127
Pests 159
  action threshold for **159**
  chemical control of 194–5
  coexistence with natural enemy 173,
    179
  cultural control of *193*, 194
  depression by natural enemy (q)
    177, **178**
  ecology 185–6

Pests (*cont'd*)
  economic injury level (EIL) of **159**
  introduced species as 169
  population characteristics of 161–5
Pesticides
  costs of 195
  insect resistance to 195
  'insurance spraying' 161
  partition coefficient of 39
  role in eutrophication 306
  usage in USA 157
  see also Individual pesticides
Pioneer species, *see* Succession
Pheromones 191
Phenanthrene 100
Phenols 260
Phenotype 72
Phosphorus
  in eutrophic freshwaters 302–10
  legumes 259, **270**
  mycorrhizae 272, 398
  precipitation in freshwaters 306,
    309–10
  soil 256–60, 271–2
    in plant successions **259**
    reclamation 270–2
    in swamps **351**
    and toxic metals 298
Phragmites australis **353**, 354
Phycomycetes 254
Phylloxera 191
Phytoplankton in eutrophic
    freshwaters 300–9
Phytoseiulus persimilis 197
Pike, *Esox lucius* 304–5, **308**
Pine, *Pinus* 272
Pirimicarb 194
Pistia stratiodes 354, 357
Plant tissue culture 106
Plasmids **74**, 79, 95, 103
  antibiotic resistance 110, 343
  gene markers 110
  genetically engineered micro-
    organisms (GEMs) 108
  hydrocarbon degradation 99
  metal tolerance 95, 110
Pleistocene 369
Plodia interpunctella 192
Poikilotherms 36, 374
Poa trivialis 88
Pollution 30
  continuous 31
  dispersal of 31

  episodic 31
  point sources of 31
  uptake of 40
  see also Individual pollutants
Polychaeta 225, 383
Polychlorinated biphenyls (PCBs) 62,
    104–5
Polycyclic aromatic hydrocarbons
    (PAHs) 62
  smokers and 82
Polyvinyl alcohol (PVA) **267**
Poplar, *Populus* 272
Population
  effective size (N_E) 149–53
  extinction 138–53
  growth
    epidemic 165–9
    exponential **118**
    logistic 119, **120**
  stability of 121–2, 140
Population vulnerability analysis
    (PVA) 151–3
Porcellio scaber 323
Positive feedback 152, 293
Potamogeton crispus **353**
  P. gramineus **323**
Potassium
  in reclamation 272
  soil 298
Potato *192*
Predation
  distribution of attacks 176–7
  functional responses in 170–3
  numerical responses in 170, 173–9
Predation rate
  of pests
    'natural enemy ravine', **166**, 167
    release point 167
Predator–prey relationships 173–6
  stability in 175–6
Prickly pear, *Opuntia megacantha* 197
Process control 329–32
Production efficiency 37
Protein polymorphisms
  in *Drosophila* 79
  in *Gambusia* 86
  see also Allozymes
Protein production, metabolic costs 77
Protozoa
  in activated sludge 343
  see also Individual genera
Prudhoe bay, Alaska 389
Pseudoscorpions 40

Pseudomonas 258
  dehalogenation by **104**
  in effluent treatment 338–9, 343
  and hydrocarbon pollution 79, 98
  metal-tolerant 95
  P. cepacia 105
  P. flourescens 108, 110, 191
  P. putida 105
  P. syringae 110
Puccinella maritima 281
Puccinia chondrillina 181
  P. graminis 192
Pyrethrum 190
Pythium root rot, *Pythium* 194

Quantitative Structure–Activity
    Relationships (QSARs) 41–2
  log K_{OW} and **42**

r, intrinsic rate of population increase
    117
  variability (V) 139
    demographic (V_1) 141
    environmental (V_e) 141–2
Radionuclides
  compartment models of 321–3
  decay 321
  in Europe 322
  in soils 321–2
  in vegetation 322
  see also Individual elements
Rarity
  of the British flora *203*
  types of 202–4
Reclamation 245, 262–78
  and decomposition 275–7
  and invertebrates 265, 276–7
  of spoils
    and acidity 273
    with alkaline wastes 268
    chemical problems with 267–75
    physical problems 265–7
    site engineering for 265
  mycorrhizae and 272, 398
  objectives in 262–3
  organic matter in **267**, 268
  soil particle size and 265
  tolerance
    microorganisms 95–6
    plants 92–4, 275
    vertebrates in 276
Recombinant technology 102–4
Recombination 77

Recruitment 122
  and age and size 129
  to catchable stock (N_C) 133–5
  and density-dependence 134
  in dynamic pool models 130–2
Red-cockaded woodpecker, *Picoides
    borealis* 150–1, 206
Red Sea 225
Red spider mite, *Tetranychus urticae*
    197
Regional ecology 390–7
Relative abundance, *see* Diversity
Relative fitness 72
Relaxation 212–16
  in birds 214
  enzymic 104
  in lizards **213**
  in mammals 214, **215**
Remediation, *see* Bioremediation
Reproductive rate (R_O) 116
Reserves
  Aston Rowant, Chilterns **282**
  Chitwan National Park, Nepal 153
  Dalby Forest, Yorkshire 238
  design of 216–20
  East African **215**, 240
  La Capelliere, France, **200**
  Myakka State, Florida **1**
  Peak District, Derbyshire **266**
  relaxation in 214, **215**
  selection of
    Analytical Hierarchy Process
      (AHP) 215–17, **236**
    Fen species scores 232, *233*
    using guilds 234, 235
    Habitat Assessment Technique
      (HAT) 235
    Habitat Evaluation Procedures
      (HEP) 234
    Habitat Suitability Index (HSI)
      234
    indicator species for 229
    indices of diversity for 228–30
    monetary evaluation of 237–8
    Rio Palenque, Ecuador 204
    service evaluation of 238
    by size 218–20
    SLOSS argument 216
    Sites of Special Scientific Interests
      (SSSIs) *231*
    Tour Du Valat, Camargue 227
    Walthamstow Marshes, London
      238

Reserve management *239*, 239–40
  and disturbance 240
  see also Management
Residence time 31, 315
  in activated sludge process 342
  soil 298, 300
  vegetation *299*
  see also Individual pollutants
Resistance, *see* Stability
Restoration 244–5, 262, 278–84
  of freshwater 245
  of grassland 281–2
  of hedgerows 279, 284
  of prairie 281
  by seeding 281–2
  of soils
    agricultural 278–9
    arid–alkaline 278–9
  of woodland 279, 284
Rhine 21
Rhinocerus 150, 153
Rhizobium 269, 272, 277
Rhododendron ponticum 165
Romanomermis culicivorax 189
Rice *192*
Richmond Park, London **156**
Risk assessment and GEMs 109
Roadside verges 387
Rotifers 344
r-selected species 54, 119, 162–4
  birth rate of 162
  characteristics of *163*
  disturbed habitats and 394
  earthworms 277
  as pests 163–4, 168
Ruderal dominants 252–4
Rudd, *Scardinius erthyopthalmus* **308**
Rumen ecology 357
Ryegrass, *Lolium* 349

Sahel 278
Salicin 190
Salicornia **200**, 207, 280
Salmonella 361
  S. typhimurium 79
Salvinia, *Salvinia rotundifolia* 354
Saprotrophs 254
Satellite remote sensing 240, 389, 390
Scale
  disturbance and 7, *394*
  ecological methodology and 6–7, 372–3

and management 8
Scirpus lacustris 353, 354
Scope for Growth (SfG) 64–8
  in community-level monitoring 381
  and lysosomal activity 66
  in microorganisms 60
  in mussels 66
  and population growth 67–8
  and tolerance 68
Screw-worm, *Cochliomyia hominivorax* 189–90
Scrobicularia plana 50, 52
Seal, *Callorhinus ursinus* 135
Seasonality 7, 34, 385
Searching efficiency (a) 173, 177, **178**
Selection and fishing intensity 135
Selection coefficient 89
Semiochemicals 190–1
Sentinel species *47*, 48–50
  lichens as 48
  marine *49*, 48–50, **51**
Set-aside, habitat restoration 279, 280
Sewage 322–48
  nature of 335
  sludge and reclamation 265, **267**
Shadow prices 237
Shrew, *Sorex araneus* 320
Silesia, Poland **368**
Site selection, *see* Reserves
Silwood Park, Berkshire **251**
Sludge age 342
Sludge disposal 348–50
Smelters
  Avonmouth 299
  Gusum **58**
  Missouri 298
Soil
  ATP 60
  CO_2 production and pollution **59**
  clay content 268
    in pollutant binding 274, 322
  decomposition and metals in 58
  development 264–5
  erosion of 278–9
  enzymic activity and metals 59–60
  invertebrates 255
  glacial 263–4
  organic matter *256*
  pesticides in 57
  processes
    decomposition 57, 254–6
    nutrient cycles 57, 254–6
    remediation 94–102

removal 283
respiration and pollution 57
skeletal 262–3, *264*
stabilizers 265, **267**
Spatial heterogeneity 385, 388–9
  see also Landscape ecology
Specialist species 80
Species 5
Species–area relationships 207–11
  equilibrium number of species (Ŝ)
    209, **217**
    and niche class 211
  relaxation and 212–15
  in woodland vascular plants 230
  see also Islands
Species equitability, *see* Diversity
Species diversity, *see* Diversity
Species introductions
  and extinction of native species
    223–4
  and grassland communities 224
Species richness, *see* diversity
Spoils
  chemical problems 267–75
  physical problems 265–7
Sphaerotilus 344
Stability 17–21
  adjustment **19**, 20, 311
  amplitude **19**, 20
  community 20–1, 224–5, 376–7
    and guilds 376–7
    species number/connectance (SC)
      376
    and succession **244**
  cyclical change 21
  and diversity 224, 375–8
  elasticity **19**, 20, 247, 312, 377
  ecosystem 287, 294–5, 373–8, 393–4,
    394
    in eutrophic freshwaters 305,
      310
  homeorhesis 18–20, **19**
  homeostasis 18–20
  inertial **19**, 311, 322
  oceans 373–4
  microcosm 309
  population 17–20, 121–2, 137
    and extinction 140
    and predator–prey relationships
      175–6
    stable limit cycles and 121
  resistance 20
  in stochastic models 20

terrestrial ecosystems 373–4
trophic structure and 21
types of 18–20
Staphylococcus aureus 95
Steady-state, *see* Equilibrium
Stenotopic species 80
Stigeoclonium 338
St Johnswort (Klamath weed),
    Hypericum perforatum 185
Strawberries 110
Streams 387
Stress 21–4, 77–9
  and abundance biomass comparison
    (ABC) 55
  and ecosystem collapse *11*, 371
  genetic variation and 78–9
  hierarchy theory and 23
  protein production and 61
  response at different levels *11*, *18*,
    22–3, 29
  see also Disturbance
Stress-tolerators 165, 234, 253–4
Strip-mining **242**
Subbaromyces 338
Succession 245
  allogenic 243
  in anaerobic digesters 347–8
  autogenic 243
  competition in 250
  cyclical change and 247
  disturbance and 245–7
  energetics and 400
  facilitation in 247–50
  in heathlands 247
  inhibition in 247–50
  K-selected plants in 252, 260
  in landfill 361–2
  late successional plants 250
  mechanisms of change 245–6
  nutrient availability during 256–62
  of old field communities 250
  pioneer plants in 250
  primary 243, 247, 263
  r-selected plants in 252
  secondary 243, 247, 262
  in temperate forests 248
  tolerance in 247–8
Sudan 266
Sullem Voe Oil terminal **65**
Sulphur 273
  radioactive 321
  see also Acid deposition
Supertramps, *see* Incidence functions

Surplus yield models 124–9
Synergism 32
Synonomes 190
Synoptic population growth **166**
Syrphidae (Diptera) 183
Systems ecology, *see* Models

2, 4, 5-T **104**, 105
Tannins 260
Taenia 348
TCA cycle 104
Teleonemia scrupulosa 169
Temperate forest
  niche differentiation in 383
  soil processes 255, 297
  Vermont 299
  Walker Branch 298, *299*
Thiobacillus
  in biological filtration 339
  in reclamation 273
  in sludge treatment 350
Threshold concentrations, *see*
      Dose-response
Thymus praecox 203, 240
Tilapia, *Tilapia* **356**
Tobacco *192*
Tolerance 34
  costs
    to *Agrostis* 92
    to *Amaranthus retroflexus* 92
    to *Senecio vulgaris* 92
  metal
    in grasses 88–91
  in *Nereis* 85
  as a pollution indicator 84–8
Tomato 106
Toxic effects, chronic and acute 44
Toxicity test 4–6, 311
Tragopon dubius 250
Tramps, *see* Incidence functions
Transduction **75**
Transformation **75**, 77
Transgenic organisms 106
Transposition **76**
Transposons **76**, 79, **103**
Tricking filters, *see* Biological
      filtration
Trioxys pallidus 177
Trophic levels
  pollutant mobility between 318–23
  problems of definition 318
Trophic structure
  biomanipulation of 307–8

Tropical forests 382–4
  diversity
    and area 203–4
    endemism 204
  global atmospheric carbon and **400**
  niche differentiation in 382
  nutrient cycling in 225, 384
  plant–animal associations 224
  key resources in 224
  Pleistocene refugia 383–4
  soil processes 255–6
    nutrients in 278
    tree loss and 278
  tree genera 383
Trout, brook *Salvelinus fontinalis*
      63
  rainbow *Salmo gairdneri* 63
Tryptophan 190
Tundra
  Alaskan 389–90, *391–2*
  ecology 289
Tunicates **370**
Turtles, Kemp's ridley, *Lepidochelys
      kempi* **139**
Typha 234
  T. latifolia **353**
Typhus 361

Ulothrix 338
United Nations Environment
      Programme (UNEP) 397
Univoltine species 116

Vaccinium 390
Variation
  genetic 72–9
    and stress 78
  neutral 72
Verbascum thapsus 250
Verticillium lecanii **187**
Vibrio 339
Vigna unquiculator 192
Viruses
  in pest control 186, **187**
  in transduction **75**
Volatile fatty acids (VFAs) 346, 361–2,
      364
Vole, *Microtus agrestis* **320**
Vorticella 344

Wadden Sea 55
Watersheds as units of conservation
      218

Weeds
control by pathogens 189
habitat characteristics of 164
habitat restoration and 280, 284
in Guyana 163–4
population characteristics of 164
West Indies 214
Wetlands 232–8
Camargue **200**, 227
Delaware 235
in effluent treatment 238, 350
Fenlands 228, 232–4
Florida **1**, 238, 350–2
Iowa 211
Mississippi 224
Prairie wetlands 235
Swedish 235
Whales
Fin, *Balaenoptera* 120, 135, 137, **138**
Humpback, *Megaptera* **114**
Wheat 194
Willow, *Salix* 190, 272
Woodland
Bedfordshire 230
transfer for restoration 284
Woodlice
in composting 365
in landfill sites 360

in soil development 255
see also Individual genera
Wytham Wood 282

Xanthotoxins 190
Xenobiotics 95
Xenorhabdus 189

Yield (Y) 123
economic **128**
fishery
effort and 126, **127**, 132–3
recruitment and 129
forests 137

Zinc
in crabs 45, 63
and metallothionein (Mt) 63
in molluscs 41, 52, 61
release from sludge 350
soil 298–300
spoil 59
mobility of 349
tolerance to
in grasses *88*, **90**
in *Nereis* 85
Zooglea ramigera 338, 342
Zooplankton 300–9